Analysis of Microarray Gene Expression Data

ANALYSIS OF MICROARRAY GENE EXPRESSION DATA

MEI-LING TING LEE

Channing Laboratory
Department of Medicine
Brigham and Women's Hospital
and
Harvard Medical School
Boston, Massachusetts, USA

Kluwer Academic Publishers
Boston/Dordrecht/London

Distributors for North, Central and South America:
Kluwer Academic Publishers
101 Philip Drive
Assinippi Park
Norwell, Massachusetts 02061 USA
Telephone (781) 871-6600
Fax (781) 871-6528
E-Mail <kluwer@wkap.com>

Distributors for all other countries:
Kluwer Academic Publishers Group
Post Office Box 322
3300 AH Dordrecht, THE NETHERLANDS
Telephone 31 78 6576 000
Fax 31 78 6576 474
E-Mail <orderdept@wkap.nl>

 Electronic Services <http://www.wkap.nl>

Library of Congress Cataloging-in-Publication

Lee, Mei-Ling Ting
Analysis of Microarray Gene Expression Data
ISBN HB   0-7923-7087-2
ISBN PB   1-4020-7789-0
ISBN E-book   1-4020-7788-2

Contents

List of Figures

List of Tables

Preface

I thank Jeff Sklar and his former research team at the Brigham and Women's Hospital for introducing me to microarray technology in 1999. At that time, statistical methodology for analyzing microarray data was a new research field that needed much development and published reports in the literature were sparse. It took me several weeks at Frank Kuo's laboratory observing the procedures and details before I began to understand how gene expression is measured in microarray experiments. During the past few years, statistical models and methods for microarray data have been studied by many investigators. There is still, however, much room for improvement. Hence I thought it might be a useful contribution if I published a synthesis of what I have learned.

I thank David Beier, Mason Freeman, Cynthia Morton, and Rus Yukhananov for providing datasets for illustrations in the book. I thank Harry Björkbacka for contributing the chapter on microarray technologies and for providing insightful comments for the chapter on DNA, proteins, and gene expression. I thank Ming-Hui Chen, Frank Kuo, Weining Lu, and Pi-Wen Tsai for providing helpful comments on draft chapters of this book. I am especially grateful to Alex Whitmore for his many constructive comments, contributions, encouragement, and his tireless efforts in reading preliminary drafts. I thank Paul Guttry, Jaylyn Olivo, and Nancy Voynow of the Editorial Office at Brigham and Women's Hospital for proofreading the manuscript. Some errors might remain in the book, but their number would be greater without the help I have received. In order to write this book, I have worked in my office during every weekend and holidays for the past three years. I thank my family for their understanding. This project is supported in part by National Institutes of Health grants CA89756, HG02510 and HL72358.

Thanksgiving in Boston, 2003

This book is dedicated
to my parents.

I

GENOME PROBING
USING MICROARRAYS

Chapter 1

INTRODUCTION

After genome sequencing, microarray technology has emerged as a widely used platform for genomic studies in the life sciences. DNA microarrays have orderly arrangements of nucleic acid spots at high density. Many research studies have demonstrated the general usefulness of genome probing using microarrays. While genomics aims to give biologists an inventory of all genes used to assemble life forms, microarray technology provides high-throughput measurements in molecular biology, yields information for the reconstruction of complex gene control networks, and offers a panoramic view of the consequences of controlling gene transcriptions.

Microrray technology provides a systematic way to survey deoxyribonucleic acid (DNA) and ribonucleic acid (RNA) variation. DNA microarray chips have revolutionized genetic research in the same way that silicone chips revolutionized the computer industry and applications[1]. Microarray technologies allow the transcription levels of thousands of genes to be measured simultaneously. Understanding biological systems with thousands of genes will require organizing similar parts by their properties. Methods to group genes with similar expression patterns have proved useful in identifying genes that contribute to common functions or genes that are likely to be co-regulated[2]. The hypothesis that many human diseases may be accompanied by specific changes in gene expression has generated much interest in gene expression monitoring at the genome level using arrays. Microarray gene expression studies open up fresh avenues of cancer class discovery and class prediction[3]. Microarrays can be used in the determination of prognosis in histologically similar tumors with variable tendency to recur or spread and in the classification of tumors of uncertain histotype or tissue origin. Large-scale gene expression analysis can also increase the depth of diagnostic and drug-effect profiling. By comparing gene expression in normal and abnormal cells, microarrays can accelerate the discovery of key biological processes for therapeutic targeting. Gene-expression profiling using microarrays permits a simultaneous analysis of multiple markers[4]. Microarrays can be used to screen for polymorphisms within the population that may protect against or predispose to disease[5]. High-throughput microarrays can be used as a screen for early detection of disseminated tumor cells in peripheral blood[6]. DNA microarrays have been used to identify two classes of familial *BRCAx* breast cancers that differ in their expression of a large number of genes[7]. Arrays have also been used in comparing expression profiles of chronic lymphocytic leukemia specimens with and without immunoglobulin gene mutations[8], among many other applications.

While simultaneous measurement of thousands of gene expression levels provides a potential source of profound knowledge, success of the microarray technology depends on the precision of the measurements and on the integration of computational tools for data mining, visualization, and statistical modeling. With this technology, the expression of thousands of genes are measured in parallel, and the data obtained from image analysis are inherently noisy. In Chapter 4 we show that there are many sources of variation in microarray experiments. To obtain reliable gene expression data, experimental procedures need to be rigorously controlled to minimize noise and extraneous variation. Internal controls are needed on the arrays to allow for possible errors such as imperfect hybridization and repetitive sequences.

With the abundance of data produced from microarray studies, Lander (1996) pointed out that the greatest challenge is analytical. The impact of microarray technology on biology will depend heavily on data mining and statistical analysis. A sophisticated data-mining and analytical tool is needed to correlate all of the data obtained from the arrays, to group them in a meaningful way, and to perform statistical analysis in order to investigate hypotheses of interest. Experimental design and statistical methods provide powerful analytical tools to biologists for the study of living systems. Through statistical analysis and the graphical display of clustering and classification results, microarray experiments allow biologists to assimilate and explore the data in a natural and intuitive manner.

The challenge to statisticians is the nature of the microarray data. Instead of having large numbers of sample observations for a few variables, microarray data usually involve thousands of gene variables but few specimen samples. Microarray experiments raise numerous statistical and computational questions in diverse areas such as image processing, cluster analysis, machine learning, discriminant analysis, principal component analysis, multidimensional scaling, analysis of variance models, random effects models, multiplicative models, multiple testing, models with measurement errors, models to handle missing values, mixture models, Bayesian methods, and sample size and power determination.

The analysis of microarray data is a research field that is evolving. The contribution of this book is to provide general readers with an integrated presentation of various topics on analyzing microarray data. With the modest aim of providing an introduction to microarray technology and exploring different analytical methods that have been considered, the book is organized into four parts. We begin in Part I by providing the needed background knowledge about array technology. In order to familiarize readers with the necessary genetic terminology, basic genetic concepts are briefly reviewed in Chapter 2. In Chapter 3, we introduce different types of microarray platforms and review the basic procedures involved in microarray experiments. Array data measured from different platforms are discussed in Chapter 3. Important problems involving array data variation, background correction, and normalization are discussed in Chapters 4 to 6. The main focus of this book begins with Part II, where we provide a systematic presentation of statistical issues and methods for analyzing microarray gene expression data. We give a basic description of different types of experimental designs and discuss the advantages and disadvantages of some common designs used in microarray studies. The methods of analysis of variance for fixed effects and random effects models are presented in Chapter 10. Because of the large number of genes involved, the issue of multiple testing discussed in Chapter 11 should not be ignored in analyzing array data. The useful technique of permutation tests is considered in Chapter 12. Bayesian methods for analyzing array data are explored in Chapter 13. We present power and sample size considerations for microarray studies in Chapter 14. Related research topics not described in this book that require further investigation are also mentioned. Although clustering methods were used in analyzing array data in earlier literature, we present clustering methods after statistical methods, in part because clustering using model-based normalization might provide more accurate results. Various unsupervised learning methods are discussed in Part III. In Part IV we consider supervised learning methods.

Notes

[1] Lander, E.S. (1996). *Science*, **274**, 536-539.

[2] Spellman, P.T., Sherlock, G., Zhang, M.Q., Iyer, V.R., Anders, K., Eisen, M.B., Brown, P.O., Botstein, D. and Futcher, B. (1998). *Molecular Biology of the Cell*, **9**, 3273-329.

[3] Golub, T.R., Slonim, D.K., Tamayo, P., *et al.* (1999). *Science*, **286**, 531-537.

[4] Welsh, J.B., Zarrinkar, P.P., Sapinoso, L.M., *et al.* (2001). *Proceedings of the National Academy of Sciences*, USA, **98**, 1176-1181.

[5] Harcia, J.G. (1999). *Nature Genetics*, **21**, 42-47.

[6] Martin, K.J., Graner, E., Li, Y., Price, L.M., Kritzman, B.M., Fournier, M.V., Rhei, E., and Pardee, A.B. (2001). *Proceedings of the National Academy of Sciences, USA*, **98**, 2646-2651.

[7] Hedenfalk, I., Ringner, M., Ben-Dor, A., Yakhini, Z., Chen, Y., Chebil, G., Ach, R., Loman, N., Olsson, H., Meltzer, P., Borg, A., Trent, J. (2003). *Proceedings of the National Academy of Sciences, USA*, **100**, 2532-2537.

[8] Rosenwald, A., Alizadeh, A.A., Widhopf, G., *et al.* (2001) *Journal of Experimental Medicine*, **194**, 1639-1648.

Chapter 2

DNA, RNA, PROTEIN, AND GENE EXPRESSION

This chapter provides a basic understanding of the background of microarray experiments for readers who are not familiar with molecular biology. Further reading is encouraged, as an understanding of the underlying biology is required to make correct decisions in statistical analysis. For a more detailed description of genes and genetic analysis, see Lewin[1] (2000), Cooper[2] (2000), Griffiths et al.[3] (1999), Lehninger et al.[4] (2000), Griffiths et al.[5] (2000), Alberts et al.[6] (2002), and Dale and Schantz[7] (2002). Readers with a good knowledge of biology are advised to skip directly to Chapter 3 on Microarray Technologies or Chapter 4 on Inherent Variability in Microarray Data.

2.1. The Molecules of Life

A *cell* is the minimal unit of life. There are a multitude of specific chemical transformations that not only provide the energy needed by a cell, but also coordinate all of the events and activities within that cell. The life process involves a wide array of molecules ranging from water to small organic compounds (e.g., fatty acids and sugars), and macromolecules (DNA, proteins, and polysaccharides) that define the structure of the cells. Macromolecules control and govern most of the activities of life. *Deoxyribonucleic acid* (DNA) molecules store information about the structure of macromolecules, allowing them to be made precisely according to cells' specifications and needs.

DNA is a very stable molecule that forms the "blueprint" of an organism. The DNA structure encodes information as a sequence of chemically linked molecules that can be read by the cellular machinery and guides the construction of the linear arrangements of protein building blocks, which eventually fold to form functional proteins. Molecular biology deals with how information is stored and converted to all the components and interactions that make up a living organism. Each cell contains a complete copy of its genetic material in the form of DNA molecules. The DNA can be copied and passed on to the cell's progeny through a mechanism called *replication*. The genetic information can also be copied as a transportable "working copy" composed of *ribonucleic acid* (RNA) molecules, which are closely related to DNA. This process is called *transcription*. The RNA is transferred to a machinery that synthesizes *protein* molecules based on the information carried by the RNA. This process is called *translation*. The process sequence is illustrated by the following chart.

$$\text{DNA} \xrightarrow{\text{transcription}} \text{RNA} \xrightarrow{\text{translation}} \text{protein}$$

What has just been described is the *central dogma* of molecular biology that formulates how information is stored and converted to all the components and interactions that build up a living organism.

Proteins are the most functionally versatile of the life molecules. Being the "work horses" or "machines" of a cell, proteins catalyze an extraordinarily wide variety of chemical reactions and also serve as the building blocks of cellular structures. They are the building blocks of muscles, skin, and hair, as well as the enzymes that catalyze and control all chemical reactions in an organism, ranging from food digestion to nerve impulses and the components that are responsible for DNA replication, transcription, and translation.

In the following sections we will discuss the building blocks and the higher-order structure of the macromolecules of life.

2.2.   Genes

Genes are the units of the DNA sequence that control the identifiable hereditary traits of an organism. A *gene* can be defined as a segment of DNA that specifies a functional RNA. The total set of genes carried by an individual or a cell is called its *genome*. The genome defines the genetic construction of an organism or cell, or the *genotype*. The *phenotype*, on the other hand, is the total set of characteristics displayed by an organism under a particular set of environmental factors. The outward appearance of an organism (phenotype) may or may not directly reflect the genes that are present (genotype). Today the complete genome sequences of several species are known, including several bacteria, yeasts, and humans. With microarray technology we can study the expression of all the genes in an organism simultaneously. Such genome-wide studies will help to uncover and decipher cellular processes from a completely new perspective.

2.3.   DNA

Except for some viruses, the genetic material of all known organisms consists of one or more long molecules of deoxyribonucleic acid (DNA). The chemical components of the DNA molecule dictate the inherent properties of a species. DNA is made up of chains of chemical building blocks called *nucleotides.* Each nucleotide consists of a phosphate group, a deoxyribose sugar molecule, and one of four different nitrogenous bases usually referred to by their initial letters: *guanine* (G), *cytosine* (C), *adenine* (A), or *thymine* (T). Genetic information is encoded in DNA by the sequence of these nucleotides. The information stored in the sequence of nucleotides in terms of the four nitrogenous bases is analogous to a long word in a four-letter alphabet.

The carbons in the deoxyribose sugar group of a nucleotide are assigned numbers followed by a prime symbol $(1', 2',$ etc.$)$. In DNA, the nucleotides are connected to each other via a link of the $5'$ hydroxyl phosphate group of one pentose ring of the deoxyribose sugar to the $3'$ OH group of the next pentose ring. The chemical connections between the repeating sugar and phosphate groups are called *phosphodiester bonds.* With one $5'$ end and the other $3'$ end, each chain is said to have polarity. It is conventional to write nucleic acid sequences in the $5' \rightarrow 3'$ direction. DNA forms a double helix of two intertwined chains (strands) of nucleotides. The two polynucleotide chains run in opposite directions; that is, one strand runs in the $5' \rightarrow 3'$ direction, while the other strand runs in the $3' \rightarrow 5'$ direction.

It was proposed in the now classic manuscript by Watson and Crick[8] in 1953 that the two nucleotide chains are held together by *hydrogen bonds* that form between the nitrogenous bases. The polarity of the double helix requires specific hydrogen bonding between the bases so that they fit together. Guanine preferentially hydrogen-bonds with cytosine, and adenine can bond preferentially with thymine. That is, G pairs only with C, and A pairs only with T. These matching base pairs are referred to as *complementary.* For example, a short segment with ten nucleotides might be of the form

$$5' - ATGCCCTGAC - 3'$$
$$3' - TACGGGACTG - 5'$$

The specific base pairing of DNA is the mechanism by which encoded information can be transferred from generation to generation with very little alteration. When the DNA is copied by the processes of replication and transcription, the double helical structure of the DNA is opened up and a copy is made based on the specificity of the base pairing. Nucleotides are rejected if they do not have the correct base pairing (i.e., are unmatched). Even though these copying mechanisms proceed with high fidelity, mistakes are sometimes made through the insertion of a non-matching nucleotide, thus creating a *mutation*. Random mutation is one of the foundations for evolution, since it can introduce variations in the genetic code over time.

The genome of an organism is made up of one or more long molecules of DNA that are organized into *chromosomes*. A chromosome consists of an uninterrupted length of double-stranded DNA that contains many genes. For most bacteria and fungi, the cells contain only one copy of the genetic material, and these organisms are called *haploid*. In higher organisms, two copies of each chromosome and its component genes are present; these organisms are called *diploid*. Human cells contain two sets of 23 chromosomes, for a total of 46. During reproduction 23 chromosomes from the father and 23 from the mother are combined and mixed to make a new set of 46 chromosomes in the progeny. The chromosome pairs in the diploid cells may not be identical and may contain variants of the same gene. The variations have been created by random *mutation* over time. A mutation consists of a change in the sequence of base pairs (bp) in DNA. A mutation in a coding sequence may change the sequence of amino acids in the protein. Gene variants like this that occupy the same position or *locus* on a chromosome are called *alleles*. A gene may have multiple alleles. This mixing of gene variants explains how traits like eye color can be inherited from either the mother or the father. Two chromosomes with the same gene components are said to be *homologous*.

The unit of replication is the chromosome. When a cell divides, all the chromosomes are replicated. When a chromosome is replicated, all its genes are replicated. In addition to containing information about a protein component or function, genes also contain regulatory elements that determine when and where the gene in question needs to be transcribed and translated.

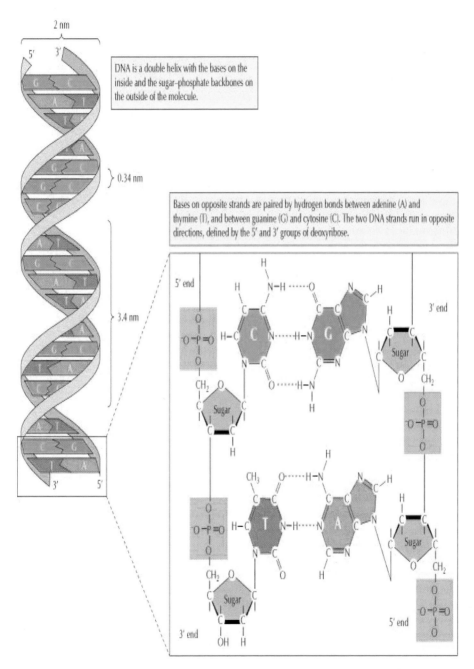

Figure 2.1. The Double-helix Structure of DNA. Source: The Cell - A Molecular Approach, by Cooper, G.M., Copyright 2000, Sinauer Associates Inc., Sunderland, Massachusetts. Reprinted with permission from Sinauer Associates Inc.

2.4.   RNA

As was introduced in section 2.1, the core biochemical flow of genetic information can be summarized as the process of RNA synthesis (transcription) and the process of protein synthesis (translation). The first step in making a protein is to copy, or transcribe, the information encoded in the DNA of the genes into a single-stranded molecule called ribonucleic acid (RNA). Since this process is similar to the process of copying written words, the synthesis of RNA from DNA is called *transcription*. The DNA is said to be transcribed into RNA, and the RNA is called a *transcript*. The nucleotides of RNA contain the sugar ribose, while the nucleotides of DNA contain deoxyribose that has one more oxygen. Furthermore, instead of thymine, RNA contains *uracil* (*U*), a base that has hydrogen-bonding properties identical to those of thymine. Hence the RNA bases are *G*, *C*, *A*, and *U*. RNA is less stable than DNA. RNA synthesis requires the RNA polymerase enzyme complex that binds to a specific sequence at one end of a gene (the *promoter*) and separates the two strands of DNA. It moves along the gene, maintaining the separated strand "bubble", and uses only *one* of the separated strands as a template, synthesizing an ever-growing tail of polymerized nucleotides that eventually becomes the full-length transcript. Hence, RNA is a single-stranded nucleotide chain, not a double helix. Since RNA is always synthesized in the $5' \to 3'$ direction, the addition of ribonucleotides by RNA polymerase is at the $3'$ end of the growing chain.

There are two general classes of RNAs. Those that take part in the process of decoding genes into proteins are referred to as "*informational* RNAs" called *messenger* RNA (mRNA). In the other class, the RNA itself is the final functional product. These RNAs are referred to as "*functional* RNAs". Functional RNAs are the *transfer* RNAs (tRNA) and the *ribosomal* RNA (rRNA), which are both part of the intricate protein synthesis machinery that translates the informational mRNA into protein.

Figure 2.2 shows that the sequence of messenger RNA is complementary to the sequence of the bottom strand of DNA and is identical to the top strand of DNA, except for the replacement of *T* with *U*. A messenger RNA includes a sequence of nucleotides that corresponds to the sequence of amino acids in the protein. This part of the nucleic acid is called the *coding region*. Because mRNA is an exact copy of the DNA coding regions, mRNA analysis can be used to identify *polymorphisms* in coding regions of DNA. A polymorphism is a DNA region for which nucleotide sequence variants exist in a population of organisms. Such variations can sometimes explain the occurrence of a disease or enzyme deficiency within a population. Hence, a considerable effort has been put into trying to identify such variations. Microarray technology can be used both in the identification of polymorphisms and in the diagnosis of polymorphism-related disease. Organisms whose cells have a membrane-bound nucleus are called *eukaryotes*. For example, animals and plants are eukaryotes. In eukaryotic cells, the initial pre-mRNA transcription product can be many times longer than needed for translation into protein. At the 5′ end of a eukaryotic gene, there is a regulatory region to which various proteins bind, causing the gene to be transcribed at the right time and in the right amount. A region at the 3′ end of the gene contains a sequence encoding the termination of transcription. In the genes of many eukaryotes, the protein-encoding sequence is interrupted by varying numbers of segments called *introns*. The coding sequence segments interrupted by the introns are called *exons*. Introns are removed in the *splicing process* to generate the final mature mRNA ready to be translated by the protein synthesis machinery.

An example of DNA with two base-paired strands

5' ATGCGGACCTGACATGCCGTTAGAGAC 3'

3' TACGCCTGGACTGTACGGCAATCTCTG 5'

RNA is synthesized from one strand of DNA
In the 5' to 3' direction

5' AUGCGGACCUGACAUGCCGUUAGAGAC 3'

Figure 2.2. RNA is synthesized by using the top strand of DNA in the 5′ to 3′ direction as a template for complementary base pairing

2.5. The Genetic Code

The sequence of nucleotides in DNA is important not because of its structure, but because it *codes* for the sequence of amino acids that dictate the structure of a protein with a defined function, be it structural or catalytic. The relationship between a sequence of DNA and the sequence of the corresponding protein is called the *genetic code*. The genetic code is read in groups of three nucleotides, or *codons*, each of which represents one amino acid. Because each position in the three-nucleotide codon could be one of the four bases A, C, G, and T, there are a total of $4 \times 4 \times 4 = 64$ possible different codons, each representing an amino acid or a signal to terminate translation. As there are only 20 common amino acids, several different codons can code for the same amino acid (the genetic code is said to be degenerate due to this many-to-one relationship). Since the genetic code is read in non-overlapping triplets, there are three possible ways of translating any nucleotide sequence into a protein, depending on the starting point. These are called *reading frames*. A reading frame that starts with a special initiation codon (AUG-methionine) and extends through a series of codons representing amino acids until it ends at one of three termination codons (UAA, UAG, UGA) can potentially be translated into a protein and is called an *open reading frame* (ORF). A long open reading frame is unlikely to exist by chance. The identification of a lengthy open reading frame is strong evidence that the sequence is translated into protein in that frame. An open reading frame for which no protein product has been identified is sometimes called an *unidentified reading frame* (URF).

2.6.   Proteins

The primary structure of a *protein* is a linear chain of building blocks called *amino acids*. There are 20 amino acids that commonly occur in proteins. These amino acids are linked together by covalent bonds called *peptide bonds*. A peptide bond is formed through a condensation reaction during which one water molecule is removed. Because of the manner in which the peptide bond forms, a polypeptide chain always has an amino (NH_2) end and a carboxyl (COOH) end. This primary chain is coiled and folded to form a functional protein. Proteins are the most important determinants of the properties of the cells and organisms. The biological role of most genes is to encode, or carry, information for the composition of proteins. This composition, together with the timing and amount of each protein produced, determines the structure and physiology of an organism, i.e., the phenotype.

Because the process of reading the mRNA sequence and converting it into an amino acid sequence is like converting one language into another, the process of protein synthesis is called *translation*. The four-letter alphabet of the genes is translated into the 20-amino-acid alphabet of proteins in ribosomes. Ribosomes are big complexes of several proteins and *ribosomal* RNA (rRNA). The rRNA functions to guide mRNA into a correct starting position by binding to special sequences present in the beginning of all mRNAs. The translation of the genetic code into a protein is achieved with the help of *transfer* RNAs (tRNAs). The tRNAs contain a trinucleotide sequence complementary to the codon called the *anticodon*. Each species of tRNA molecules is charged with a specific amino acid in an enzymatic reaction, hence coupling a certain amino acid to a certain anticodon nucleotide triplet on the tRNA molecule. In essence, the translation of the DNA code to protein amino acids is done in this enzymatic coupling step. The ribosome subsequently aligns the mRNA codon with the matching tRNA anticodon, and if the base pairing matches, the amino acid carried by the tRNA is attached to the growing chain of amino acids to form a polypeptide chain. Hence, the specific base pairing of the nucleotides once again ensures that the correct information is transferred. When the ribosome reaches a stop codon, it releases the polypeptide chain, which then folds into the defined three-dimensional structure of a protein. Proteins must often undergo post-translational modifications to become active. These modifications can, for instance, be cleavages of the polypeptide chain at predefined sites or binding of additional molecules like lipids, sugars, or co-factors that assist in catalysis of chemical reactions.

2.7.   Gene Expression and Microarrays

Gene expression is the process by which mRNA, and eventually protein, is synthesized from the DNA template of each gene. The first stage of this process is *transcription*, when an RNA copy of one strand of the DNA is produced. In eukaryotes it is followed by RNA *splicing*, during which the introns are cut out of the primary transcript and a mature mRNA is made. As part of the maturation process, a tail of adenine nucleotides is added to the 3' end of the mRNA. This poly *A* tail can vary greatly in length and is believed to stabilize the mRNA molecule. Transcription and splicing of RNA occur in the nucleus. The next stage of gene expression is the *translation* of the mRNA into protein. This occurs in the cytoplasm. In the process of gene expression, RNA provides not only the essential substrate (mRNA) but also components of the protein synthesis apparatus (tRNA, rRNA).

Some protein-encoding genes are transcribed more or less constantly; they are sometimes called *housekeeping genes* and are always needed for basic reactions. Other genes may be rendered unreadable or, to suit the functions of the organism, readable only at particular moments and under particular external conditions. The signal that masks or unmasks a gene may come from outside the cell; for example, from a nutrient or a hormone. Special *regulatory sequences* in the DNA dictate whether a gene will respond to the signals, and they in turn affect the transcription of the protein-encoding gene. Understanding which genes are expressed under which condition gives invaluable information about the biological processes in the cell. The power of microarray technology lies in its ability to measure the expression of thousands of genes simultaneously.

2.8.   Complementary DNA (cDNA)

Complementary DNA (cDNA) is used in recombinant DNA technology. cDNA is complementary to a given mRNA and is usually made by the enzyme *reverse transcriptase,* first discovered in retroviruses. *Reverse transcription* allows a mature mRNA to be retrieved as cDNA without the interruption of non-coding introns. The coexistence of mRNA and cDNA establishes the general principle that information in the form of either type of nucleic acid sequence can be converted into the other type. In microarray technology the process of reverse transcription is frequently used to incorporate fluorescent dyes into cDNA complementary to the mRNA transcripts.

$$RNA \xrightarrow{\text{reverse transcription}} cDNA$$

2.9.   Nucleic Acid Hybridization

The specific base pairing of nucleic acids is the foundation of microarray technology. The specific pairing of an artificial DNA sequence probe with its biological counterpart allows for exact identification of the sought-after unique sequence or gene.

Because of the base-pairing arrangments, the two strands of DNA can separate and re-form very quickly under physiological conditions that disrupt the hydrogen bonds between the bases but are much too mild to pose any threat to the covalent bonds in the backbone of the DNA. The process of strand separation is called *denaturation* or melting. Because of the complementarity of the base pairs, the two separated complementary strands can be re-formed into a double helix (the two strands are then said to be annealed). This process is called *renaturation.* The technique of renaturation can be extended to allow any two complementary nucleic acid sequences to anneal with each other to form a duplex structure.

Hybridization is the biochemical method on which DNA microarray technology is based. Nucleic acid sequences can be compared in terms of complementarity that is determined by the rules for base pairing. In a perfect duplex of DNA, the strands are precisely complementary. It is possible to measure complementarity because the denaturation of DNA is reversible under appropriate conditions. Detecting and identifying nucleic acid (DNA, mRNA) with a labeled cDNA probe that is complementary to it is an application of nucleic acid hybridization. DNA microarrays utilize hybridization reactions between single-stranded fluorescent dye-labeled nucleic acids to be interrogated and single-stranded sequences immobilized on the chip surface. The next chapter will discuss the microarray technology in detail.

Notes

[1] Lewin, B. (2000). *Genes, VII*, Oxford University Press, New York.

[2] Cooper, G.M. (2000). *The Cell - A Molecular Approach*, 2nd ed., Sinauer Associates Inc, Sunderland, Massachusetts.

[3] Griffiths, A.J.F., Gelbart, W.M., Miller, J.H., and Lewontin, R.C.. (1999). *Modern Genetic Analysis*, Freeman, New York.

[4] Lehninger, A.L., Nelson, D.L., and Cox, M.M. (2000) *Principles of Biochemistry*, Worth Publishing, 3rd ed.

[5] Griffiths, A.J.F., Miller, J.H., Suzuki, D.T., Lewontin, R.C., and Gelbart, W.M. (2000). *An Introduction for Genetic Analysis*, Freeman, New York.

[6] Alberts, B., Johnson, A., Lewis, J., Raff, M., Roberts, K., Walter, P. (2002). *Molecular Biology of the Cell*, Garland Publishing, 4th ed.

[7] Dale, J.W. and von Schantz, M. (2002). *From Genes to Genomes: Concepts and Applications of DNA Technology*. John Wiley and Sons, Ltd, England.

[8] Watson, J.D., and Crick, F.H.C. (1953). *Nature*, **171**, 737-738.

Chapter 3

MICROARRAY TECHNOLOGY

This chapter is contributed by Harry Björkbacka, Ph.D.
Lipid Metabolism Unit, Massachusetts General Hospital, Boston

Microarray technology allows measurement of the levels of thousands of different RNA molecules at a given point in the life of an organism, tissue, or cell. Comparisons of the levels of RNA molecules can be used to decipher the thousands of processes going on simultaneously in living organisms. Also, comparing healthy and diseased cells can yield vital information on the causes of diseases. Microarrays have been successfully applied to several biological problems and, as arrays become more easily available to researchers, the popularity of these kinds of experiments will increase. The demand for good statistical analysis regimens and tools tailored for microarray data analysis will increase as the popularity of microarrays grows. The future will likely bring many new microarray applications, each with its own demands for specialized statistical analysis.

In order to analyze any experimental data correctly, it is fundamental to understand the experiments that generated the data. Microarray experiments contain many steps, each with its individual noise and variation. The final result may be affected by any of the steps in the process. Good experimental design and careful statistical analysis are required for successful interpretation of microarray data. This chapter will review the most commonly used microarray technology platforms, pointing out their strengths and weaknesses. More in-depth descriptions of some specific microarray protocols can be found in pioneering research articles by Shalon et al.[1], Lockhart et al.[2], Lander[3], Lipschutz et al.[4], Brown and Botstein[5], Eisen and Brown[6], Southern et al.[7], Bowtell[8], Cheung et al.[9], and books by Schena[10], Baldi and Hatfield[11], Bowtell and Sambrook[12], and Grigorenko[13], among others.

3.1. Transcriptional Profiling

With today's advances in microarray technology, large sets of gene expression data can be created. Such catalogues are called *gene expression* or *transcriptional profiles*, and the process of gathering the data is called profiling. Transcriptional profiling can be either sequencing- or hybridization-based.

3.1.1. Sequencing-based Transcriptional Profiling

Sequencing-based approaches include sequencing of complementary DNA (cDNA) libraries and serial analysis of gene expression (SAGE). Libraries of cDNA are created by reverse transcription of mRNAs expressed in a tissue or a cell type under some treatment or condition. Individual cDNA clones are created by recombinant DNA technology. Sequencing the cDNA reveals the identity of the clone either as a known sequence in a public database or as a novel unknown sequence. The number of clones with the same unique sequence is related to the expression levels in the mRNA pool used to create the library.

The basic idea of SAGE is to generate short cDNA "sequence tags" from a pool of mRNA, combining them and sequencing several "tags" at a time. A *sequence tag* is a stretch of 10-14 *base pairs* (bp) that contains sufficient information to uniquely identify a transcript, provided that the tag is obtained from a unique position within each transcript. The tags are created by special restriction enzymes that cut DNA at specific sequences. First, the cDNAs are cut with an enzyme at a short recognition sequence, and then with another enzyme that cuts about 20 base pairs away from the same recognition sequence. The tags are also created such that they can be cloned and sequenced easily. The number of times a particular tag is observed in the sequencing data indicates the expression level of the corresponding transcript. An overview of SAGE is given in Figure 3.1.

Biotin-labeled double-stranded cDNA is cleaved with a restriction endonuclease (anchoring enzyme). The 3′ ends of the resulting cDNA fragments are then purified using streptavidin-coated magnetic beads, and the resulting cDNA fragments are divided into two populations, each of which is ligated to a different linker (1 or 2 in the figure) containing a type-IIS restriction endonuclease (tagging enzyme) recognition sequence.

Such enzymes cleave DNA up to 20 bp away from their recognition site. Digestion of the two cDNA populations thus results in the generation of a short sequence consisting of the linker and a short portion of its adjacent cDNA. Following the creation of blunt ends, the two populations are ligated to each other and total cDNA is amplified by polymerase chain reaction (PCR), resulting in the generation of products with two tags (a ditag) orientated tail-to-tail, with an anchoring enzyme recognition site at either end. Following cleavage at each anchoring enzyme recognition sequence and concatenization of ditags via this site, products are cloned and individual clones consisting of at least 25 tags (25-75) are selected for sequencing.

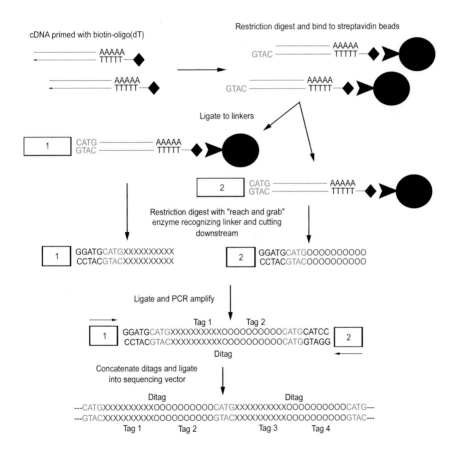

Figure 3.1. Sequencing-based transcriptional profiling and serial analysis of gene expression (SAGE)

3.1.2.   Hybridization-based Transcriptional Profiling

Microarray technology has evolved from Ed Southern's insight that labeled nucleic acid molecules could be used to identify nucleic acid molecules attached to a solid support. Hybridization methods, such as Southern and Northern blots, colony hybridizations, and dot blots, have long been used to identify and quantify nucleic acids in biological samples. These methods traditionally attempt to identify and measure only one gene or transcript at a time.

Hybridization methods have evolved from these early membrane-based, radioactive detection embodiments to highly parallel quantitative methods using fluorescence detection. Some key innovations have made it possible to develop techniques that analyze hundreds or thousands of hybridizations in parallel. The first was the use of non-porous solid supports, such as nylon filters or glass slides, which facilitate miniaturization. The second was the development of methods for spatial synthesis and robotic spotting of oligonucleotides and cDNAs on a very small scale. These methods have made it possible to generate arrays with very high densities of DNA, allowing tens of thousands of genes to be represented in areas smaller than standard glass microscope slides. In fact, today it is technically possible to generate arrays of probes representing all the genes of a genome on a single slide. Finally, improvements in fluorescent labeling of nucleic acids, fluorescent-based detection, and image processing have improved the accuracy of microarrays.

Before describing the process of generating and using microarrays in more detail, a clarification of the nomenclature is needed. At least two nomenclature systems currently exist in the literature for referring to DNA hybridization partners. There is no general consensus on the usage of the terms *probe* and *target*, and researchers have used these two terms interchangeably in a number of publications. With respect to the nucleic acids whose entwining represents the hybridization reaction, the identity of one is defined as it is tethered to the solid phase, making up the microarray itself. The identity of the other is revealed by hybridization. *Nature Genetics*[14] and Duggan *et al.*[15] adopted the nomenclature that the tethered nucleic acids spotted on the array are the probes, and the fluor-tagged cDNAs from a complex mRNA mixture extracted from cells are the targets. This book will follow this nomenclature.

3.2.   Microarray Technological Platforms

Microarrays allow large numbers of DNA clones with known sequences to be immobilized as an array of detection units (*probes*), while the pool of RNAs to be examined (*targets*) is fluorescently labeled and then hybridized to the detectors. There are three main microarray technological platforms, namely spotted cDNA arrays, spotted oligonucleotide arrays, and *in-situ* oligonucleotide arrays (e.g., Affymetrix GeneChip Arrays). The differences between these three platforms lie in the way the arrays are produced and the types of probes used (see Figure 3.2).

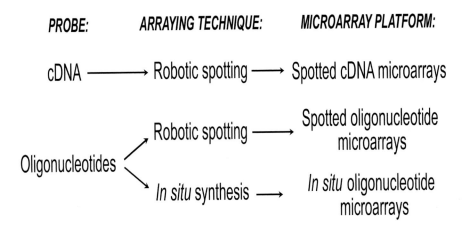

Figure 3.2.   The three common microarray platforms are distinguished by the probe type and arraying technique used in manufacturing.

In spotted cDNA arrays, Figure 3.3, full-length cDNA clones or *expressed sequence tag* (EST) libraries are robotically spotted and immobilized on the support. Many laboratories already have cDNA libraries, so generation of these arrays requires only investment in the robotic equipment to spot, or array, the cDNA. Spotted cDNA arrays have an advantage over other types of arrays in that unknown sequences can be spotted. Thus, for organisms for which no or only limited genome sequence information is available, spotted cDNA microarrays are the only choice for genome-wide transcriptional profiling.

Spotted oligonucleotide arrays, Figure 3.4, are very similar to spotted cDNA arrays, except that synthetic oligonucleotides (abbreviated oligos) instead of cDNA are used as probes. In fact, the same robotics can be used to manufacture both types of arrays. When sequence information is available, oligonucleotide probes of 20-70 nucleotides can be designed and synthesized. Using designed oligonucleotide probes gives better control over what part of the gene will be utilized for hybridization. The oligonucleotides can, for instance, be designed to optimally differentiate between highly similar transcripts that might cross-hybridize on a cDNA array.

In-situ oligonucleotide arrays, Figure 3.5, were developed by Fodor *et al.*[16] and Affymetrix, Inc. *In-situ* oligonucleotide arrays use a combination of photolithography and solid-phase oligonucleotide chemistry to synthesize short oligonucleotide probes (25-mer oligos) directly on the solid support surface. The number of oligonucleotides (50,000 probes per 1.28 square centimeters) on a chip manufactured by this method vastly exceeds what can be achieved by spotting solution robotically. Affymetrix Inc. has chosen to utilize this advantage to construct an array with several oligonucleotide probes and cross-hybridization controls for each target gene. However, the researcher has little, if any, control over what probes are used on pre-manufactured arrays like the Affymetrix GeneChip arrays. On the other hand, comparison of results between different laboratories is facilitated by the use of products from a common manufacturer.

For *in-situ* oligonucleotide arrays, the test and reference samples (or the treatment and control samples) are hybridized separately on different chips. In contrast, for either spotted cDNA arrays or spotted oligonucleotide arrays, a test and a reference sample labeled with two different fluorescent dyes are commonly simultaneously hybridized on the same arrays. This difference affects how microarray data generated with single-color or two-color arrays are analyzed (see section 3.8).

3.3. Probe Selection and Synthesis

For large-scale gene expression studies, the first step is to select and prepare the specific hybridization detectors (probes).

cDNA microarray technology provides great flexibility in the choice and production of ordered arrays, since the probes can be created from cDNA libraries. A *cDNA library* is a set of plasmid vectors with inserted mRNA segments turned into cDNA by reverse transcription, usually harbored in bacterial clones. *Plasmids* are independently replicating small extrachromosomal DNA molecules. The probe to be used on a cDNA array can be amplified from such a cDNA library by PCR (Figure 3.3). PCR allows amplification of a specific segment of template DNA between two sequence-specific hybridized oligonucleotide primers. Many laboratories have already created cDNA libraries for a wide variety of organisms, cells, and tissues. For organisms whose genomes have been fully sequenced, one can *amplify* every known and predicted open reading frame (ORF) in the genome using reverse transcription PCR (RT-PCR) and sequence-specific primers. In organisms with smaller genomes and infrequent introns, such as yeast and prokaryotic microbes, purified total genomic DNA serves as a template, and sequence-specific oligonucleotides are used as primers.

When dealing with large genomes and genes with frequent introns, such as those of the human and mouse, cloned expressed sequence tags (EST), individual full-length cDNA clones, or collections of partially sequenced cDNAs corresponding to each of these transcripts can be used as the source of gene-specific detector probes in an array. Many methods are available for recovering purified cDNA from the PCR amplification reaction. A simple method is to prepare purified template cDNAs from the bacterial colonies that harbor them and follow-up with ethanol precipitation, gel filtration, or both, to prepare relatively pure cDNA for printing. The choice of template source and PCR strategy vary with the organism being studied.

Synthesized oligonucleotides can also be used as probes in spotted microarrays (Figure 3.4). Genes of interest are chosen from public sequence databases including GeneBank, dbEST, and UniGene. Many variables have to be considered in selecting the sequence of the oligonucleotide to be made. First, the length of the oligonucleotide has to be chosen. The longer the oligonucleotide, the more specific it will be. However, longer oligonucleotides are more costly and more difficult to make. Today several commercial oligonucleotide sets are available for mouse, human, and other organisms, varying in probe length between 30 and 70 nucleotides. Second, the probes must be selected so that they are specific for their target genes. If similarities exist between probes on the same microarray, they can cross-hybridize to more than one gene target, making the results hard to analyze. Third, all oligonucleotides must have similar hybridization properties. Usually all the probes are designed so that their melting temperature is within 1-2 degrees Celsius and that they have a similar content of G and C nucleotide base pairs. Several probe selection algorithms have been developed, but so far no consensus exists on the most effective design principles. For instance, there are several algorithms just for calculating oligonucleotide melting temperatures. Other considerations that may go into designing the probes are self-hybridization properties (palindromic sequences) and synthesis efficiency of certain sequences. The location of the probe along the message may also be important.

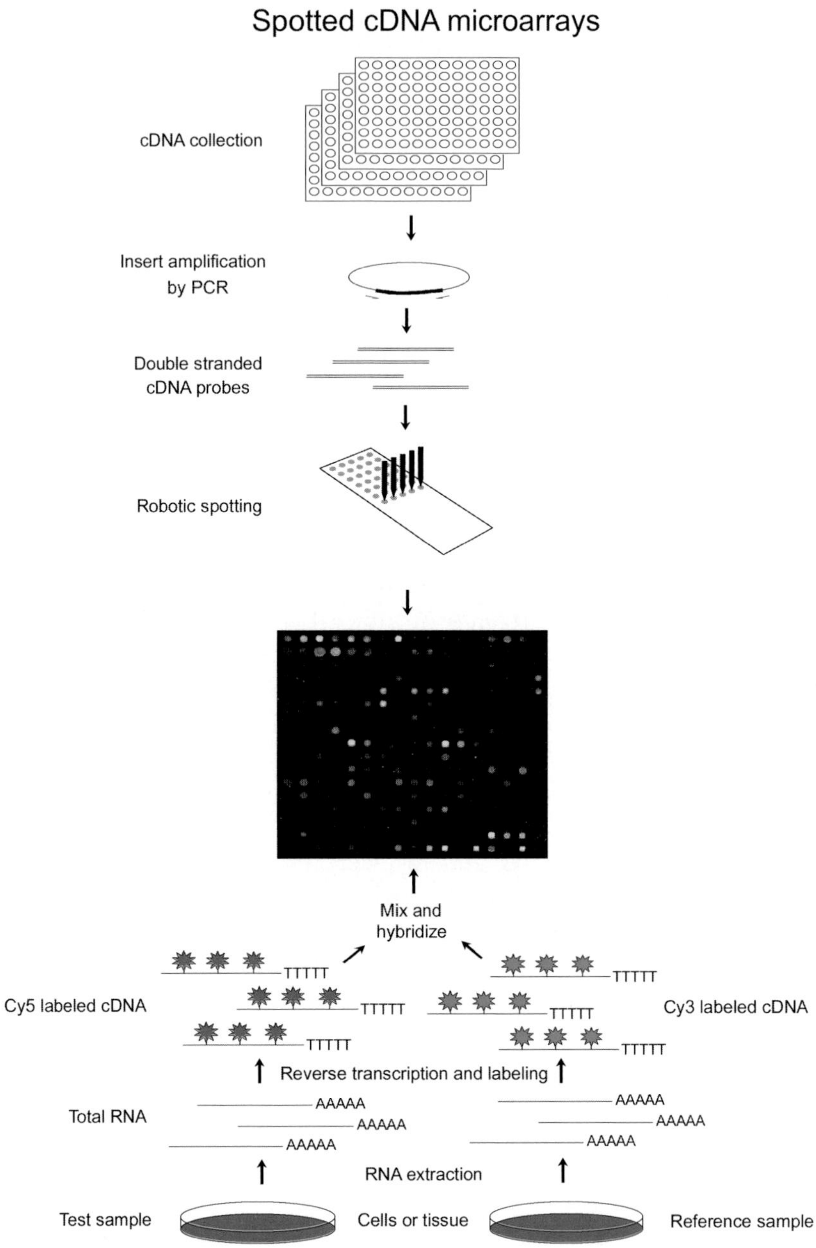

Figure 3.3.   Spotted cDNA microarrays

Spotted oligonucleotide microarrays

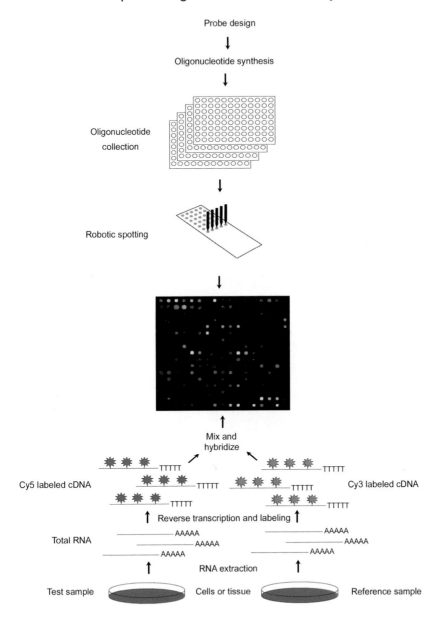

Figure 3.4. Spotted oligonucleotide microarrays

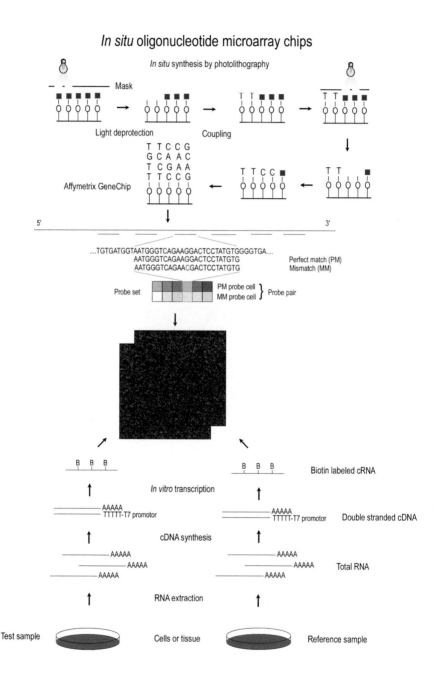

Figure 3.5. In-situ oligonucleotide microarrays

The probe can be selected to hybridize to the 3′-*untranslated region (UTR)*, 3′ or 5′ coding sequences. Another matter to consider is the complementarity of the probes. Before designing oligonucleotide probes for spotted arrays, a labeling strategy must be chosen to assure that probes and targets will hybridize. The mRNA is the sense strand of the message, and when converting the mRNA to single-stranded labeled cDNA, the cDNA will be antisense and then only hybridize specifically to sense strand oligonucleotide probes (see section 3.5 on Target Labeling). The probes in spotted cDNA arrays are double-stranded and can hybridize either to a labeled anti-sense or to a sense strand target.

The probes for *in-situ* oligonucleotide arrays are synthesized directly on the surface of the support. The probes are 20 nucleotides long and are organized as perfect-match versus mismatch pairs, with the mismatch probe acting as a control for hybridization specificity. *Perfect-match* (PM) probes are designed to be complementary to the target sequence. *Mismatch* (MM) probes are designed to be complementary to the target sequence except for one base mismatch at the central position that has equal distance from either ends of the probe. Mismatch probes serve as controls for cross-hybridization. A *probe cell* is a single square feature on an array containing a PM or MM probe. The size of the feature can vary depending on the array type. Each probe cell contains millions of probe molecules. A *probe pair* consist of two probe cells: a PM and its corresponding MM. On the array, a probe pair is arranged with the PM cell directly adjacent to the MM cell. When the probes are designed, several probe pairs are selected to represent each transcript. A *probe set* designed to detect one transcript usually consists of 16-20 probe pairs.

DNA chips may contain 100,000 different oligos in a 4-cm^2 area, and each probe cell has approximately 10^7 oligo molecules. By designing oligos that span an entire exon using a register of one nucleotide change between adjacent probe cells and a window of 25 nucleotides at a time, it is possible to utilize DNA chip technology for DNA resequencing. In this resequencing strategy, the midpoint nucleotide (number 13 in a 25-mer) is synthesized as a *G, C, A, or T*. Using PCR products and hybridization conditions that discriminate between perfect and single base-pair mismatch duplexes, it is possible to read the sequence across the target DNA based on the most intense signal in each set of four oligo probes. Affymetrix is also producing *in-situ* oligonucleotide arrays that contain probes corresponding to every known *single nucleotide polymorphism* (SNP) in the human genome[17]. A SNP is a single base-pair site within the genome at which more than one of the four possible base pairs is commonly found in natural populations. SNPs can be inherited markers of a disease.

3.4.   Array Manufacturing

The *in-situ* oligonucleotide arrays developed by Affymetrix, Inc. are made using *photolithography*, in which light is passed through holes in a mask (Figures 3.5). The light deprotects chemically reactive groups in locations on the support specified by the holes in the mask. A nucleotide can subsequently be coupled to the activated group. By varying the location of the holes in the mask and the nucleotides coupled in each step, oligonucleotide probes can be synthesized *in situ* at a very high density.

In spotted cDNA arrays and spotted oligonucleotide arrays, the probe is deposited as a solution on the surface of the support and then attached (Figures 3.3 and 3.4). Several different choices of support materials are available, ranging from plastic polymers to glass. The surface of the popular glass microscope slides are usually coated with chemicals to reduce the background fluorescence and nonspecific binding of the labeled target. This can be done by the manufacturer or in-house before arraying the probes. Surface coatings can be relatively simple poly-lysine coatings or more complex 3-D molecular matrix layers.

There are two basic methods for spotting or arraying cDNA and oligonucleotides onto the support. Contact spotting relies on pins to pick up solution by capillary action and deposit drops upon contact. The first cDNA microarrays were created in this fashion in Patrick Brown's laboratory at Stanford University. Non-contact spotting is done by variations of the inkjet technology, in which small drops of solution can be sprayed with high precision onto a surface. Inkjet printers have been used to spot both cDNAs and oligos (Packard Piezoelectric dispensing system). They also have been used to spot free nucleotides to synthesize oligonucleotides *in situ* on the solid support (Agilent Technologies SurePrint Inkjet technology). After spotting, the probes have to be attached covalently to the surface of the support. The method of choice for cDNAs is ultra-violet (UV) light cross-linking, which forms attachments at random sites along the probe molecule. Oligonucleotides may also be UV cross-linked, but also more specifically attached by a linker molecule. Adding a linker molecule to the end of the oligonucleotide in the synthesis allows the oligonucleotides to be specifically attached at one end to the surface of the support material. The support material surface must have the appropriate reactive groups to allow attachment of oligonucleotides via a linker molecule. Since not all surface coatings have reactive groups that can be used to couple oligonucleotides covalently, the choice of coating depends on what attachment method will be used. If linkers are used, the oligonucleotides must be allowed to react to form the covalent bonds, and the remaining unreacted reactive groups on the surface must be blocked before the slide is ready for use.

3.5.   Target Labeling

A wide variety of target labeling methods are available today. Figure 3.6 describes three commonly used labeling regimens. RNA can be labeled through reverse transcription (RT) incorporating modified nucleotides (nt), either directly tagged with a fluorescent marker or later chemically attached to the modification. A very different strategy involves amplification of the RNA message by running *in vitro* transcription (IVT) from a promoter sequence incorporated into the cDNA in the RT reaction. Several cRNA transcripts can be created from each cDNA in the IVT reaction amplifying the original message. Biotin-modified nucleotides incorporated in the IVT step later serve as handles for fluorescent dye tagging.

Most methods involve reverse transcription of the mRNA in some fashion. The choice of target labeling strategy for both spotted arrays and *in-situ* oligonucleotide arrays must ensure that the probes and the labeled targets are complementary. In the simplest labeling strategy, a fluorescent dye-modified nucleotide is incorporated in the reverse transcription of mRNA to cDNA. The reverse transcription needs to be primed by a hybridized short oligonucleotide primer. The primer can be either an oligo-dT (usually 12-18 nucleotides long) that anneals to the poly-A tail of eukaryotic mRNAs or a random primer (typically 6 nucleotides long) that will initiate reverse transcription at random sites along the mRNA. The number of nucleotides modified by fluorescent groups must be balanced to avoid incorporating dye molecules too frequently (which will quench the fluorescence signal) yet frequently enough to provide a good signal.

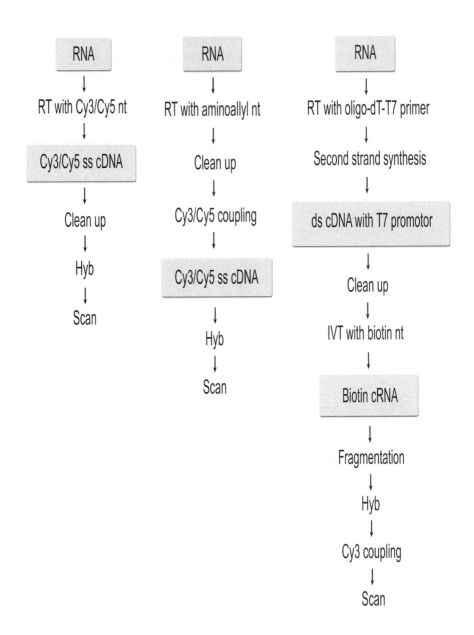

Figure 3.6. Three commonly used labeling regimes: direct labeling (left), aminoallyl labeling (middle), and IVT labeling (right). RT: reverse transcription; ss: single-stranded; ds: double-stranded; nt: nucleotide; IVT: in vitro transcription; Hyb: hybridize.

For both spotted cDNA arrays and spotted oligonucleotide arrays, usually two samples (a test and a reference) are hybridized to each slide to compensate for slide differences in the spotting process (Figures 3.3 and 3.4). The test and reference mRNA samples (for instance, a diseased and a healthy tissue sample) are reverse-transcribed and labeled with two different fluorescent dyes in separate test tubes and later pooled for simultaneous hybridization to the microarray slide probes. The most popular choices for fluorescent dyes are the Cy3 (green) and Cy5 (red) dyes. Several options exist, however, and systems have been developed to allow for detection of multiple dye-labeled hybridized targets[18]. The emission spectra of all the dyes used have to be sufficiently separated in wavelength to allow for individual detection. A drawback of using dye-modified nucleotides in the reverse transcription reaction is that different dye molecules can be incorporated with different efficiency into the cDNA. To overcome this obstacle, labeling methods have been developed in which nucleotides modified with an aminoallyl group are incorporated into both the test and the reference sample in different test tubes. The dye can later be attached to the test and reference cDNA with aminoallyl groups in a chemical reaction that presumably is less sensitive to the structural bias of the dye molecule.

Traditionally, 50-100 μg of total RNA (mRNA, rRNA, and tRNA) has been needed to obtain good signals. With the development of better reverse transcription enzymes and more efficient labeling procedures, the need for total RNA is down to about 10-20 μg. Some biological samples are so small that such RNA quantities cannot be extracted. Hence, PCR or *in vitro* transcription-based methods for amplifying the mRNA have been developed. With PCR, any nucleotide sequence can be hugely amplified, but a frequent concern with such methods is that the relative abundance of different mRNA species can be skewed in the amplification process and might not accurately reflect the levels in the original sample. In general, the *in vitro* transcription amplification methods are thought to amplify mRNAs with less bias. In *in vitro* transcription, the reverse transcription reaction is primed by an oligo-dT primer with a promoter sequence extension. After the reverse transcription, the double-stranded cDNA is generated and *complementary RNA (cRNA)* can be synthesized by transcription from the *promoter sequence* in a test tube. Several cRNA molecules can be transcribed from each double-stranded cDNA template.

In-situ oligonucleotide arrays are manufactured with good precision, allowing for comparison between arrays that have been hybridized with a single labeled target sample (Figure 3.5). *In vitro* transcription labeling is the method of choice for labeling targets for *in-situ* oligonucleotide arrays (Figure 3.5). For GeneChip arrays from Affymetrix, biotin-modified nucleotides are incorporated into the cRNA, which is then fragmented to approximately equal lengths to give more uniform hybridization properties. The biotin modifications are used as a handle for dye attachment. Methods have also been developed in which mRNA can be chemically labeled directly without the need of conversion to cDNA (PerkinElmer Life Sciences). Methods that attach more dye per transcript have also been developed, including systems in which the dye is enzymatically attached to pre-hybridized reverse-transcribed targets (PerkinElmer Life Sciences) and systems in which dye-oligonucleotide multimers (dendrimers) are hybridized to reverse-transcribed targets modified with a universal oligonucleotide extension (Genisphere).

3.6.   Hybridization

Hybridization of the labeled target to the probes on a microarray is performed by adding the targets dissolved in hybridization buffer to the slide within a confined space, followed by incubation for a given amount of time at a certain temperature. The hybridization can, for instance, be performed under a microscope slide cover slip or within a chamber that limits the volume. Volumes are kept small to reduce the time of hybridization. Automated hybridization stations have been developed that agitate the hybridization solution over the slide and allow for better control of hybridization conditions, which gives lower backgrounds and better reproducibility. The hybridization conditions need to be set so as to promote the specific hybridizations between the target and individual probes and limit *nonspecific hybridizations* to the support itself or other probes. This is achieved mainly by varying the temperature and the ionic strength of the hybridization buffer. The temperature needs to be lower than the melting temperature of the probes but sufficiently high to reduce nonspecific hybridizations. The salt concentration, pH, and other characteristics of the buffer may also promote specific hybridizations. It may be advantageous to add competing DNA like salmon sperm DNA, Cot-1 DNA (enriched with mammalian repetitive sequences), and poly-A DNA (to block nonspecific hybridization to poly A regions). After hybridization for anywhere from several hours to overnight, the hybridization solution is discarded and the slides are subjected to washes of varying ionic strength to remove nonspecifically bound targets with increasing stringency. After the wash regimen, the slide is dried and is ready to be scanned. The dyes used are degraded over time by exposure to light, so hybridized slides and labeled target solutions need to be stored in the dark.

3.7. Scanning and Image Analysis

After hybridization, an image of the array with hybridized fluorescent dyes must be acquired. Microarray scanners have confocal lasers or other light sources to produce light at the wavelength that excites the fluorescent dyes. The fluorescence emission intensity of the dyes is captured in high-resolution monochrome images acquired for each fluorescent dye. The scanner software then displays a composite colored image for multi-dye hybridizations. The goal is to measure, for each spot on the array, the relative amount of fluorescence from each dye hybridized with its target. Next, the probe spots have to be identified and the fluorescence intensities quantified from the high-resolution image.

The process of scanning spotted cDNA microarray images can be separated into three tasks, including *gridding*, *segmentation*, and *intensity extraction*. Gridding or addressing is the process of assigning coordinates to each of the spots. A grid is placed at the approximate location of the spots, and an algorithm is used to fine-tune the location of the spots and classify pixels as part of a spot or the background. Also, the grid file typically contains information about the identity of the individual spots.

The *segmentation* procedure allows the classification of pixels either as foreground (i.e., as corresponding to a spot of interest) or as background. Segmentation algorithms designed to detect spot boundaries include edge detection algorithms, histogram methods, fixed-circle, and adaptive-circle segmentation. The location of the spots can be manually corrected to account for effects such as dust particles missed by the algorithm. Once the locations of the spots have been decided, the fluorescence intensities of all the pixels within each spot and outside it (i.e., the local background) are calculated and a result file is reported, in a step called *intensity extraction*. This includes calculating, for each spot on the array, foreground fluorescence intensities, background fluorescence intensities, locations, averages, medians, standard deviations, and possibly quality measures. These result files contain all the raw fluorescence data available for analysis of an microarray experiment.

Few microarray images are perfect. Image analyis is confounded by many factors such as the compensation for the non-zero fluorescent background observed on most arrays and blemishes on the slides. Common problems in spot shape and appearance such as comet tails or donut holes are often observed. Locally high background and weak signals are also problematic phenomena in array images.

3.8.   Microarray Data

Microarray data have two basic qualities, biological significance and statistical significance. The biological significance tells how much the expression of a gene is influenced by the condition under study, i.e., the expression ratio or the fold change. The biological significance is what researchers are interested in. The statistical significance tells how trustworthy the biological significance is. Because of the many sources of variability in microarray experiments, the statistical analysis is crucial for successful interpretation of the biological phenomena under study. To perform meaningful statistical analysis of microarray data, one must understand the format of the microarray data and how to interpret the data. The microarray data format for *in-situ* oligonucleotide arrays is very different compared to that of spotted microarrays, and each data format has its own analysis and interpretation requirements.

3.8.1.   Spotted Array Data

Gene expression data obtained from either spotted cDNA arrays or from spotted oligonucleotide arrays are similar in format. There are several software packages, both commercial and freeware, available for image analysis and quantification of spotted microarrays. The formats of the result files vary greatly. Typically, spreadsheets are used to report raw array data, including the location of the spots, gene identity, mean and median of the pixel intensities within a spot, and local background intensities.

The simultaneous hybridization of two specimen samples labeled with Cy3 (green) and Cy5 (red) dyes has special analysis requirements. The two dyes have different properties and light sensitivities. The Cy5 dye is much more readily broken down, with the result that even if equal amounts of target are hybridized, the fluorescence emission from the Cy5 dye is lower. Hence, the fluorescence signals from the two dye channels have to be normalized in order to calculate correct expression ratios. Normalization not only corrects for different dye properties but also for concentration differences between the co-hybridized test and reference samples. Normalization methods or algorithms must be selected with reference to the kind and degree of systematic bias that is present.

In practice, expression data obtained from either spotted cDNA arrays or spotted oligonucleotide arrays are often reported as an expression ratio. The *expression ratio* is simply the normalized ratio of the fluorescence intensity of the test sample and the reference sample for a certain gene. Often intensities of both the test and reference samples are background corrected (subtracted) before calculating the normalized ratio. Genes that are upregulated two-fold have an expression ratio of two, and genes that are downregulated two-fold have an expression ratio of 0.5. The expression ratios are often expressed as log-2 ratios. A gene that is upregulated two-fold will have a log-2 expression ratio of 1 and a gene downregulated two-fold will have a log-2 ratio of -1.

It may not be possible to calculate an expression ratio for all genes on an array. A gene may be unexpressed in the reference sample but highly induced in the test sample. Calculation of an expression ratio in such a case would need division by zero. These genes are usually of high interest, for instance in diseased versus normal gene expression. To be able to get at least an approximate expression ratio, even though it will be an underestimate of the true ratio, one often has to set a lowest allowed level of signal. This lowest allowed signal can be set to an arbitrary value or to the level of noise, for instance, some specified number of standard deviations of the background signal. It may be advantageous to discard or attribute low-intensity signals with confidence scores to filter out unreliable intensity readings. Since the image processing has to be highly automated to increase the throughput of analysis, methods like outlier detection, which improve the data quality after quantification, may have considerable impact on the reliability of microarray data.

3.8.2. In-situ Oligonucleotide Array Data

The GeneChip Absolute Analysis[19] calculates a variety of metrics using hybridization intensities measured by the scanner. GeneChip probe arrays are scanned at high pixel resolution. In the case of a higher density probe array, it creates 8 pixels x 8 pixels (on average) for every probe cell, or a total of 64 pixels per probe cell (viewable in the .DAT file). A single intensity value for every probe cell, representative of the hybridization level of its target, is derived. The bordering pixels are excluded. The remaining pixel intensity distribution is calculated, and the intensity value associated with 75% of the distribution is used as the *Average Intensity* of the probe cell. The *Average Intensities* for all probe cells are saved in the .CEL file. Some metrics utilize intensity data from the entire probe array and are used for *Background* and *Noise* calculations. Other metrics compare the intensities of the sequence-specific PM probe cells with their control MM probe cells for each probe set[20]. The use of the PM minus MM differences averaged across a set of probes is aimed at reducing the contribution of background and cross-hybridization while increasing the quantitative accuracy and reproducibility of the measurements.

The GeneChip software provides both absolute analysis and comparison analysis algorithms. The software calculates an average intensity value for every probe cell. Then, the background is calculated and substracted from the intensities of all probe cells. The noise is calculated by determining the degree of pixel-to-pixel variation within the same probe cells used to calculate the background. The numbers of positive and negative probe pairs are determined for every probe set. A *Positive* probe pair is one in which the intensity of the sequence-specific PM probe cell is significantly higher than the intensity of the control MM probe cell. A *Negative* probe pair is one in which the intensity of the MM probe cell is significantly higher than the intensity of the PM probe cell. Log average ratios and average differences are calculated directly using probe cell intensities. The log average ratio is derived from the ratio of PM probe cell intensity to that of the control MM. The average difference for each probe set (the average of the differences between every PM probe cell and its control MM probe cell) is directly related to the level of expression of the transcript. By examining the positive fraction, the positive/negative ratio, and the log average ratio, a decision matrix may be employed to determine whether a transcript is *Present* (P), *Marginal* (M), or *Absent* (A; undetected). The Absolute call is displayed in the .CHP file in the GeneChip software.

GeneChip Comparison Analysis performs additional calculations on data from two separate probe array experiments in order to compare gene expression levels between two samples. The analysis employs normalization by scaling techniques to minimize differences in overall signal intensities between the two arrays, allowing for more reliable detection of biologically relevant changes in the samples. The Comparison Analysis begins with the user designating an Absolute Analysis of one probe array experiment as the source of Baseline data and a second probe array experiment as the source of Experimental data to be compared to the Baseline. The results are used in a decision matrix to derive a Difference Call. Since the average difference of a transcript is directly related to its expression level, an estimate of the Fold Change of the transcript between the baseline and the experimental samples is also calculated.

Although Affymetrix has developed prediction rules to guide the selection of probe sequences with high specificity and sensitivity[21], inevitably there remain some probes that hybridize to one or more nontarget genes. In the standard analysis[22], the mean and standard deviation (SD) of the PM-MM differences of a probe set in one array are computed after excluding the maximum and the minimum. If a difference is more than 3 SD from the mean, a probe pair is marked as an outlier in this array and is discarded when calculating average differences of both the baseline and the experimental arrays. One drawback to this approach is that a probe with a large response might well be the most informative but may be consistently discarded. Furthermore, if one wants to compare many arrays at the same time, this method tends to exclude too many probes. Li and Wong[23] (2001) show that even after making use of the control information provided by the MM intensity, the information on expression provided by the different probes for the same gene are still highly variable. They note that the between-array variation in PM−MM differences can be substantial and that the variation due to probe effects is larger than the variation due to arrays. In addition, human inspection and manual masking of image artifacts is currently time-consuming. Li and Wong (2001) use a multiplicative model to detect and handle cross-hybridizing probes, image contamination, and outliers from other causes.

3.9. So I Have My Microarray Data - What's Next?

The remainder of this book will focus on statistical analysis of microarray data. Microarray technology has been used mostly as an exploratory or survey methodology, where the most interesting findings are verified by other more precise methods. The goal of the statistical analysis then becomes to limit the number of false positives and to validate clusters or groups of genes that were found based on their expression pattern. Microarray experiments may never produce precise quantitative results for all measured transcripts. Instead, the power of microarray technology lies in its potential to survey the entire transcriptome in one experiment. With sound statistical analysis, patterns emerging from such a holistic view may give insights that traditional experiments never can.

3.9.1. Confirming Microarray Results

Microarrays are often used to find candidate genes for further studies. Because of the cost of microarrays, many investigators cannot afford thorough experimental designs with the numbers of replicates needed for a good statistical analysis. It then becomes imperative to verify the microarray result before spending a lot of time and money on follow-up experiments. Even with sound statistical analysis it may prove advantageous to first verify the most interesting genes coming from the analysis with an independent method. Two methods of choice are Northern blot analysis and quantitative real-time PCR.

3.9.2.   Northern Blot Analysis

Specific RNA sequences can be detected by blotting and hybridization analysis using techniques very similar to those originally developed by Ed Southern for DNA. Different RNA molecules can be size-separated or fractionated by gel electrophoresis. The gel electrophoresis is run under denaturing conditions to keep the single-stranded RNA molecules denatured, thus allowing good separation. The RNA to be separated is loaded into a gel-like solid support consisting of an agarose polymer. Upon application of a current, the RNA molecules will migrate in the gel due to their phosphate backbone negative charge. A long molecule will migrate slowly through the agarose gel, while a short molecule will migrate quickly. After allowing the RNA molecules to separate for a while, the electrophoresis is stopped and the fractionated RNA is transferred from the agarose gel to a membrane support (Northern blotting) by capillary transfer. After transfer the RNA is immobilized on the membrane. The membrane can now be probed with radioactively labeled single-stranded cDNA probes complementary to the gene of interest. The amount of radioactivity hybridized to a test and reference sample separated on the same gel is proportional to the relative expression of the gene of interest in the test and reference samples. For more information on the Northern blot technology, see Ausubel *et al.*[24] (1993) and Sambrook *et al.*[25] (1989).

Chu *et al.*[26] (1998) compared Northern blot assay of gene expression and corresponding microarray data of sporulation in budding yeast. As shown in Figure 3.7, the authors found that the pattern of expression as assayed by Northern analysis was very similar to that determined by microarray analysis.

3.9.3.   Reverse-transcription PCR and Quantitative Real-time RT-PCR

The powerful amplification of nucleic acids achieved by the PCR methodology makes it an excellent method to use to detect low-abundance nucleic acids. Since RNA cannot serve as a template for PCR, the first step to quantify RNA is to reverse transcribe the RNA sample into cDNA. *Reverse transcription PCR* (RT-PCR), not to be confused with *quantitative real-time RT-PCR* (QRT-PCR), can be used for semi-quantitative assays of RNA abundance, simply by sampling the PCR mixture followed by gel quantification of the amplified product. Yet, as the name implies, this is at best only semi-quantitative. The advent of fluorescence techniques applied to the reverse transcription PCR methodology, together with instrumentation capable of combining amplification, detection, and quantification, has revolutionized the possibilities of nucleic acid quantification.

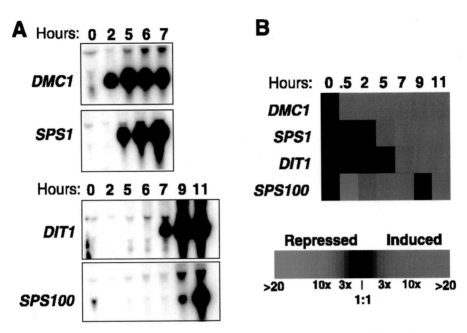

Figure 3.7. Northern blot assay of gene expression and corresponding microarray data. (A) Samples from the indicated time points were assayed by Northern analysis (19). Genes were chosen to be representative of the four previously identified temporal classes. DMC1, SPS1, DIT1, and SPS100 belong to the early, middle, mid-late, and late classes, respectively (1, 2). (B) RNA samples from the same time course as in (A) were analyzed by microarray analysis. Data are graphically displayed with color to represent the quantitative changes. Increases in mRNA (relative to pre-sporulation levels) are shown as shades of red, and decreases in mRNA levels are represented by shades of green. Source: Science, volume 282 (1998). Reprinted with permission.

In quantitative real-time RT-PCR (QRT-PCR) the amplification can be monitored by fluorescence in "real-time" in every cycle of amplification. There is already a wide variety of specialized methods and reagents utilized in QRT-PCR. The most commonly used chemistries include 5′ nuclease assays using TaqMan probes, molecular beacons, and SYBR green I intercalating dyes, all providing means to monitor the accumulation of PCR product in each cycle of the PCR reaction. The fluorescence values recorded during every cycle represent the amount of PCR product amplified to that point in the amplification reaction. The point at which a statistically significant signal can be recorded above background is called the threshold cycle. The more template present at the beginning, the fewer PCR cycles that are required to reach the threshold cycle. The threshold cycle is inversely proportional to the logarithm of the starting amount of template in the PCR reaction. By constructing standard curves of known amounts of starting template, unknown samples can be quantified very accurately. See Ginzinger[27] (2002) for a review of gene quantification using quantitative real-time PCR and Bustin[28] (2002) for a review of reverse transcription PCR.

Acknowledgement: Harry Björkbacka is supported by the Swedish Research Council.

Notes

[1] Schalon, D., Smith, S.J., Brown, P.O. (1996). *Genome Research*, **6**, 639-645.

[2] Lockhart, D.J., Dong, H., Bryne, M.C., Follettie, M.T., *et al.* (1996). *Nature Biotechnology*, **14**, 1675-1680.

[3] Lander, E.S. (1996). *Science*, **274**, 536-539.

[4] Lipshutz, R.J., Fodor, S.P., Gingeras, T.R., Lockhart, D.J. (1999). *Nature Genetics*, **21**, 20-24.

[5] Brown, P.O., and Botstein, D. (1999). *Nature Genetics*, **21**, 33-37.

[6] Eisen, M.B., Brown, P.O. (1999). *Methods in Enzymology*, **303**, 179-205.

[7] Southern, E., Mir, K., Shchepinov, M. (1999). *Nature Genetics*, **21**, 5-9.

[8] Bowtell, D. (1999). *Nature Genetics*, **21**, 25-32.

[9] Cheung, V.G., Morley, M., Aguilar, F., Massimi, A., Kucherlapati, R., Childs, G. (1999). *Nature Genetics*, **21**, 15-19.

[10] Schena, M. (2000). *DNA Microarrays*. Oxford University Press, New York.

[11] Baldi, P., Hatfield, G.W. (2002). *DNA Microarrays and Gene Expression*, Cambridge University Press, Cambridge, U.K.

[12] Bowtell, D., Sambrook, J. (Editors) (2002). *DNA Microarrays: A Molecular Cloning Manual.*, Cold Spring Harbor Laboratory.

[13] Grigorenko, E.V. Editor. (2002). *DNA Arrays: Technologies and Experimental Strategies*, CRC Press, New York.

[14] Phimister, B. (1999), *Nature Genetics*, **21**, 1.

[15] Duggan, D.J., Bittner, M., Chen, Y., Meltzer, P., Trent, J.M. (1999). *Nature Genetics*, **21**,10-14.

[16] Fodor, S.P., Rava, R.P., Huang, X.C., Pease, A.C., Holmes, C.P., Adams, C.L. (1993). *Nature*, **364**, 555-556.

[17] Miesfeld, R. L. (1999), *Applied Molecular Genetics*, Wiley-Liss, New York, 238-243.

[18] Hessner, M.J., Wang, X., Hulse, K., Meyer, L., Wu, Y., Nye, S., Guo, S.-W., Ghosh, S. (2003). *Nucleic Acids Research*, **31**, No. 4, e14.

[19] Affymetrix (2001). *Affymetrix Microarray Suite 5.0 User Guide*, Affymetrix Inc., Santa Clara, CA.

[20] Affymetrix GeneChip, 700228 rev.2.

[21] Lockhart, D., Dong, H., Byrne, M., Follettie, M., Gallo, M., Chee, M., Mittmann, M., Wang, C., Kobayashi, M., Horton, H., *et al.* (1996). *Nature Biotechnology*, **14**, 1675-1680.

[22] Wodicka, L., Dong, H., Mittmann, M., Ho, M., and Lockhart, D. (1997). *Nature Biotechnology*, **15**, 1359-1367.

[23] Li, C., Wong, W.H. (2001). *Proceedings of the National Academy of Sciences, USA*, **98**, 31-36.

[24] Ausubel, F.M. et al. (editors) (1993). *Current Protocols in Molecular Biology*, John Wiley and Sons, Inc., New York.

[25] Sambrook, J., Fritsch, E.F. and Maniatis T. (1989). *Molecular Cloning: A Laboratory Manual*, 2nd edition, Cold Spring Harbor Laboratory Press.

[26] Chu, S., DeRisi, J., Eisen, M., Mulholland, J., Botstein, D., Brown, P.O., Herskowitz, I. (1998). *Science*, **282**, 699-705.

[27] Ginzinger, D.G. (2002). *Experimental Hematology*, **30**, 503-512.

[28] Bustin, S.A. (2002). *Journal of Molecular Endocrinology*, **29**, 23-39.

Chapter 4

INHERENT VARIABILITY
IN MICROARRAY DATA

This chapter begins with a statistical characterization of the genetic populations that are investigated in microarray studies. It then goes on to examine the nature of microarray gene expression data that are generated from these populations and their sources of variability and error.

4.1.   Genetic Populations

The target objects in a microarray study might be cell lines (such as yeast cells), biological specimens (such as tumor tissues), or biological systems under varying experimental conditions (such as organisms exposed to different levels of a toxin). For expository convenience, we refer to these target objects as either biological specimens or experimental conditions. These phrases may be replaced by equivalent terms, such as cell lines, tissues, and so on, when discussing particular applications of microarray technology. The index set of biological specimens or experimental conditions under investigation is denoted by $\mathcal{N} = \{1, 2, \ldots, N\}$, where N denotes the total number of sample specimens or conditions in the study.

Each biological specimen under investigation defines a distinct biological population with a particular set of genes. The union of these gene sets for all N specimens forms the *population gene set* for the study, denoted by $\mathcal{G}_P$. Thus, the genes of any specimen in the study will be found in the set $\mathcal{G}_P$. The word 'gene' is used here to encompass a wide range of real genes, DNA strands, and other biologically coded objects that can bind to or hybridize with probes mounted on an array slide.

The *designated gene set* in a microarray study refers to the set of probes that are spotted on the array. The designated gene set on the array is denoted here by the index set $\mathcal{G}_A = \{g : g = 1, 2, \ldots, G\}$, where G is the total number of probes in the designated set. In practical applications of microarray technology, the designated gene set for a microarray study often differs from the true population set $\mathcal{G}_P$, possibly to a great extent. The difference between the population gene set $\mathcal{G}_P$ and the designated gene set $\mathcal{G}_A$ on the array will usually be unknown, with set $\mathcal{G}_A$ containing genes that may not be in $\mathcal{G}_P$, and vice versa. In addition to genes in the population gene set $\mathcal{G}_P$, the designated set $\mathcal{G}_A$ may contain foreign genes, gene replicates, cDNA fragments, other gene-like objects, and even blanks. Foreign genes are genes that are alien to the biological specimens under study but are purposely included in the microarray for reasons of control, calibration, and monitoring of the array results. The presence of replicates implies that not all spots represent genetically distinct probes. Replication offers the benefits of improving statistical precision and diagnostic checking.

To characterize the biological population, we consider the genetic makeup of the population in relative terms. The genetic material in each biological specimen has different concentrations of the distinct genes in the designated set $\mathcal{G}_A$ on the array. We shall denote the concentration of genetic material in specimen n that is attributable to gene g by ζ_{gn}. By definition, $\zeta_{gn} \geq 0$ for each gene g and specimen n. If $\zeta_{gn} = 0$, then gene g is not present in biological specimen n. In microarray studies, the values of ζ_{gn} are quantities of interest and they are invariably unknown. Therefore, they must be estimated. More often, ratios of the form $\zeta_{gn'}/\zeta_{gn}$ for pairs of distinct experimental conditions n' and n are of interest. When these ratios are not unity, they indicate that gene g is up- or down-regulated under one experimental condition relative to the other. Observe that the concentrations ζ_{gn} convey no information about genes that are in the population gene set $\mathcal{G}_P$ but are missing from the designated set $\mathcal{G}_A$ on the array.

A key design consideration in microarray studies is the composition and size of the designated gene set $\mathcal{G}_A$. The number of genes G in the gene set may be in the hundreds or in the thousands. The number of these genes that are expected to be differentially expressed between different biological specimens or experimental conditions may be very small, numbering perhaps in the dozens. One of the scientist's challenges is to cast the net for designated genes widely enough to include all genes that are implicated in a particular scientific question but narrowly enough that spurious or chance indications of differential expression are not numerous. In a later chapter we re-examine this issue in the context of sample size and power issues for microarray studies.

Expression levels for genes are dictated physiologically by a combination of genetic factors such as promoters, enhancers, and splice sites and a large number of environmental factors, including temperature, stress, and light, which can lead to changes in the levels of hormones and other signalling substances that affect gene expression levels[1]. Like mRNA levels, protein levels also change in response to physiological factors and changes in environment. Unlike DNA and RNA, however, protein activity is not measured by hybridization but rather by binding of proteins to labelled oligonucleotides[2].

4.2.   Variability in Gene Expression Levels

A microarray study extracts a sample of genetic material from each specimen and through a laboratory and measurement routine obtains a reading of genetic intensity for each gene in the designated gene set $\mathcal{G}_A$. Many factors influence the relationship between these readings and the genetic population concentrations ζ_{gn}, making it difficult to estimate the concentrations from the microarray data. Indeed, the purpose of this book is to address this exact problem.

4.2.1.   Variability Due to Specimen Sampling

Sampling of genetic material for microarray studies generally takes place at several levels. Any discussion of variability must be clear about which sources of variability are under consideration. The first level of sampling is usually encountered when the biological specimens are selected in some broad context. For example, a uterine tumor of a particular host patient may be selected as the specimen. If there are several such tumors in the host, however, then the one chosen for the study is a sample from among all tumors present in the host and, indeed, may be viewed as a sample of all future tumors that might develop in the host. In subsequent discussion, the selected specimen will usually be taken as the starting point of the statistical investigation. Inferences will relate to this biological specimen and not to more general populations that are farther back in the selection process. This definition of a biological specimen is a convenient operating definition. It is not meant to suggest that inferences farther back to more general populations are not of interest. In some applications, in fact, study design and analysis may need to relate to a very general level of definition. For example, it is conceivable that a uterine tumor study might take the relevant tumor population as all those that are extant in U.S. females of a given race in a specified age bracket at a given point in time. The connection between such a general population and the microarray results for a small sample of uterine tumors taken from a few women in this population is very distant and tenuous, spanning sampling variability introduced at several intervening stages of sample selection. In order to generalize the findings to the general population of women, the microarray studies should be based on a *random sample* taken from the general patient population. To reiterate, the following discussion and analysis will assume that the immediate biological specimens in the laboratory are the target biological populations of interest for statistical inferences.

Even after the biological specimen is in hand, variability in expression measurement can arise from many factors. One of the first sources of variability encountered in many microarray studies is that produced by selecting the sample of genetic material from the population of interest. For example, a sample of tumor tissue must be selected from a patient's tumor in a microarray study of uterine cancer. It is clear that the sample material may vary to the extent that the tumor is not a uniform biological object. The genetic composition of the sample may differ depending on whether the core biopsy is taken, for instance, from the peripheral zone or the transition zone of the tumor. The microarray study design must take sampling variability of this kind into account. For example, the design might call for taking a systematic selection of several samples from different parts of the tumor.

4.2.2.   Variability Due to Cell Cycle Regulation

In a microarray study based on samples from yeast cultures synchronized by three independent methods, Spellman *et al.*[3] (1998) created a catalog of 800 yeast genes whose transcript levels vary periodically within the cell cycle. They note that, in addition to random fluctuations in the data, cross-hybridization between genes whose DNA sequences are similar can produce false positives when only one of the genes is actually cycle-regulated. False positives can also occur when an unregulated gene overlaps the mRNA for a cell cycle-regulated gene; the cDNA corresponding to the regulated gene would hybridize with the unregulated gene's DNA.

4.2.3.   Experimental Variability

Microarray experiments can encounter technical problems at any step. Few microarray images are perfect. Various systematic errors in microarray measurements may exist because of the preparation of probes, targets, and arrays as well as the procedures of image analysis. Variations can also be laboratory- or system-dependent. Schuchhardt *et al.*[4] (2000) present a systematic study of sources of distortion and noise in microarray studies involving spotting techniques. Investigations by Wang *et al.*[5] (2001) and by Yang *et al.*[6] (2001) also show how numerous the potential sources are and how significant each might be in application. Listed below are some general sources of variation that are common to microarray experiments.

Preparation of mRNA.
As we mentioned earlier, the fluor-tagged cDNAs from a complex mRNA mixture extracted from cells are often termed targets. Depending upon tissue and sensitivity to RNA degradation, targets may be different from sample to sample.

Reverse transcription.
Reverse transcription to cDNA may result in DNA species of varying lengths.

PCR amplification.
Clones are subject to PCR amplification. The amplication is difficult to quantify and may fail completely.

Pin geometry and transport volume.
Pin geometry can produce systematic variation. Pins have different characteristics and surface properties. The amount of transported target volume can fluctuate randomly even for the same pin of the arrayer.

Slide heterogeneity.
The fraction of target cDNA that is chemically linked to the slide surface from the droplet is unknown. Furthermore, the target may be distributed unequally over the slide or the hybridization reaction may perform differently in different parts of the slide.

Fluorescence labeling.
Depending upon nucleotide composition, radioactive (fluorescence) labeling may also fluctuate.

Hybridization reaction.
The efficiency of the hybridization reaction is influenced by a number of experimental parameters, notably temperature, time, buffering conditions and the overall number of target molecules used for hybridization. Cross-hybridization within a gene family may also occur.

Nonspecific hybridization is an error source that cannot be completely excluded.

Non-specific background and over-shining.
Non-specific radiation (fluorescence) and signals from neighboring spots can be present.

Image processing and data acquisition.
Results from image analysis can be affected by background noise and overshining from neighboring spots, non-linear transmission characteristics, and the procedures used in handling saturation effects and variations in spot shapes.

In general, the exact gene expression levels are unknown, and particular strategies have to be developed to quantify systematic errors in microarray experiments. A variety of correction methods can be found in the literature, including comparison of duplicated spots to quantify the variability for the same array and the same pin; analysis of control spots to quantify the variability from pin to pin and variations across the filter; checking the reproducibility on different filters; analysis of empty background spots for non-specific noise and overshining; and use of dilution series of the target[7]. Brazma *et al.*[8] (2001) proposed the Minimum Information about a Microarray Experiment (MIAME) as a standard for recording and reporting microarray-based gene expression data. Any single microarray output is subject to substantial variability even under the relatively controlled conditions of an experiment. It is advisable to consider appropriate experimental designs and perform multiple stages of quality control before hybridizing valuable experimental samples.

4.3.   Test the Variability by Replication

Replication serves to sharpen the precision of statistical inferences drawn from microarray studies. Replication also helps to evaluate whether data from arrays are uniform in some appropriate statistical sense, providing a quality check on the scientific investigation.

4.3.1.   Duplicated Spots

To check the consistency of microarray experiments and to study the effects of variability and replication on the reliability of cDNA microarray findings, Lee, Kuo, Whitmore, and Sklar[9] (2000) conducted a study to investigate whether the locations of cDNA spots on slides may produce variation in measurements of transcriptions. A single human tissue sample formed the target biological sample. Only output from channel 1 (green) contained expression readings for the target tissue. Output from channel 2 (red) contained noise alone. The study design consisted of 288 genes, each printed at three locations on the same slide. By comparing the signals from these triplicates, Lee *et al.* evaluate the minimum variability that is likely to be inherent in a microarray system and learn more about the reproducibility of the array process and the outcome of analysis. The experiment was designed so that 32 of the 288 genes would be expected to be highly expressed because of Alu repeats that should cross-hybridize to similar sequences widely distributed among expressed and unexpressed portions of the human genome. Results based on individual replicates, however, show that there are 55, 36, and 58 highly expressed genes in replicates 1, 2, and 3, respectively. On the other hand, we will show in later chapters that by applying appropriate statistical methods one can pool the readings from the three replicates and obtain more accurate analytical results such that only 2 of the 288 genes are incorrectly classified as expressed. As a result, a minimum of three replicates is recommended in a microarray study. This replication test data set is used to demonstrate a number of points later in this text.

4.3.2. Multiple Arrays and Biological Replications

There are different levels of biological replication. Depending upon the nature of the microarray study, the sampling units may consider biological replicates, e.g., from inbred species. If inbred species are not readily available, species of the same strain can be used. In applications with human tissues, variation between individuals could be much larger than other sources of variation, and hence sampling from a large number of individuals will be necessary. When multiple specimens from the same individuals are available, they can sometimes be considered replicates. For tumor tissues, however, even multiple specimens from the same tumor can have considerable variation. Moreover, if RNA samples are divided into multiple plates for hybridization, multiple arrays can be made based on each RNA specimen. These replications can be very useful in testing the reproducibility of the findings based on array data. Baggerly *et al.*[10] (2001) propose parametric models for log ratios obtained from replications within a sample (within a channel) and replications between samples (between arrays and/or between channels). Wang *et al.*[11] (2003) proposed quantitative quality control measures for microarray experiments.

Notes

[1] Schena, M., Editor. (2000). *DNA Microarrays*, Oxford University Press, New York.

[2] Bulyk, M.L., Huang, X., Choo, Y., Church, G.M., (2001), *Proceedings of the National Academy of Sciences, USA*, **98**, 7158-7163.

[3] Spellman, P.T., Sherlock, G., Zhang, M.Q., Iyer, V.R., Anders, K. , Eisen, M.B., Brown, P.O., Botstein, D. and Futcher, B. (1998). *Molecular Biology of the Cell*, **9**, 3273-329.

[4] Schuchhardt, J., Dieter, B., Arif, M., *et al.* (2000), *Nucleic Acids Research*.

[5] Wang, X., Ghosh, S., and Guo, S. (2001). *Nucleic Acids Research*, **29**, No. 15, e75.

[6] Yang, Y.H., Dudoit, S., Luu, P., and Speed, T.P. (2001). In Bittner, M.L., Chen, Y., Dorsel, A.N., and Dougherty, E.R. (eds), *Microarrays: Optical Technologies and Informatics*, SPIE Society for Optical Engineering, San Jose, CA.

[7] Herzel, H., Beule, D., Kielbasa, S., Korbel, J. (2000), *Chaos*, **11**, 1-3.

[8] Brazma, A., Hingamp, P., Quackenbush, J., Sherlock, G., Spellman, P., Stoeckert C., Aach, J., Ansorge, W., Ball, C.A., Causton, H.C., Gaasterland, T., Glenisson, P., Holstege, F.C.P., Kim, I.F., Markowitz, V., Matese, J.C., Parkinson, H., Robinson, A., Sarkans, U., Schulze-Kremer, S., Steward, J., Taylor, R., Vilo, J., and Vingron, M. (2001). *Nature Genetics*, **29**, 365-371.

[9] Lee, M.-L.T., Kuo, F.C., Whitmore, G.A., and Sklar, J., (2000). *Proceedings of the National Academy of Sciences, USA*, **97**, 9834-9839.

[10] Baggerly, K.A., Coombes, K.R., Hess, K.R., Stivers, D.N., Abruzzo, L.V., and Zhang, W. (2001). *Journal of Computational Biology*, **8**, 639-659.

[11] Wang, X., Hessner, M.J., Pati, N., and Ghosh, S. (2003). *Bioinformatics*, **19**, 1341-1347.

Chapter 5

BACKGROUND NOISE

The reading or measurement for gene expression is usually a fluorescence intensity measurement or some other quantitative marker. The reading may have undergone various adjustments within the instrument system, such as calibration. Thus, any description of gene expression data must be accompanied by an explanation of how the values are produced by the instrument system. The expression measurements will invariably contain a component that represents background noise. This chapter considers the nature of this noise and methods for taking it into account.

5.1.   Pixel-by-pixel Analysis of Individual Spots

In cDNA arrays, gene expression corresponds to the fluorescence intensity of an image that is measured pixel by pixel. The discussion begins by considering an analysis at the pixel level. To demonstrate a pixel-by-pixel analysis, we cite a study on budding yeast *Saccharomyces cerevisiae* considered by Brown *et al.*[1] (2001). In this study, the array compared two different strains (A and B) of wild-type *S. cerevisiae*. Following methods described in Schena *et al.*[2] and DeRisi *et al.*[3], mRNA from strain A was copied to cDNA and labeled with Cy3 (green), and mRNA from strain B was copied to cDNA and labeled with Cy5 (red). The cDNAs were hybridized to a spotted DNA microarray. The array carried about 6,200 genes. Images of the hybridized arrays were obtained from a modified fluorescence microscope that scans the slide at several wavelengths.

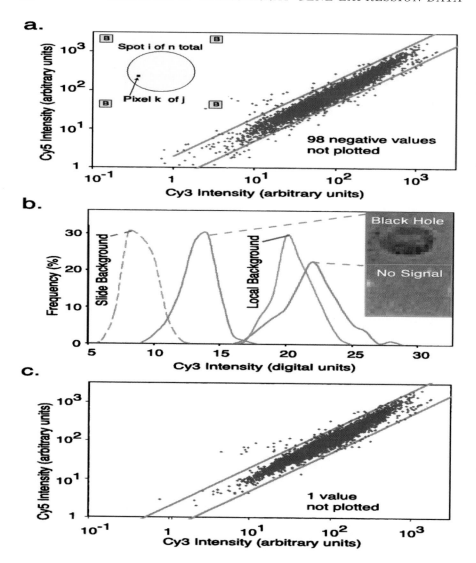

Figure 5.1. Methods for determining local background.
(a) Gene expression graph of Cy5 vs. Cy3 intensity for 6,200 yeast genes with local background used to offset signal intensities. Not shown are 98 spots with negative intensities. (Inset) Local background was determined by averaging red and green intensities over 16 pixels at each of the four corners (marked B) of a rectangular region surrounding each spot. (b) Distribution of Cy5 intensities for a region of the slide away from the hybridization area (dotted blue line), a spot with negative intensity (a "black hole," red line), local background surrounding the black hole (solid blue line), and a nearby low-intensity spot (green line). Each distribution is derived from over 150 sampled pixels. (c) Gene expression graph as in (b) but with best-fit background derived by Brown et al. (2001). Only one (obviously scratched) spot has negative intensity and is not shown. Source: Proceedings of the National Academy of Sciences (PNAS), volume 98, Copyright (2001) National Academy of Sciences, U.S.A. Reprinted with permission from PNAS.

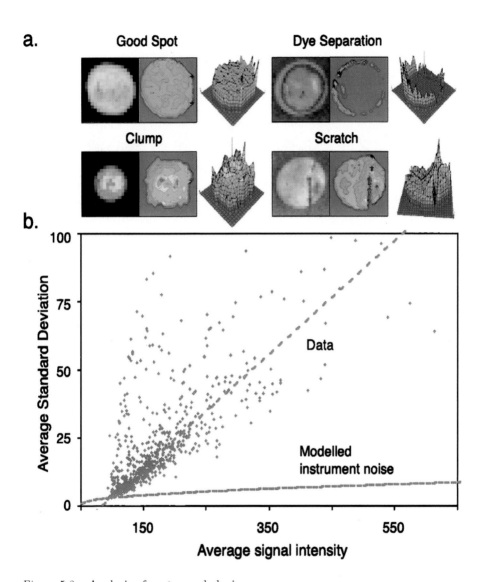

Figure 5.2. Analysis of spot morphologies.
(a) A gallery of spots from cDNA-based microarrays and their corresponding red-green ratio is graphed three-dimensionally. Clockwise from the top left: a high-quality spot, a spot exhibiting dye separation, a scratched spot, and a clumped spot (b) relationship between average signal intensity and the standard deviation of the pixel-by-pixel intensities for all 6,200 spots. The red dotted line shows the trend-line; the green-dotted line shows expected instrument noise based on photon counting statistics. Source: Proceedings of the National Academy of Sciences (PNAS), volume 98, Copyright (2001) National Academy of Sciences, U.S.A. Reprinted with permission from PNAS.

The fluorescence signal from a spot, having an approximate diameter of 100μm, is registered on roughly 200 pixels. Thus, each spot provides a large number of red and green intensity measurements. Brown *et al.* determine the amount of real and background fluorescence at each spot on the array. The mean intensity for a spot, averaged over all pixels, consists of (1) fluorescent signal from cDNA hybridized to the spot, (2) background fluorescence arising from non-specific binding, and (3) unwanted contributions from noise and proton counting error.

The authors determined the background arising from non-specific binding by measuring local background intensity at the four corners of a rectangular region of interest surrounding the spot (Figure 5.1). They found that local background was an inexact measure of non-specific fluorescence for microarray spots. When absolute intensities were compared on a pixel-by-pixel basis for a spot that (a) looked like a "black hole," (b) a region surrounding the black hole (the local background), (c) a weakly fluorescent spot, and (d) a point on the slide outside the hybridization area, the observed intensity distributions showed significant overlap. The center of the black hole was clearly less intense than the local background, and the weakly fluorescent spot was only slightly brighter. The problem of negative intensities arises not because the spot is incorrectly located during image segmentation but rather because a more-fluorescent probe is actually binding to the area surrounding the spot than to the spot itself.

To explore the intensity variations systematically, the authors plotted the standard deviation of the pixel-by-pixel intensities for each spot (averaged across both channels) against the spot's average signal intensity. Figure 5.2 shows that the standard deviations in the pixel-by-pixel intensity distributions are large in absolute magnitude and rise linearly as the signal intensity increases. In contrast, measurement noise, including photon counting noise, rises only as the square root of the signal.

Because the microarrays analyzed in the authors' study did not contain hybridization standards, they derive a computational method to determine background levels from the experimental spots themselves and conclude that pixel-by-pixel information for microarray images can be used to formulate measures that assess the accuracy with which an array has been sampled.

5.2.   General Models for Background Noise

Instead of pixel-by-pixel analysis, we discuss in this section some general models for handling microarray background noise.

The full data set consists of gene expression measurements w_{gn} for all genes $g = 1, \ldots, G \in \mathcal{G}_A$ and for all specimens $n = 1, \ldots, N \in \mathcal{N}$. The w_{gn} are usually arranged in a $G \times N$ matrix of values, with genes corresponding to rows and specimen samples corresponding to columns. Different microarray platforms yield different types of expression measurements. The exact relationship of measurement w_{gn} to the true concentration ζ_{gn} of gene g in specimen n depends on the technology, imbedded adjustments that have been used, and, importantly, on the background noise that is present. We defer the discussion of missing and saturated intensity values to later chapters.

5.2.1.   Additive Background Noise

A plausible statistical description for the gene expression reading w_{gn} is that of a sum of two components, one representing latent (i.e., unobserved) true gene expression x_{gn} and the other representing background noise B_{gn}, as follows.

$$w_{gn} = x_{gn} + B_{gn} \tag{5.1}$$

The true unknown gene expression component x_{gn} is assumed to increase proportionally with ζ_{gn}, the true concentration of gene g in specimen n. Thus, $x_{gn} = 0$ if $\zeta_{gn} = 0$; in other words, the x_{gn} component is absent from w_{gn} if the specimen n does not contain gene g. The background noise component B_{gn} is assumed to be mathematically independent of ζ_{gn}. The assumption is also made that x_{gn} and B_{gn} are additive and probabilistically independent. Both components are taken as nonnegative. Observe that additive noise gives an incorrect indication of genetic material being present even when none is truly there (a potential false positive).

When gene g is absent from specimen n, then the reading w_{gn} is a pure observation of background noise, i.e., $w_{gn} = B_{gn}$. If all background noise values constitute a random sample from a common noise distribution, then genes that are known to be absent from the specimen yield a sample of observations from the noise distribution and can be used to infer its form and parameter values. A later discussion of ANOVA models for gene expression data exploits this observation. Figure 5.3 displays a histogram of gene expression readings for a data set consisting of 864 cDNA spots that is discussed in detail later. The left-hand portion of the histogram largely reflects the noise distribution.

5.2.2.   Correction for Background Noise

The nature of noise and its correction depends on the microarray technology being used. Data from spotted arrays provide an uncorrected measure of expression intensity w_{gn}, as well as an estimate of the background noise B_{gn}. If this background estimate is denoted by $\hat{B}_{gn}$, then a background-corrected expression reading $w_{gn}^{(c)}$ is provided by

$$w_{gn}^{(c)} = w_{gn} - \hat{B}_{gn} = (x_{gn} + B_{gn}) - \hat{B}_{gn} \tag{5.2}$$

The ScanAlyze system, for example, outputs measures CH1I and CH2I, which are uncorrected mean pixel intensities for the array spot for two fluorescent hybridizations[4] and also produces background corrections CH1B and CH2B for the same channels.

The correction procedure (5.2) may yield negative values for gene expression, especially where component x_{gn} is small (or zero) or the background estimate $\hat{B}_{gn}$ is large. The following variant of the background-corrected reading is obtained when negative values from (5.2) are set to zero.

$$w_{gn}^{(c)} = \max\{w_{gn} - \hat{B}_{gn}, 0\} \tag{5.3}$$

If $w_{gn}^{(c)}$ is zero (or negative), the inference is generally made that $\zeta_{gn} = 0$, i.e., that gene g is absent from specimen n.

The simple subtraction of a background correction provided by the instrument software does not necessarily produce a better reading for gene expression. It must never be forgotten that $\hat{B}_{gn}$ is only an estimate of B_{gn}, and a bad estimate may be worse than having no estimate at all.

The difference between the true and estimated noise, i.e.,

$$B_{gn} - \hat{B}_{gn}, \tag{5.4}$$

appears in the background-corrected expression reading (5.2). It is hoped that the true and estimated background noises are exchangeable random variables, meaning that if $g(b, \hat{b})$ is their joint probability density function (p.d.f.), then $g(b, \hat{b}) = g(\hat{b}, b)$ for all outcome values b and $\hat{b}$. This condition would imply that B_{gn} and $\hat{B}_{gn}$ share the same mean and variance and that their difference $B_{gn} - \hat{B}_{gn}$ has a mean of zero, i.e., $E(B_{gn} - \hat{B}_{gn}) = 0$. One would also expect that the difference has a smaller variance than the original background noise, i.e.,

$\sigma^2(B_{gn} - \hat{B}_{gn}) < \sigma^2(B_{gn})$. If the estimate $\hat{B}_{gn}$ imitates the first two moments of the true background noise B_{gn} in the sense of having the same mean and variance, then it can be shown that $\hat{B}_{gn}$ and B_{gn} are moderately correlated if the corrected intensity measure $w_{gn}^{(c)}$ is to be less variable than the original intensity reading w_{gn}. Specifically, the true background noise and its estimate must have a correlation coefficient exceeding 0.5 if the background correction is to be helpful in this sense.

Diagnostic checks are available to verify whether background noise correction has been effective. The checks are not conclusive but can be helpful. The most effective checks involve examining genes suspected of having no true gene expression. These will be genes for which w_{gn} is small. This judgment can be made using a histogram such as that in Figure 5.3. One can select a cutoff point at which it appears that the bulk of genes found to the left of the cutoff mainly reflect simple noise with no true gene expression. For genes to the left of the cutoff, the observed expression level w_{gn} will consist largely of the pure background noise B_{gn}. For these genes, the B_{gn} and their estimates $\hat{B}_{gn}$ should be close if the background estimates are reliable. This proximity can be checked in a plot of w_{gn} against $\hat{B}_{gn}$ for w_{gn} smaller than the cutoff.

As background correction is never perfect, model (5.1) is also valid for background-corrected data because each reading retains two components. The component reflecting latent gene expression, namely x_{gn}, is present. Also, the difference $B_{gn} - \hat{B}_{gn}$ remains as a residual background noise component. Now, however, the residual background component may assume both positive and negative values. Of course, if the background correction has been effective, the residual background component will tend to have a smaller magnitude than the original background noise component.

In the remainder of this chapter and those that follow, the context will make it clear whether the gene expression readings under discussion are background-corrected. We will only use superscript (c) where there is a need for explicit notation.

5.2.3. Example: Replication Test Data Set

The replication test data set introduced in Lee *et al.* (2000) will be used to illustrate some of the concepts and methods in this section. As was mentioned in section 4.3, the data set was produced from a small experiment that aimed to study the effects of variability and replication on the reliability of cDNA microarray findings. The study design consisted of 288 genes, each printed at three locations on the same slide. A single human tissue sample formed the target biological sample. The experiment was designed so that 32 of the 288 genes would be expected to be highly expressed because of Alu repeats[5] that should cross-hybridize to similar sequences widely distributed among expressed and unexpressed portions of the human genome. Only output from channel 1 (green) contained expression readings for the target tissue. Output from channel 2 (red) contained noise alone. Counting the $288(3) = 864$ spots as the designated gene set, we have $G = 864$. There is only one biological specimen, so $N = 1$. Note that the designated gene set $\mathcal{G}_A$ on the array contains triplicates of 288 distinct genes. This data set is used to demonstrate a number of points in the discussion that follows and also later in the text.

The readings of gene expression w_{g1} for the 864 cDNA spots are displayed in a histogram in Figure 5.3. Here index $n = 1$ because only one specimen is under consideration. The histogram shows the gene expression data in terms of their common logarithms (i.e., logarithms to base 10). The gene expression data are the output denoted as CH1I in the ScanAlyze system, which are the uncorrected mean pixel intensities of spots for the green fluorescent hybridization (Eisen, 1999). Observe the large concentration of small readings that generally correspond to noise. A scattering of larger readings also appear that are mainly associated with gene probes that should be highly expressed. The logarithmic scale amplifies the detail in the lower range of the data.

The noise component of the data appears unimodal and roughly symmetrically distributed. The distribution of the smaller number of expressed genes stands out as a separate distribution to the right end of the histogram scale. Thus, the distribution pattern of the data looks very much like a mixture of gene expression readings for a large number of unexpressed genes (noise alone) and a smaller number of genes expressed to varying degrees.

Taking the Replication Test Data Set as a case example, a log-value of 3.8 is used as a cutoff for unexpressed genes (based on a rough visual judgment for Figure 5.3). It is found that 761 of the 864 spots lie below this cutoff level. As the experiment involved 256 gene triplicates that should be unexpressed, the count $3(256) = 768$ closely matches this count of 761. Among the 761 spots, a plot of w_{gn} against $\hat{B}_{gn}$ shows a fairly clear linear relationship but one that does not follow the line of identity. The scatter plot and line of identity appear in Figure 5.4. The correlation coefficient for w_{gn} and $\hat{B}_{gn}$ is 0.773. What is a little more surprising is that only 9 of the background-corrected readings $w_{gn}^{(c)}$ for the 761 unexpressed genes are negative. If the background noise

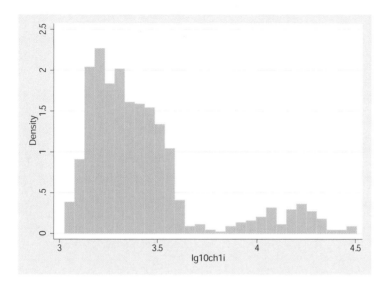

Figure 5.3.   Histogram of log-expression data for the Replication Test Data Set (using common logarithms)

estimates are unbiased estimates of their true values, then about half of the unexpressed genes would be expected to have negative values for $w_{gn}^{(c)}$.

The plot shows clearly that the values of w_{gn} for the 761 spots are consistently larger than their counterpart background noise estimates $\hat{B}_{gn}$ (i.e., they lie above the line of identity). This fact is also indicated by a comparison of their mean values; 2250 being the mean for w_{gn} and 1722 for $\hat{B}_{gn}$. Thus, the results of the diagnostic check suggest that background correction for this case example is moderately successful at best.

The case example is convenient because there is a clear separation of expressed and unexpressed genes (Figure 5.3). In many applications, however, the separation is not as clear, and the set of unexpressed genes required for the diagnostic test will not be so easy to discern.

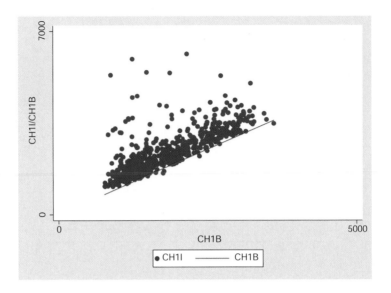

Figure 5.4. Scatter plot of expression intensities and background estimates for 761 genes classified as unexpressed. The plot also shows the line of identity.

5.2.4.   Noise Models for GeneChip Arrays

On the basis of the difference measure PM−MM from Affymetrix GeneChip arrays, Li and Wong[6] (2001) introduce the following noise model

$$PM_{ij} - MM_{ij} = \theta_i\phi_j + \epsilon_{ij}, \tag{5.5}$$

where θ_i is the fitted expression index of sample i, $i = 1,\ldots,n_s$, $\epsilon_{ij} \sim N(0,\sigma^2)$ is the error term, and ϕ_j is the sensitivity of probe j , $j = 1,\ldots,n_p$, with the constraint $\sum_j \phi_j = n_p$ in order that the solution be unique.

Šášik, Calvo, and Corbeil[7] (2002) point out that, under the Li and Wong model in (5.5), the expression index θ_i can be negative. Also, genes with negative θ can still be classified as present (as can genes with negative average difference in the Affymetrix Microarray suite[8] (5.0)). Hence, they propose a model that does not suffer from the above problems. It is based on an assumption similar to that underlying the Li-Wong model, that fluorescent intensity of a PM probe (properly adjusted for background B) is directly proportional to the concentration ζ_i of the transcript on sample i, that is,

$$PM_{ij} - B \sim \phi_j\zeta_i \tag{5.6}$$

Taking the logarithm (to base 2) of both sides, the multiplicative model in (5.6) can be written as an additive model on the logarithmic scale

$$\log_2(\text{PM}_{ij} - B) \sim \log_2 \phi_j + \log_2 \zeta_i \tag{5.7}$$

Defining $\eta_{ij} \equiv \log_2(\text{PM}_{ij} - B)$, $\psi_j \equiv \log_2 \phi_j$, and $\gamma_i \equiv \log_2 \zeta_i$, then the model can be written as

$$\eta_{ij} = \psi_j + \gamma_i + \epsilon_{ij} \tag{5.8}$$

where $\epsilon_{ij} \sim N(0, \sigma^2)$. There is one such equation for each transcript, and they allow for the possibility that the variance σ^2 of the error term is different for each transcript. In model (5.8), regardless of the fitted values of ψ_j and γ_i, the probe sensitivities ϕ_j and concentrations ζ_i can never be negative, because $\phi_j \equiv 2^{\psi_j} > 0$ and $\zeta_i \equiv 2^{\gamma_i} > 0$.

5.2.5.   Elusive Nature of Background Noise

The preceding sections have discussed the nature of background noise and statistical models for its correction. It was noted that background correction must be checked to ensure that it has been beneficial. Several investigations have shown that the subject is elusive for both spotted arrays and *in-situ* oligonucleotide arrays.

Many researchers have chosen to use foreground intensities from spotted arrays without background correction because of concerns regarding the effectiveness of subtracting the machine-generated background noise measure.

One of several problems with background measurement is image segmentation. From a statistical perspective, image segmentation is a data-dependent procedure. Segmentation in spotted microarrays may yield biased intensity readings because array spots tend to be centered on those regions of the image with the brightest fluorescence. This bias is illustrated by the case example that considered the analysis of background correction for the Replicate Test Data Set in section 5.2.3. Figure 5.4 revealed that the background intensity calculated by the ScanAlyze system in this illustration appeared to be too small, on average, leading to background-corrected intensities that were predominantly positive for genes that were known to be unexpressed.

The difficulty of separating signal from noise has also been noted in *in-situ* oligonucleotide microarrays. Naef *et al.*[9] (2003), for example, note that, whereas conventional wisdom views the PM probes as carrying the signal while MM probes serve as non-specific controls, their experience shows that the MM probes actually track the signal. They state that "the MMs should be viewed as a set of average lower affinity probes". They then go on to develop a method to exploit the signal content of the MM probes. Along a similar line, Irizarry *et al.*[10] (2003) remark that "Recent results ... suggest that subtracting MM as a way of correcting for non-specific binding is not always appropriate". They cite two sources for this remark and go on to say that "until a better solution is proposed, simply ignoring these values is preferable."

Notes

[1] Brown, C.S., Goodwin, P.C., and Sorger, P.K. (2001). *Proceedings of the National Academy of Sciences, USA*, **98**, 8944-8949.

[2] Schena, M., Shalon, D., Davis, R.W., and Brown, P.O. (1995), *Science*, **270**, 467-470.

[3] DeRisi, J.L. , Iyer, V.R., and Brown, P.O. (1997), *Science* **278**, 680-686.

[4] Eisen, M.B. (1999). *ScanAlyze User Manual*, Version 2.32; Stanford University: Stanford, CA.

[5] Lee, M.-L.T., Kuo, F.C., Whitmore, G.A., Sklar, J. (2000). *Proceedings of the National Academy of Sciences, USA*, **97**, 9834-9839.

[6] Li, C., Wong, W.H. (2001). *Proceedings of the National Academy of Sciences, USA*, **98**, 31-36.

[7] Sasik, R., Calvo, E., Corbeil, J., (2002), *Bioinformatics*, **18**, 1633-1640.

[8] Affymetrix (2002). http://www.affymetrix.com/products/25mer-content.html.

[9] Naef, F., Socci, N. D. and Magnasco, M. (2003). *Bioinformatics*, **19**(2), 178-184.

[10] Irizarry, R.A., Bolstad, B.M., Collin, F., Cope, L.M., Hobbs, B. and Speed, T.P. (2003). *Nucleic Acids Research*, **31**(4), 1-8.

Chapter 6

TRANSFORMATION AND NORMALIZATION

6.1. Data Transformations

Gene expression readings w_{gn} will not necessarily have desired statistical properties without being transformed. Desirable properties may include normality or constant variance, for example. Here we discuss transformations of the data and related issues that have proven useful in analyzing microarray data. We shall let $y_{gn} = h(w_{gn})$ denote the transformed value where h is a selected mathematical function. We will limit our attention to monotonic increasing transformations, i.e., those for which transformed measurements y increase with observed measurements w.

6.1.1. Logarithmic Transformation

By far the most common transformation applied to microarray readings is the logarithmic transformation

$$y_{gn} = \log w_{gn}. \qquad (6.1)$$

The base of the logarithm may be 2, 10, or the natural logarithmic constant e. The choice of base is largely a matter of convenient interpretation.

There are several issues that arise with the use of a logarithmic transformation. If the readings are initially corrected for background noise as described in (5.2) or (5.3), then the *background-corrected transformed value*, which we denote by

$$y_{gn}^{(c)} = \log w_{gn}^{(c)}, \qquad (6.2)$$

will be undefined if $w_{gn}^{(c)}$ is zero or negative. Thus, the transformation is reserved for readings where intensity exceeds background. This fact will affect the interpretation of statistical models and findings.

A logarithmic transformation is used for microarray data because it tends to provide values that are approximately normally distributed and for which conventional linear regression and ANOVA models are appropriate. If the transformed data are approximately normal, the implication is that the raw readings are lognormally distributed. The multiplicative version of the central limit theorem can be used as a partial justification for the assumption of a lognormal form for the components of additive background noise model (5.1). For example, both the latent expression component x_{gn} and noise component B_{gn} of the microarray reading w_{gn} may be generated as the result of many independent multiplicative random effects that produce the final values of these components. As it happens, however, even if both x_{gn} and B_{gn} are independent and lognormally distributed, their sum will not have this kind of distribution. Thus, the log-transformed values y_{gn} can only be approximately normal in this situation.

The final test of whether log-intensity readings are appropriate for statistical analysis of microarray data is whether reliable and useful results can be derived. Various diagnostic methods are available for checking for major departures from normality and other model failures. For example, normal probability plots can be used to provide a check on normality. Some of these diagnostic methods will be demonstrated in case studies taken up later. As already noted, the log-transformation is widely used and has been well defended by investigators in many studies. Thus, it is a key transformation for microarray gene expression data. Figure 5.3 shows the distribution of such data after a logarithmic transformation.

6.1.2.   Square Root Transformation

It is reasonable to expect that intensity readings in microarrays will be proportional to the number of occurrences of fundamental molecular events such as hybridizations. The constant of proportionality would be the quantum of fluorescence, radiation, or other signal produced by a single fundamental event. The fundamental events would fall into two categories: a homogenoeus category representing hybridizations that count as true expression for a gene and a heterogenous category of events (such as non-specific binding) that represent noise. The latter category contains events that are not of scientific interest but nonetheless are being counted by the microarray instrument. A plausible model for the number of each type of event is a Poisson distribution. As the gene expression and noise components in model (5.1) are assumed to be independent and the sum of independent Poisson variables remains Poisson distributed, it follows that the gene expression reading w_{gn} will be proportional to a Poisson random variable. The Poisson model implies that the variance of w_{gn} is proportional to its mean value. Microarray data often have variability that tends to increase with the level of the reading. If this relation is one in which the variance increases in proportion to the mean, then the preceding Poisson model is plausible.

For microarray data that follow this Poisson model, the square root function

$$y_{gn} = \sqrt{w_{gn}} \tag{6.3}$$

is a variance-stabilizing transformation. Thus, the transformed values y_{gn} will tend to have a constant variance. For the Replication Test Data Set, the histogram of square-root transformed data is not much different in appearance than that shown for log-transformed data in Figure 5.3.

6.1.3.   Box-Cox Transformation Family

The two preceding transformations are two members of the Box-Cox family of transformations. This family is defined as follows.

$$y_{gn} = \frac{w_{gn}^d - 1}{d} \tag{6.4}$$

The square root transformations correspond to parameter d being $1/2$. The logarithmic transformations correspond to the limit of equation (6.4) when d approaches 0. The case where $d = 1$ corresponds to taking no transformation at all, except for a change in origin.

The Box-Cox family provides a range of transformations that may be examined to see which value of d yields transformed values with the desired statistical properties.

6.1.4.   Affine Transformation

It is sometimes desirable or necessary to shift the origin of the intensity measurement scale by some fixed amount, say a. Thus, if w_{gn} is the intensity reading, then $a + w_{gn}$ is the adjusted reading. We shall refer to a as a *shift parameter* or *offset* and to the transformation as an *affine transformation*, meaning a shift away from the origin.

The affine transformation may be combined with other transformations such as the logarithmic transformation, in which case we have

$$y_{gn} = \log(a + w_{gn}).$$

As the logarithmic transformation requires positive arguments, the affine transformation may be used with the logarithmic transformation where intensity readings have been background-corrected and some readings are negative. In this case, the offset parameter a would be chosen so that $a + w_{gn}^{(c)}$ is positive for all readings.

Kerr *et al.*[1] (2002) have found that applying opposing affine transformations to the two color intensities on a spot of a cDNA array may improve the correspondence of intensity readings. Denoting the two color readings for a spot g on the nth array by $w_{gn}^{(R)}$ and $w_{gn}^{(G)}$, a plot of $\log(w_{gn}^{(R)})$ against $\log(w_{gn}^{(G)})$ for all g usually shows a curvilinear scatter of points, indicating that the two color fluors measure gene expression differently. The authors show that the following affine transformation may eliminate the nonlinearity in the scatter plot and put the two colors on an equal footing.

$$y_{gn} = \log \ \left(\frac{w_{gn}^{(R)} - a_n}{w_{gn}^{(G)} + a_n} \right) \tag{6.5}$$

In this case, the offset parameter a_n is chosen separately for each array n and then applied across all genes on the array. The red (R) and green (G) color readings can be reversed in calculating this log-ratio of intensities. Observe that the offset parameter tends to keep the average intensity of the combined colors roughly unchanged but shifts one color reading relative to the other. The shift is most pronounced for low intensities. The authors propose that parameter a_n be chosen to minimize the sum of the absolute deviations of the y_{gn} observations from their median value for all genes g. This fitting criterion has the effect of giving a nearly horizontal scatter in an MA plot for the two transformed intensities. See section 6.2.4 for a discussion of such plots.

The effect of the affine adjustment on the transformed expression intensities may need to be explored in different applications. For example, with a transformation $y = \log(a + w)$, the first derivative dy/da equals $1/(a + w)$ for natural logarithms. This fact tells us that a small change in the shift parameter a will have a small impact on large expression values because $1/(a + w)$ declines rapidly with increasing w but will be large when $a + w$ is close to zero.

6.1.5. The Generalized-log Transformation

Extending a model introduced by Rocke and Lorenzato[2] (1995) and Rocke and Durbin[3] (2001), Durbin *et al.*[4] (2002) propose a two-component error model for gene expression data from microarrays. Let w denote the measured raw expression level, μ the true expression level, and b the mean background noise (mean intensity of unexpressed genes). They demonstrate that the measured expression levels from microarray data can be modelled as

$$w = b + \mu e^\eta + \epsilon \tag{6.6}$$

where $\eta \sim N(0, \sigma_\eta^2)$ represents the proportional error that always exists, but is noticeable mainly for highly expressed genes, and $\epsilon \sim N(0, \sigma_\epsilon^2)$ represents the error for the background noise. Random variables η and ϵ are taken as independent.

Under this model, the variance of the measured intensity w at true level μ is given by

$$\mathrm{Var}(w) = \mu^2 \, S_\eta{}^2 + \sigma_\epsilon{}^2 \tag{6.7}$$

where $S_\eta \equiv \sqrt{e^{\sigma_\eta^2} \left(e^{\sigma_\eta^2} - 1\right)}$ denotes the approximate relative standard deviation (RSD) of w for high levels of expression.

At low expression levels (i.e., μ close to 0), the measured expression can be written as

$$w \approx b + \epsilon \tag{6.8}$$

implying that the measured expression in (6.8) is approximately normally distributed with mean b and constant variance $\sigma_\epsilon{}^2$. When μ is large, the measured expression may be modelled by

$$w \approx \mu \, e^\eta \tag{6.9}$$

On the log scale, (6.9) can be written as

$$\ln(w) \approx \ln(\mu) + \eta \tag{6.10}$$

which implies that $\ln(w)$ has constant variance for μ sufficiently large and that $\ln(w)$ is distributed approximately as a normal random variable with mean $\ln(\mu)$ and variance σ_η^2. All terms in (6.6) play a significant role only when the expression level μ falls between these two extremes.

The measured expression w is then distributed as a linear combination of a normal and a lognormal random variable and has variance (6.7). Therefore, the distribution of the measurement error changes depending on μ, i.e., the variance changes with the mean in a non-linear fashion.

Durbin *et al.* (2002) use a mathematical procedure called the delta-method to derive the following transformation for microarray data that stabilizes the asymptotic variance over the full range of the measured data.

$$y = h(w) = \ln[\, w - b + \sqrt{(w - b)^2 + c}\,] \qquad (6.11)$$

Here c is the constant

$$c = \frac{\sigma_\epsilon^2}{\exp(\sigma_\eta^2) - 1}. \qquad (6.12)$$

The transformation in (6.11) has the general form reported by the authors, but the expression for constant c in terms of the model parameters differs from that obtained in the cited paper. The two versions of the transformation differ little when σ_η^2 is small. This transformation stabilizes the asymptotic variance of data distributed according to model (6.6). For a large value of w, the transformation (6.11) is approximately the natural logarithm. At w near zero, the transformation (6.11) is approximately linear. This transformation was considered by Hawkins[5] (2002) in the context of another application.

Geller *et al.*[6] (2003) show that data from Affymetrix GeneChips conform to the same two-component model in Durbin *et al.* (2002). Huber *et al.*[7] (2002) also consider a family of transformations that is related to the generalized-log family.

6.2.   Data Normalization

The laboratory preparation of each biological specimen n on a microarray slide introduces an arbitrary scale or dilution factor that is common to gene expression readings w_{gn} for all genes g. Analysts usually correct the readings for this scale factor and other variations using a process called *normalization*. The purpose of normalization is to minimize extraneous variation in the measured gene expression levels of hybridized mRNA samples so that biological differences (differential expression) can be more easily distinguished.

Normalization is commonly based on the expression levels of all genes on an array. Assuming that most genes are similarly expressed in both the test and the reference sample and that the number of upregulated genes largely matches the number of downregulated genes, normalization factors can be based on the total fluorescence, expression ratios, or regression analysis. Total fluorescence normalization assumes that approximately the same total amount of test and reference sample has hybridized, and thus the total fluorescence of both dyes used should be the same on the array. A normalization factor calculated from the ratio of the total fluorescence of the dyes can be used to re-scale the intensity of each gene on the array.

Practical experience has shown that, in addition to array effects, other extraneous sources of variation may be present that cloud differential gene expression if not taken into account. These sources include many mentioned in Chapter 4, such as effects of dye color, pin tips, and spatial anomalies on slides. For example, an examination of two-color cDNA data sets shows that fluorescence of an array spot varies with the amount of hybridization in a different way for each dye color, i.e., depending on whether the fluor is red or green. These systematic color differences need to be taken into account. Pin-tip differences represent a potential source of variation. Spots are printed on the array by pins that are configured in a particular pattern for the experiment. As the pins are robotically controlled, the print pattern is reproduced across the arrays. To quote Yang et al.[8] (2002), pin-tip effects may result from "slight differences in the length or in the opening of the tips, and deformation after many hours of printing." Some systematic variation may be associated with regions of the slide surface. Pin-tip differences are one regular source of spatial variation, but others may also be present. A slide may need to be partitioned into regions and adjustments carried out for each region.

It will be demonstrated that normalization should encompass all material sources of extraneous variation. In essence, normalization gives approximately the same 'average' level of gene expression across all genes for each combination of experimental conditions. Sometimes analysts also wish to scale expression readings for each gene g so that the 'average' level of gene expression is the same across all conditions for each gene. The normalization procedures differ with respect to which kind of average is used and what sources of variability are taken into account. Several authors have given systematic discussions of normalization. See, for example, Yang et al (2002), among others.

6.2.1.   Normalization Across G Genes

In this section we give a general framework for normalization based on the Box-Cox family of transformations. For any transformed intensity $y_{gn} = h(w_{gn})$ in the form of (6.4), the *normalized value* is given by the centered differences

$$y_{gn} - \overline{y}_{+n} = h(w_{gn}) - \frac{1}{G} \sum_{g=1}^{G} h(w_{gn}) \qquad (6.13)$$

where $\overline{y}_{+n}$ is the arithmetic mean value of the transformed intensity y_{gn} over all genes g. From equation (6.13) and the definition of a mean, these transformed normalized readings necessarily sum to zero for each experimental condition, as follows.

$$\sum_{g=1}^{G} (y_{gn} - \overline{y}_{+n}) = 0 \qquad (6.14)$$

To relate the transformed mean $\overline{y}_{+n}$ to a measure of central location on the original scale of intensity (i.e., on the w scale), the inverse transformation h^{-1} is applied as follows, giving the measure $\hat{w}_{+n}$:

$$\hat{w}_{+n} = h^{-1}(\overline{y}_{+n}) = h^{-1}\left[\frac{1}{G} \sum_{g=1}^{G} h(w_{gn}) \right] \qquad (6.15)$$

For the special case of the identity transformation where $d = 1$ in (6.4), the normalized values of the transformed readings are simply the centered differences

$$y_{gn} - \overline{y}_{+n} = w_{gn} - \hat{w}_{+n} = w_{gn} - \overline{w}_{+n},$$

where $\hat{w}_{+n} = \overline{w}_{+n}$ is the arithmetic mean of the w_{gn} intensity readings.

For the limiting case where d approaches 0, the normalized values of the transformed readings take the logarithmic form

$$y_{gn} - \overline{y}_{+n} = \log w_{gn} - \log \hat{w}_{+n} = \log(w_{gn}/\hat{w}_{+n}),$$

where $\hat{w}_{+n}$ is the geometric mean of the intensity reading w_{gn} across genes. Often one considers the *scale-normalized intensity readings* defined by

$$w_{gn}/\hat{w}_{+n}$$

Observe that, for any given array n, if all gene expression readings w_{gn} are multiplied by an arbitrary positive constant, the scale-normalized readings would be unchanged for any of the Box-Cox transformations. Thus, the ratios eliminate any scaling factor that cuts across all genes within any experimental condition.

6.2.2. Example: Mouse Juvenile Cystic Kidney Data Set

A microarray data set introduced in Lee *et al.*[9] (2002b) will be used to illustrate normalization of data and also concepts and methods in following sections. This microarray data set was collected to investigate differential gene expression in kidney tissue from wild-type and mutant mice with juvenile cystic kidneys. Autosomal dominant polycystic kidney disease (ADPKD) is one of the most common monogenic diseases, characterized by the presence of multiple fluid-filled epithelial cysts in the kidneys. Animal models of polycystic kidney disease (PKD) are an ideal resource for investigating the perturbations of gene expression that occur in this disorder. Mouse mutants have the advantage of being maintained in inbred genetic strains, so variation of gene expression due to genetic background is minimized. Of particular significance is the opportunity to investigate differential expression that may be common to different models of PKD. These may represent molecular events that occur not as a result of a specific mutation but as a consequence of a more general injury to tubular integrity. As such, they may suggest pathways of molecular events that can be helpful for understanding the fundamental defect that occurs in PKD. Furthermore, they may suggest avenues of therapeutic intervention that can be investigated as a means to ameliorate disease progression.

The experimental design in this study was chosen by the biological scientists in advance of any consideration of statistical issues. It is not ideal from a statistical viewpoint but is not atypical as far as microarray studies are concerned. In this design, eight readings were gathered for each gene in four microarray pairs, according to the pattern set out in Table 6.1. *Array* in Table 6.1 refers to the four microarray pairs (Arrays 1 to 4). *Channel* refers to whether the expression reading comes from the Cy3 green fluor channel (Channel 1) or the Cy5 red fluor channel (Channel 2). *Type* refers to kidney tissues from two mouse species, mutant (Type 1) and wild-type (Type 2), as described in the next section. The inclusion of a color channel effect in the design was prompted by the scientists' concern that gene expression profiles might differ by channel – a concern that turned out to be justified. The experimental design is not

	Channel 1 Cy3 (Green)	Channel 2 Cy5 (Red)
Array 1	mutant	mutant
Array 2	mutant	wild-type
Array 3	wild-type	mutant
Array 4	wild-type	wild-type

Table 6.1. Experimental design for the Mouse Juvenile Cystic Kidney Data Set

mutually orthogonal in the three factors *Array, Channel,* and *Type.* In particular, each array necessarily generates expression readings on both color channels. In this particular design, *Channel* is orthogonal to *Array* and *Type*, but the latter two factors are partially confounded.

The data set in this experiment contains *ScanAlyze*[10] cDNA gene expression data for 1,728 genes. All 1,728 cDNA clones were generated in-house. Approximately 1,152 murine cDNA clones came from the murine brain cDNA library and 576 from the rat brain cDNA library. All clone sequences were verified by single-pass 3' sequencing. Each cDNA insert was amplified by PCR and was printed in duplicate on treated glass microscope slides using a Genetics Microsystems 417 arrayer in linear pattern. Averages of the duplicated readings were used in the analysis.

In the notation of this book, the design in this study involves $G = 1,728$ genes and $N = 8$ experimental conditions. We now illustrate normalization for the Mouse Juvenile Cystic Kidney Data Set. The systematic effects of array and channel (dye color) will be removed in this normalization. Table 6.2 gives the geometric mean expression levels for the 1,728 genes for each array-channel combination in the microarray study. The expression levels are not background-corrected. The geometric mean corresponds to $\hat{w}_{+n}$ for the logarithmic transformation (Box-Cox parameter $d = 0$). To illustrate a normalized reading, we note that the raw expression reading for gene g_{137} for the mutant tissue in array 2 on the green channel is 2820. Thus, its scale-normalized reading for the log-transformation is $2820/3307 = 0.8527$. The transformed normalized reading is $\log(2820) - \log(3307) = \log(0.8527) = -0.1593$, using natural logarithms. Scale-normalized readings and transformed normalized readings for other genes in the data set would be calculated accordingly. The identity in (6.14) will hold for the transformed normalized readings of the 1,728 genes under each of the eight experimental conditions in this study.

	Channel 1 Cy3 (Green)	Geometric Mean	Channel 2 Cy5 (Red)	Geometric Mean
Array 1	mutant	2813	mutant	2042
Array 2	mutant	3307	wild-type	2034
Array 3	wild-type	2040	mutant	2593
Array 4	wild-type	1815	wild-type	1797

Table 6.2. Geometric mean expression levels $\hat{w}_{+n}$ for each combination n of array, channel and mouse type in the Mouse Juvenile Cystic Kidney Data Set

Looking at the geometric means across the experimental conditions, we note that there is considerable variation in the mean values, the largest being almost double the smallest. Mean gene expression levels for mutant tissue tend to be larger than for wild-type tissue. Also, there is a suggestion that green levels are generally higher than red levels. This last observation is consistent with the remark in section 3.8 about differences in the Cy3 and Cy5 dyes.

6.2.3. Normalization Across G Genes and N Samples

Where normalization is to be carried out across both genes and experimental conditions, the *normalized value* for transformed intensity reading $y_{gn} = h(w_{gn})$ is of the form

$$
y_{gn} - \overline{y}_{+n} - \overline{y}_{g+} + \overline{y}_{++}
$$
$$
= h(w_{gn}) - \frac{1}{G} \sum_{g=1}^{G} h(w_{gn}) - \frac{1}{N} \sum_{n=1}^{N} h(w_{gn}) + \frac{1}{GN} \sum_{n=1}^{N} \sum_{g=1}^{G} h(w_{gn})
$$

Here $\overline{y}_{+n}$, $\overline{y}_{g+}$ and $\overline{y}_{++}$ denote arithmetic means of transformed intensity readings taken over genes, conditions, and over all genes and conditions, respectively. It is clear that the transformed normalized readings in this case will sum to zero across both genes and experimental conditions. As we will show later, these transformed normalized readings give a rough picture of *differential gene expression* because they represent differences in gene expression across experimental conditions for each gene.

6.2.4.   Color Effects and MA Plots

Before continuing with the discussion of normalization, it is necessary to digress briefly to consider color effects and the relation of differential expression to average intensity levels. This discussion lays the groundwork for understanding their role in normalization.

To start the discussion of color effects, we again denote the red and green intensity readings by $w^{(R)}$ and $w^{(G)}$. We suppress the subscripts for gene and experimental condition for the moment. To show that fluorescence of an array spot varies with the amount of hybridization in a different way for each color, we might plot $\log w^{(R)}$ against $\log w^{(G)}$. Many other authors have used similar graphs to demonstrate this point.

In contrast, Dudoit et al.[11] (2000) and Yang et al. (2002) propose a plot of the log intensity ratio

$$M = \log(w^{(R)}/w^{(G)}) = \log w^{(R)} - \log w^{(G)} \qquad (6.16)$$

against the average log-intensity

$$A = \log \sqrt{w^{(R)} w^{(G)}} = \frac{1}{2}(\log w^{(R)} + \log w^{(G)}). \qquad (6.17)$$

A plot of M versus A, referred to as an MA plot, amounts to a 45° rotation of the $(\log w^{(R)}, \log w^{(G)})$ coordinate system, followed by scaling of the coordinates. Difference M captures the color difference on a log-scale and A represents the average log-intensity for the two colors.

To demonstrate an MA plot and the color asymmetry, we consider data from the Mouse Juvenile Cystic Kidney Data Set. Figure 6.1 shows the difference M in normalized log-intensity for the red and green channels plotted against the average normalized log intensity A for all genes. The normalization is that in (6.14), with $h(w) = \log(w)$, which involves centering of the log-readings on each color channel. The raw data correspond to CH1I and CH2I for Array 1 (mutant tissue) in Table 6.1 and, hence, are not background-corrected.

The graph shows the zero lines for M and A (the log-averages). The graph also shows the smooth values of M as a LOWESS[12]-fitted function of A. The scatter plot of the points is concave upward. Since the gene expression readings for both colors are coming from the same mutant tissue, the curvature in the smooth fitted function illustrates the difference in intensity response between the two colors as a function of the level of expression (as measured by average A). Note that the red color tends to be relatively more intense at lower and higher intensity levels

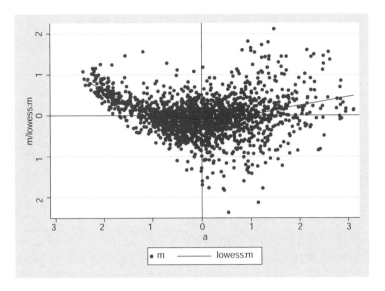

Figure 6.1. MA plot.
The difference M in normalized log-intensity for the red and green channels plotted against the average normalized log intensity A for all genes. The intensity data are from Array 1 (mutant tissue) of the Mouse Juvenile Cystic Kidney Data Set and are not background-corrected.

than green, producing the concave upward pattern in the scatter plot and smooth fitted function.

Locally weighted regression (LOWESS) (Cleveland 1979, 1981) is a method for smoothing scatterplots in which the fitted value at x_k is the value of a line fitted to the data using weighted least squares where the weight for point (x_i, y_i) is large if x_i is close to x_k and small if x_i is not close to x_k. A robust fitting procedure guards against outliers distorting the smoothed points[13].

The color asymmetry demonstrated in Figure 6.1 has led to the use of microarray study designs in which arrays are produced in pairs with the colors in one array reversed relative to the colors in the second array in order to compensate for the color differences. These are called *reversed-color designs* and are discussed in more detail in Chapter 9. Thus, a reversed-color experiment copes with the color asymmetry by obtaining two gene expression readings for each gene g under each experimental condition n, one from each color channel.

The design layout in Table 6.3 shows the structure of a reversed-color design for two experimental conditions c_1 and c_2 for a single gene. Letting $\overline{y}_1$ and $\overline{y}_2$ denote the average normalized transformed intensities

	Green dye d_1	Red dye d_2
Array a_1	sample c_1	sample c_2
Array a_2	sample c_2	sample c_1

Table 6.3.   Reversed-color design

for experimental conditions c_1 and c_2, respectively, an MA plot for this design can be constructed using the following definitions of M and A.

$$M = \overline{y}_2 - \overline{y}_1, \quad A = \frac{1}{2}\left(\overline{y}_1 + \overline{y}_2\right)$$

Note that M captures the differential expression for the two experimental conditions and A measures the mean intensity (both on the transformed scale). As $\overline{y}_1$ and $\overline{y}_2$ are averages over both arrays and colors, the effects of these two factors are neutralized. Thus, the MA plot in this case will show a relationship between differential expression and average intensity that is free of color bias. Figure 6.2 shows this plot for the Mouse Juvenile Cystic Kidney Data Set. The plot uses the data only from Arrays 2 and 3 in Table 6.1, as these contain an imbedded reversed-color design. Because the data are normalized, the MA plot has axes centered on zero.

A LOWESS function has been fitted to the plot. Note its approximate linearity and the relatively uniform scatter of points about the smooth function. Also, note the tendency for differential expression to decline with intensity. The correlation coefficient for M and A across all genes is -0.51. The figure shows that there is a relationship between differential expression and average intensity and that it is not solely a color phenomenon.

6.2.5.   Normalization Based on LOWESS Function

Yang *et al.* (2002) suggest a normalization for the color effect that uses LOWESS smoothing. The approach is referred to as *intensity-dependent normalization*. In this approach, a LOWESS function $\ell(A)$ of the average intensity A is fitted to the scatter plot and then is used to normalize the log-intensity ratio M by computing the difference $M - \ell(A)$. More specifically, the authors show that color differences may vary by the pin tip of the arrayer and therefore suggest that the preceding normalization be carried out separately for each pin tip, in essence, giving normalized log-intensity ratios of the form $M_p - \ell_p(A)$, where p is an index for the

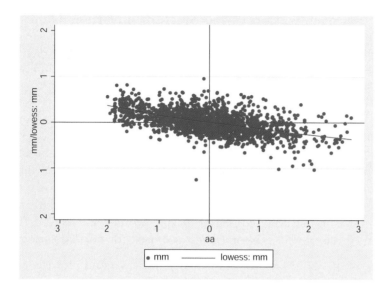

Figure 6.2. MA plot with color reversal.
The difference M in normalized log-intensity for wild-type and mutant tissues plotted against the average normalized log intensity A for all genes. The intensity data are from Arrays 2 and 3 of the Mouse Juvenile Cystic Kidney Data Set and are not background-corrected.

pin tip and $\ell_p(A)$ is the LOWESS function fitted to the data from the pth pin tip.

Instead of using the entire set of genes to fit the LOWESS curve for normalization, Tseng *et al.*[14] (2001) fit the LOWESS curve to a selected set of rank-invariant genes before conducting nonlinear normalization (see subsection 6.2.6 for more details). The curve is extrapolated to genes with the highest and lowest intensities, as these are excluded from the rank-invariant gene set by definition.

The need for nonlinear normalization is also noted by Schadt *et al.*[15] (2001) in analyzing data from *in-situ* oligonucleotide microarrays.

Because hybridization is often not uniform over the array, Wernisch *et al.*[16] (2003) considered two-dimensional LOWESS normalization. They fit a two-dimensional LOWESS surface to log-ratio values of the array set. Once a surface has been fitted, log-ratio values can be corrected by subtracting the surface values.

6.2.6.   Normalization Based on Rank-invariant Genes

Consider a cDNA spotted array experiment in which mRNAs for two specimens are separately labelled with red and green fluors and co-hybridized on the same array. One may wish to base normalization not on the entire designated gene set $\mathcal{G}_A$ but on a subset of genes that are not differentially expressed in the two specimens. This idea is pursued by Tseng *et al.* (2001) who propose a rank-invariant method to select a subset of genes as a basis for normalization, which might be called a *rank-invariant gene set*. If the ranks of the red and green intensities of a gene g are similar and if the rank of its average intensity on both channels is not extremely low or high, then one might conclude that the gene is not differentially expressed in the two specimens and therefore can be included in the rank-invariant set used for normalization.

Tseng *et al.* (2001) choose threshold numbers d and l for the following two criteria and select all genes g for the invariant set that satisfy them.

$$|\text{Rank}(w_g{}^{(R)}) - \text{Rank}(w_g{}^{(G)})| < d \qquad (6.18)$$

$$l < \text{Rank}\{(w_g{}^{(R)} + w_g{}^{(G)})/2\} < K - l \qquad (6.19)$$

Here $w_g{}^{(G)}$, and $w_g{}^{(R)}$ denote the green and red intensity levels of gene g, respectively, and $\text{Rank}(w)$ denotes the rank of intensity w among all K genes.

6.2.7.   Normalization Based on a Sample Pool

Yang *et al.* (2002) propose normalization methods that are based on robust local regression and account for intensity and spatial dependence in dye biases. They use a novel set of controls with minimal sample-specific bias over a large intensity range, referred to as a microarray sample pool.

A microarray sample pool is made from a randomly picked non-normalized plasmid library generated from mouse cerebellum. They compared different normalization methods, including within-slide normalization using the majority of genes on the microarray, and within-slide normalization using a microarray sample pool.

6.2.8.   Global Normalization Using ANOVA Models

Many authors, including Kerr, Martin, and Churchill (2000c)[17], Lee *et al.* (2000, 2002a), Wolfinger *et al.*[18] (2001) consider global normalization by incorporating appropriate main effects and/or interaction effects for normalizing factors into analysis of variance (ANOVA) models. Details of normalization by ANOVA models is discussed in detail in Chapter 10.

6.2.9.   Other Normalization Issues

Several variations have been proposed for normalizations that do not employ the whole gene set for setting the normalization adjustment factors. The basic idea is to use a subset of genes containing genes characterized by constant or known gene expression across the experimental conditions. The rank-invariant gene set introduced in Section 6.2.6 is one such variant. Other similar strategies have been proposed, as illustrated by the following.

1 A set of so-called *housekeeping genes* that are known to have relatively constant expression across experimental conditions. However, the use of housekeeping genes for normalization carries a risk, since very few genes have constant unvarying expression under all conditions or across multiple tissues.

2 A set of synthetic or alien DNA sequences, acting as *control genes*, that are spotted on the arrays but are not expected to hybridize to the specimens in the study. The addition of external controls is a good way to verify labeling and hybridization efficiencies, but usually the added control genes are too few to be utilized efficiently for normalization purposes.

3 A set of spots forming a *dilution series* for genes. The dilutions should be mirrored in measured intensities that reflect the fixed dilution ratios.

Normalization calculations may employ so-called robust methods that tend to be insensitive to unusual readings in the data set. Robust normalization may involve using a median, for instance, rather than a mean as a measure of location. The use of rank methods, such as rank-invariant gene subsets, is another robust option.

Notes

[1] Kerr, M.K., Afshari, C.A., Bennett, L., Bushel, P., Martinez, J., Walker, N.J., and Churchill, G.A. (2002). *Statistica Sinica*, **12**, 203-218.

[2] Rocke, D.M., Lorenzato, S. (1995). *Technometrics*, **37**, 176-184.

[3] Rocke, D.M., Durbin, B. (2001). *Journal of Computational Biology*, **8**, 557-569.

[4] Durbin, B.P., Hardin, J.S., Hawkins, D.M., and Rocke, D.M. (2002). *Bioinformatics*, **18**, S105-S110.

[5] Hawkins, D.M. (2002). *Statistics in Medicine*, **21**, 1913-1935.

[6] Geller, S.C., Gregg, J.P., Hagerman P., Rocke, D.M. (2003). http://handel.cipic.ucdavis.edu/˜dmrocke/gellernormpaper.pdf, (in press at *Bioinformatics*).

[7] Huber, W., von Heydebreck, A., Sültmann, H., Poustka, A., and Vingron, M. (2002). *Bioinformatics*, **18**, *Suppl. S-96-S104*.

[8] Yang, Y.H., Dudoit, S., Luu, P., Lin, D.M., Peng, V., Ngai, J., and Speed, T.P. (2002). *Nucleid Acids Research*, **30**, No.4, *e15*.

[9] Lee, M.-L.T., Lu, W., Whitmore, G.A., Beier, D. (2002b). *Journal of Biopharmaceutical Statistics*, **12(1)**, 1-19.

[10] Eisen, M.B. (1999). *ScanAlyze User Manual*, Version 2.32; Stanford University: Stanford, CA.

[11] Dudoit, S., Yang, Y.H., Callow, M.J., Speed, T.P. (2002). *Statistica Sinica*, **12**, 111-139.

[12] Cleveland, W.S. (1981). *The American Statistician*, **35**, 54.

[13] Cleveland, W.S. (1979). *Journal of the American Statistical Association*, **74**, 829-836.

[14] Tseng, G.C., Oh, M.-K., Rohlin, L., Liao, J.C., and Wong, W.H. (2001). *Nucleid Acids Research*, **29**, 2549-2557.

[15] Schadt, E., Li, C., Ellis, B., and Wong, W.H. (2001). *Journal of Cellular Biochemistry*, **84**, S37, 120-125.

[16] Wernisch, L., Kendall, S.L., Soneji, S., Wietzorrek, A., Parish, T., Hinds, J., Butcher, P.D., and Stoker, N.G. (2003). *Bioinformatics*, **19**, 53-61.

[17] Kerr, M.K., Martin, M., and Churchill, G.A. (2001c). *Journal of Computational Biology*, **7**, 819-837.

[18] Wolfinger, R.D., Gibson, G., Wolfinger, E.D., Bennett, L., Hamadeh, H., Bushel, P., Afshari, C., Paules, R. (2001). *Journal of Computational Biology*, **8**, 625-637.

Chapter 7

MISSING VALUES
IN MICROARRAY DATA

7.1.   Missing Values in Array Data

Microarray gene expression data frequently have missing values. These missing values are of concern for investigators and analysts alike. Missing values imply a loss of information. They may also imply contending with an unbalanced experimental design. Their presence should signal to investigators a departure from expected results and, hence, that there may be important issues or problems with experimental methods or conditions that require their attention.

7.1.1.   Sources of Problem

Missing values may arise in the original microarray data for various reasons, such as administrative error, defective technique, or technology failure. For example, an intended replication may be omitted, a feature of the robotic apparatus may fail, a scanner may have insufficient resolution, or an image may be corrupted.

Some researchers follow the practice of flagging readings that are suspect, and these may be converted to missing values or otherwise excluded from the analysis before proceeding. For instance, spots with dust particles, irregularities, or other bad features may be flagged manually. Spots may be flagged as 'absent' or 'feature not found' when nothing is printed in the location of a spot or if the imaging software cannot detect any fluorescence at the spot. Expression readings that are barely above the background correction (using a criterion such as less than two background standard deviations above) may also be flagged.

In two-color spotted cDNA arrays, it is not uncommon for either the reference or the treatment condition at a spot to be absent or unexpressed (especially with background correction). The resulting expression intensity ratio is undefined when the denominator is zero (being ∞) and is zero when the numerator alone is zero. Both cases pose a special problem for analysis. Various techniques can be employed in these situations. Recording them as missing is not really satisfactory. The fact that a gene may be unexpressed in the target but not in the reference, or vice versa, may be of scientific interest. These spots need to be segregated for separate investigation. The following remedies for missing values, such as imputation, would cover up a potentially interesting situation and, hence, their application would be ill-advised in this circumstance.

Missing values may also appear in a transformed data set because an undefined mathematical transformation has been applied. For example, where background correction gives corrected intensities that are negative or zero, a subsequent log-transformation will yield missing values in such cases. Care must be exercised to identify these cases as missing values.

Missing values having different causes are qualitatively different and require different remedies in the subsequent analysis and interpretation of findings. Statistical software packages handle missing values in various ways. Analysts must be aware of the differing treatments given to missing data points by their software. In the context of microarrays, some software may have an option to eliminate data for any gene that has one or more missing values under any of the experimental conditions. Other software is more selective in its treatment of missing values. A routine may only eliminate genes having one or more missing values for the experimental conditions that are under consideration for a particular analysis. One must be especially guarded about software that blindly imputes default values for missing values, such as zeros or row averages. We will return to discuss the general subject of imputation in subsection 7.4.

7.2.   Statistical Classification of Missing Data

The subject of missing values has been studied extensively by statisticians in many contexts and with varying degrees of sophistication. An authorative reference on the subject of missing values is Little and Rubin[1] (1987). Statisticians have found it helpful to classify missing values on the basis of the stochastic mechanism that produces them. The following classification system is widely used.

1. *Missing completely at random* (MCAR):
   Missing data are said to be MCAR if their missingness (i.e., the event of being missing) is independent of their own unobserved values and independent of the observed data. This kind of missing data may be viewed as arising from chance events that are unrelated to the nature of the investigation.

2. *Missing at random* (MAR):
   Missing data are said to be MAR if their missingness does not depend on their own unobserved values but does depend on the observed data. MAR represents a weakening of the assumptions of MCAR.

3. *Missing not at random* (MNAR):
   Missing data are said to be MNAR if their missingness depends on their own unobserved values. :

For an example of MCAR missingness, consider a spot that is obscured accidently by a dust particle. Its reading would be missing and classified as MCAR. At the other extreme, consider spots that show no fluorescence or that have undefined log-intensities because their background-corrected intensities are negative. Their readings will be missing and classified as MNAR because their missingness depends on the fact that their raw intensity values are zero or small.

The preceding list of categories is instructive because it leads an investigator to reflect carefully on the reasons why values are missing and the relationship that might exist between missing values and the data points that have been observed. The presence of missing values may give clues to systematic aspects of the problem.

The imputation methods that are discussed shortly are predicated on an assumed connection between the missing values and the observed data. The potential for missing values also has implications for the experimental design of microarray studies. In particular, replicates of arrays or spots on arrays provide insurance against the damaging effects of missing values. If missing values do occur by chance among a set of replicates, the observed members of the set can stand in for the missing, albeit with some loss of statistical precision.

If all intensity measurements for a given gene or condition are missing (an extreme case), then effectively the corresponding gene or condition is omitted from the study and there is no practical remedy except replicating the study to replace the missing cases. Where missing values are occasional, a few standard procedures have been proposed for handling them. We discuss these procedures in the next two subsections.

Spot	Color	
	Green	Red
1	Treatment	Control
	6800	2400
	*	*
	7600	*
2	Control	Treatment
	3500	*
	3100	6700
	2800	7200

Table 7.1. Replicated experiment with missing values marked by *. Parameters remain estimable.

7.3.   Missing Values in Replicated Designs

One faces a basic choice with missing values, namely, to proceed by omitting them from the analysis or to impute values for them in some acceptable and efficient fashion. If values are missing in a replicated experimental design, then observed replicates can stand in for the missing values. In this case, the issue is whether the parameters of the underlying statistical model can be estimated in the presence of one or more missing values. The designs with missing values are invariably unbalanced but pose no computational problem in principle provided basic estimability constaints are satisfied. Omitting missing values in a replicated design and proceeding to analyze the corresponding unbalanced experimental data set assumes, therefore, that the missingness is of the MCAR variety and, hence, that the actual fact of missingness is uninformative about the parameters of the gene expression model.

To illustrate the handling of missing values with replicated designs, consider Table 7.1, which shows a latin square design for a reverse-color cDNA experiment in which a treatment and control condition are being compared on two spots for each gene. The illustration assumes that the design is replicated with three repeats. The readings are for a single gene. Four of the 12 observations are missing and marked by an asterisk (*). An analysis of variance (ANOVA) model is fitted to the common logarithms of the data. Without giving details, the model was fitted with main effects for color, spot, and treatment. The parameters are estimable here. The analysis yields an estimate of 2.58 for the ratio of treatment to control gene expression. A confidence interval for this ratio could also be calculated in this instance.

7.4.   Imputation of Missing Values

Imputation methods replace missing values by estimates derived from the observed data, thereby converting an incomplete data set into a complete one. Various imputation methods are used, which we discuss next. These imputation methods rely on the missingness being of the MCAR type, i.e., missing completely at random.

1. *Row average:*
   A simple imputation method is to replace missing values by the average of observed values in the same row (i.e., for the same gene). Thus, readings for gene expression from observed experimental conditions are averaged to replace those that are missing for other experimental conditions. The method has little to commend it except simplicity and ease of computation, neither of which is a very compelling reason. Studies such as Troyanskaya *et al.*[2] (2001) also show that this method is not very effective. Its main weakness is that it makes no serious attempt to model the connection of the missing values to the observed data. A row average can be useful where an initial imputation is required in an iterative imputation method. Such methods generally will eventually replace the average by an improved imputed value. The SAM[3] software system offers this method as one option. For any row that has the intensity readings missing for all but one condition, however, the entire row will be replaced by this single non-missing observation and, hence, a row of identical readings is created. As a result, this row has to be eliminated from consideration because the standard deviation of the imputed intensity readings for this row is zero.

2. *k-nearest neighbors method:*
   This method matches the profile of observed data for a gene having missing values with the corresponding profiles of other genes having no missing values. A number k of these other genes having the closest matching profiles are then used to infer the missing values for the gene. To be more precise, consider a particular gene h with one or more missing values. Let $\mathcal{C} = \{1, \ldots, C\}$ denote the set of all experimental conditions in the study and let $\mathcal{G} = \{1, \ldots, G\}$ denote the set of all genes in the study. Let $\mathcal{C}_h$ be the subset of $\mathcal{C}$ for which gene h has observed values. The remaining conditions $\mathcal{C} - \mathcal{C}_h$ have missing values for gene h. Consider the subset of genes $\mathcal{G}_h$ that also have observed values for conditions $\mathcal{C}_h$. Choose the k genes among this subset that are most similar or closest to gene h in terms of their expression profiles on conditions $\mathcal{C}_h$. The average expression values for these k *nearest neighbors* on the missing conditions $\mathcal{C} - \mathcal{C}_h$ are

used as the imputed values for gene h. If the nearest neigbors also have missing values on conditions $\mathcal{C} - \mathcal{C}_h$, various backup remedies are applied. The measure of similarity or distance that is employed can vary – Euclidean distance is commonly used. The SAM software includes the k-nearest neighbors method for imputation. It employs Euclidean distance as the distance metric and uses a row average to complete the imputation if the nearest neighbor method still leaves gene h with one or more missing values.

Many variations of this method have been proposed. Some variants suggest, for example, that the imputation average be weighted by the similarity of the neighbor to the gene under consideration, with more similar neighbors being given greater weight. There is also variation in terms of the number of neighbors k to be used. A study by Troyanskaya *et al.* (2001) shows that results are adequate and relatively insensitive to values of k between 10 and 20.

3. *Regression estimate method:*
   Most commercial statistical software packages have one or more routines available for dealing with imputation. A common one involves using fitted regression values to replace missing values. The software package *Stata*, for example, offers this routine. The method works as follows. Let w_{hm} be missing for a particular gene h under condition m. As before, let $\mathcal{C}_h$ be the set of conditions for which gene h has observed values and let $\mathcal{G}_h$ be the set of genes having observed data for condition m and the conditions in $\mathcal{C}_h$. Regress w_{gm} on the expression levels in set $\mathcal{C}_h$ for all genes $g \in \mathcal{G}_h$. Use the fitted value $\hat{w}_{hm}$ from this regression as the imputed value of w_{hm}. The difference $w_{hm} - \hat{w}_{hm}$ is the imputation error in this case.

   The regression model can be applied to the original expression intensities w_{gc} as just described or to transformed values, such as log-intensities. Some checks should be made to assess the validity of the regression model for the application. For example, a logarithmic transformation of the observed expression readings may improve the applicability of the regression model. Moreover, the model must be chosen so that it does not yield invalid fitted values. For instance, a regression model that yields negative imputed values for expression on the original reading scale would be inappropriate.

4. *Principal component method:*
   This method requires that there be as many genes as experimental samples, i.e., $G \geq C$, which usually poses no difficulty as most microarray data sets have $G >> C$, i.e., the number of genes G is much

larger than the number of conditions C. This method first calculates the C principal components of the expression intensity vectors for genes, each vector being of length equal to the number of experimental conditions C. The method then resorts to regression fitting as in the previous method, but using the principal components as regressors. Each gene vector is thereby estimated by a suitable regression combination of one or more of the most important principal components. A complete microarray data set is needed to calculate principal components initially. This complete set can be obtained at the outset by using a simple imputation method, such as the row average method, to impute missing values. Subsequently, these initial imputations are replaced by imputed values provided by the first application of the principal component method. The imputation can proceed through several iterations of the principal component method until the imputations converge to stable values.

Notes

[1] Little, R.J.A. and Rubin, D.B.. (1987). *Statistical Analysis with Missing Data*, Wiley.

[2] Troyanskaya, O., Cantor, M., Sherlock, G., Brown, P., Hastie, T., Tibshirani, R., Botstein, D., Altman, R.B. (2001). *Bioinformatics*, **17 (6)**, 520-525.

[3] Chu, G., Narasimhan, B., Tibshirani, R. and Tusher, V. SAM (Significance Analysis of Microarrays): Users Guide and Technical Document (Version 1.21), Stanford University.

Chapter 8

SATURATED INTENSITY READINGS
IN MICROARRAY DATA

8.1.   Saturated Intensity Readings

Scanners used in microarray technology have an intensity *saturation threshold*. For example, in some image scanners, pixel intensities are digitized using a 16-bit gray scale, giving a range of 0 to 65535 levels of intensity. The saturation threshold is a reading of $2^{16} - 1 = 65535$. Some microarray technologies have required modification in order to address saturation problems caused by faulty design and calibration difficulties.

Microarray data need to be screened to see if intensities are at the threshold. Care must be exercised in order to identify these cases. Saturation readings are *censored observations* in the sense that the true value of w_{gc} is only known to exceed the threshold T, i.e, $w_{gc} > T$, but the precise value is unknown.

8.2.   Multiple Power-levels for Spotted Arrays

A microarray experiment was carried out in which 13,796 cDNAs for treatment and control samples from a mouse were read on the Cy3 (green) and Cy5 (red) channels, respectively. The scanning was done at six power levels, ranging from 500 to 1000 volts in 100 volt increments. Both median foreground and median background intensities were recorded separately. We refer to this data set as the Multiple Power-levels Data Set.

Figure 8.1, based on this data set, illustrates saturation. In this figure, we have plotted Cy3 foreground intensity at each voltage against Cy3 foreground intensity at 500 volts for the 13,796 genes (cDNAs). The

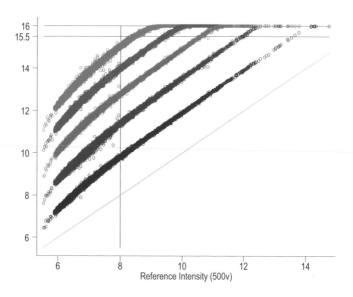

Figure 8.1. Plots of Cy3 foreground intensities at power levels of 600 (blue), 700 (purple), 800 (red), 900 (brown), and 1000 (green) volts, against readings at the reference level of 500 volts. A line of identity is shown for the 500-volt variable. Scales are logarithmic to base 2. Saturation occurs at 16 on the logarithmic scale, with partial saturation occurring at about 15.5.

plots use logarithmic scales to base 2 so the upper limit of intensity is 16. The 500-volt level is used as a reference intensity here so the plot of the 500-volt readings against themselves shows up as a line of identity. Several features of the plots are noteworthy.

1 The plots at higher voltages level off at 16 on the logarithmic scale. Spots at 16 are saturated readings, and these are more numerous at higher voltages.

2 The plots show that, as voltage increases, the foreground intensity increases almost in fixed proportion. These proportionate changes show up as roughly parallel scatter plots because of our use of logarithmic scales. The amplification of intensity from each 100-volt increment occurs at a slightly diminishing rate with voltage.

3 Some curvature appears in the scatter plot for each voltage at low intensities. The figure shows a vertical line at a log-intensity of 8 on the reference scale, which roughly separates the linear from the non-linear response segments. Each plot approaches a linear asymptote

Voltage	500	600	700	800	900	1000
Totally saturated	0	2	21	111	338	744
Partially saturated	0	6	36	129	361	790
Unsaturated	13,796	13,788	13,739	13,556	13,097	12,262

Table 8.1. Numbers of totally, partially, and unsaturated spots at each voltage in the Multiple Power-levels Data Set

as intensity increases, although the asymptote at 1000 volts is not well-defined.

4 Only a handful of genes have log-intensities below 6 on the reference scale.

5 Near the saturation threshold, the plots also show a little curvature. This curvature is presumably the result of saturation occurring for some pixels in the image but not others. We shall refer to this phenomenon later as *partial image saturation*.

6 There appears to be a greater scatter of points around the line of relationship at lower intensities than at higher intensities. Further analysis of deviations (using LOWESS smoothing) shows, however, that this greater scatter is mainly caused by the larger numbers of spots in the lower intensity range. Variability is, in fact, quite constant across the unsaturated range of readings at each voltage, with only slightly greater dispersion in the low-intensity region.

We shall use the line plotted at level 15.5 to define the onset of partial saturation. The numbers of totally saturated, partially saturated, and unsaturated spots as a function of voltage are tabulated in Table 8.1. We note for this data set that any spot that is saturated at one voltage remains saturated at higher voltages.

8.2.1. Imputing Saturated Intensity Readings

Scanning at several power or dilution levels allows saturated intensities to be estimated and also provides slightly improved precision for intensity measurement. Software such as *masliner*[1], developed by Dudley *et al.* (2002), performs this kind of analysis. We sketch the underlying logic of this kind of analysis below, although specific packages have their own customized routines and refinements. We now use the term *amplification* to describe the increased intensity at higher power levels (or in less diluted samples).

A number of methods can be used for imputing intensity values for partially and totally saturated spots. As the observations are censored, statistical methods for censored data can be used. These methods require a specification of an appropriate statistical model of intensity and a description of the censoring mechanism. In this context of microarray data, however, a satisfactory job is done by treating the saturated spots as having missing values and using imputation methods to give them values.

To demonstrate the approach, we will use the Multiple Power-levels Data Set and apply a version of regression imputation. As the regression method relies on the log-intensities at different power levels having a linear relationship with each other, we exclude readings in the region of non-linear response. As a specific criterion, we eliminate readings with log-intensities of 8 or less on the reference (500 volt) scale – see the vertical line in Figure 8.1. This cutoff does not eliminate any saturated spots. The regression imputation method is as follows:

1 Set the totally and partially saturated cases (i.e., those with logarithmic readings above 15.5) to missing values.

2 In an initial iteration, regress intensities at each voltage level on the known intensities at lower voltage levels. Observe in Table 8.1 that voltage 500 has only unsaturated spots. Thus, we begin with voltage 600 and, for each gene, regress the voltage-600 intensity on the intensity for voltage 500. The missing voltage-600 readings are given imputed values equal to the fitted values from this regression. Moving on to voltage 700, the voltage-700 intensity is regressed on the intensities for voltages 500 and 600, including the imputed values for voltage 600. Again, missing values at voltage 700 are replaced by fitted values from the regression. The procedure is repeated successively for the higher voltages, 800, 900, and 1000.

3 In a second iteration, again proceeding gene by gene, regress the intensity at each successive voltage on the intensities of all other voltages, replacing missing values by imputed values from the initial iteration. Update the imputed values for each successive regression at a higher voltage level.

Figure 8.2 shows the revised scatter plots, corresponding to Figure 8.1, but now with imputed values replacing intensities exceeding 15.5 on the logarithmic scale. The method seems to have made a reasonable adjustment for saturation and produced a roughly linear intensity scale at each voltage level as expected. The asymptotes at different voltages are

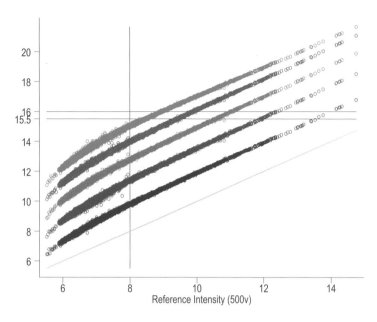

Figure 8.2. Plots of Cy3 foreground intensities at power levels of 600 (blue), 700 (purple), 800 (red), 900 (brown), and 1000 (green) volts, against the readings at the reference level of 500 volts. Partially and totally saturated readings have been replaced by imputed values derived by regression. A line of identity is shown for the 500-volt variable. Scales are logarithmic to base 2. Saturation occurs at 16 on the logarithmic scale, with partial saturation occurring at about 15.5.

roughly parallel except at the highest voltage, for which the imputation also happens to involve the greatest extent of extrapolation.

8.3.   High Intensities in Oligonucleotide Arrays

Naef *et al.*[2] (2003) present a detailed calibration study of GeneChips, in which a number of transcripts are hybridized at known concentrations. They show that high-concentration bias is a serious issue in GeneChips, which is probably related to chemical saturation in the absorption process of the target to the probes.

Their results show that conventional analysis techniques have difficulty detecting small changes in high transcript concentrations. While the variance is smallest for high concentrations, it appears that potential bias has been neglected. In comparing two conditions, they conclude that the high-intensity GeneChip microarray data are almost always very tightly scattered about the diagonal and that there are rarely genes in that region that show fold changes greater than 1.5 or 2. They show that fold changes as large as 4 in expression levels are often compressed to fold changes barely larger than 1.

In contrast to the standard view in which the perfect matches (PMs) are thought to carry the signal, while the mismatches (MMs) serve as non-specific controls, Naef *et al.* (2003) use PM and MM probes equally. Specifically, instead of the difference PM−MM, both log transformed intensities PM and MM are used in their analysis. Expecting that some MM probes will be less affected by saturation than the PM probes[3], they use the method of locally linear embeddings[4] to reduce the bias at high concentrations.

Notes

[1] Dudley, A.M., Aach, J., Steffen, M.A., and Church, G.M. (2002). *Proceedings of the National Academy of Sciences, USA*, **99**, 7554-7559.

[2] Naef, F., Socci, N.D., Magnasco, M. (2003). *Bioinformatics*, **19**; 178-184.

[3] Naef, F., Lim, D.A., Patil, N. and Magnasco, M. (2002b). *Phys. Rev. E.*, **65**, 040902.

[4] Roweis, S.T. and Saul, L.K. (2000). *Science*, **290**, 2323-2326.

II

STATISTICAL MODELS
AND ANALYSIS

Chapter 9

EXPERIMENTAL DESIGN

Historically, experimental design[1] has been used in improving agricultural products, crop yields, and livestock[2]. In the past few decades, experimental design has also played a key role in quality improvement in industrial engineering. Properly designed experiments help an investigator to avoid systematic errors at the analysis stage when comparing the effects of treatments on response measures of interest. Hence, a properly designed experiment can provide the maximum amount of information from the available data.

Microarray experiments can benefit from adopting principles of good experimental design[3]. In this chapter, we describe some of the issues and ideas that are unique to experimental design for microarray studies. The presentation in this chapter is far from exhaustive, however, and the reader is referred to some of the cited references for more information. For a good review of experimental design issues, see Cochran and Cox[4] (1992), Winer[5] (1971), Milliken and Johnson[6] (1992), and Neter, Kutner, Nachtscheim, and Wasserman[7] (1996), among others.

In this chapter, we will first classify experiments by the factors to be considered in the comparisons and by design structure. We then discuss some prototypical designs for microarrays. Some of these designs reflect the historical development of the subject (albeit a short history), and others reflect current and evolving practices.

9.1. Factors Involved in Experiments

An *experimental factor* is an explanatory variable whose influence on the response variable in the experiment is of interest. The different settings that a factor may take on in an experiment are referred to as *factor levels*. Experiments may include one or more factors controlled at their respective levels. The corresponding data analyses may be referred to as one-way, two-way, or, in general, multi-way analyses, depending on the number of factors involved in the design. In some microarray studies, the design is specified and analyzed gene by gene. In other designs, the gene set is considered as a design factor having the individual genes as its levels.

1. **One Factor:**

   A simple microarray experiment may look for changes in gene expression across a single factor of interest. Let k denote the number of factor levels. For example, an experiment involves RNAs from kidney tissues taken from mice exposed to three different concentrations of a toxin (low, medium, high). The factor of interest is toxin exposure, which has $k = 3$ levels. Samples from multiple mice are collected at each factor level (toxin exposure). A high-density oligonucleotide microarray is prepared for each tissue sample. Expression measurements are obtained from the arrays for each gene in the mouse gene set. A one-way comparison of the expression levels across the three toxin levels can be used to identify a differentially expressed gene.

2. **Two Factors:**

   In a two-way analysis, there are two factors involved in the comparison. If the first factor has k_1 levels and the second factor has k_2 levels, then a two-way factor structure consists of a total of $k_1 \times k_2$ combinations of levels. For example, if there are two types of mice, mutant and wild type, involved in the previous experiment, then, in addition to the factor for toxin exposure (with three levels), a second factor having two levels (the two types of mice) is also taken into account in the comparison.

3. **Multiple Factors:**

   When there are more than two factors involved in the experiment, the multi-way structure is often called a *factorial design*. An m-way factorial design consists of m factors having $k_1, \ldots, k_m$ levels, respectively. Together there are a total of $k_1 \times \cdots \times k_m$ combinations. If all combinations are taken into account in the design, it is called a *complete factorial design*. For example, in a mouse experiment, the multiple factors taken into account may include sex, age group, mouse type, and other characteristics.

4. Fractional Replication and Latin Square Designs:

When the number of factors involved in an experiment is large, the number of possible combinations of the factor levels required for a complete factorial design is huge. Because of the costs and labor involved in microarray experiments, a complete factorial design is often beyond the resources of investigators. It is often necessary in this situation to use only a subset of all possible combinations of factor levels. For example, consider an experiment with seven factors, each having two levels. There are $2^7 = 128$ possible combinations of factor levels in a complete factorial experiment. In a complete 2^7 design, each main effect estimate is an average over 64 combinations of the other factors and these estimates of main effects may be more precise than is needed. It is feasible to obtain good estimates of main effects from a much smaller experiment by using a carefully chosen subset of the complete factorial design. This chosen subset is called a *fractional replication* or a *fractional factorial design*.

Consider a mouse experiment that involves comparisons among two mouse types, two age groups, and two sexes using oligonucleotide arrays. A complete 2^3 factorial design involving these three 2-level factors (mouse type, age group, and sex) would involve a total of 8 possible combinations. If we can ignore interaction effects that might exist among mouse type, age group, and sex, then a half-replicate of a 2^3 factorial design can be considered. Hence, the investigator would only need 4 specimens, instead of 8, for the microarray experiment. Each half-fraction of a full 2^3 design is often called a 2^{3-1} fractional factorial design, or a 2×2 *latin square design*[8]. Specifically, the combinations included in a half replicate of a 2^3 factorial design should satisfy the following two conditions.

(a) To estimate a main effect efficiently, the two levels of each factor must appear equally often.

(b) To estimate main effects independently, each pair of main effects should be orthogonal to each other.

Thus, all four combinations of the levels of any two factors must appear equally often. As a result, there are only two distinct 2^{3-1} fractional factorial designs that satisfy these two conditions. If we index the age groups as the rows and the genders as the columns, and place mouse type in the cells, then each of these two designs can be represented as a 2×2 latin square design as illustrated in Tables 9.1 and 9.2.

	Male	Female
age group a_1	mutant	wild-type
age group a_2	wild-type	mutant

Table 9.1.   A half-replicate of a 2^3 design forms a 2×2 latin-square.

	Male	Female
age group a_1	wild-type	mutant
age group a_2	mutant	wild-type

Table 9.2.   The other half-replicate of a 2^3 design also forms a 2×2 latin-square.

In either of the two half-replicates, the main effects of any factor are orthogonal to those of the other two factors. The main effects of any factor, however, are confounded with the two-factor interactions not involving that factor. Hence, if one can assume that the two-factor and higher-order interaction effects are negligible, the design of each half-replicate allows the experimenter to estimate independently the main effects of any of the three main factors (mouse type, age group, and sex). Using a half-replicate design, therefore, saves half of the experimental resources.

In experiments where both half-replicates are available, by combining the two latin-squares, we have a complete 2^3 factorial design. The benefit of having both half-replicates is that the experimenters can estimate two-factor interaction effects. If the two 2×2 latin-squares constitute blocks in the experiment, then the blocks are confounded with the three-factor interaction effects. In a design with both latin-squares, all the two-factor interaction effects, in addition to the main effects, can be estimated independently. Two-factor interaction effects cannot be estimated if only one latin-square is used repeatedly.

9.2.   Types of Design Structures

In addition to consideration of possible factors to be included in the analysis, it is important to consider the structure of the experimental design at the planning stage.

1. Completely Randomized Design:

   To illustrate this design, consider a homogeneous population of mice. Assume that k different treatments (one factor with k levels) are of interest in a gene expression study of this population. If a random sample of mice drawn from this population are randomly assigned in fixed numbers to the treatments, then the experiment is called a *completely randomized design*. If an equal number of mice are assigned to each treatment, then the design is called *balanced*.

2. Randomized Complete Block Design:

   A blocking approach can be employed in microarray experiments to increase sensitivity and to diminish inherent variation and, thereby, to enhance the detection of significant differences. The term *blocking* has its origins in agricultural experiments. For example, crop yields are influenced by soil fertility, amount of sunlight, moisture, and other factors. To remove variation contributed by these factors when comparing yields of, say, k strains of wheat, each of the k strains would be randomly assigned for planting on k plots of the same block of land. The experiment would employ, say, b different blocks of land for this purpose. In genetic experiments, for instance, multiple RNA samples taken from the same mouse may be used for different treatments so as to eliminate possible intra-sample variation. Here treatments correspond to wheat strains, each mouse is a block and the plots in each block correspond to the RNA samples taken from the same mouse. The k treatments of interest would be randomly assigned to k RNA samples from each of b mice. Where the number of treatment conditions matches the number of experimental units in each block, as in these examples, the design is called a *randomized complete block design*.

3. Incomplete Block Design:

   Should the number of experimental units that can be accommodated in each block be less than all possible treatment conditions of interest, then the design is said to be *incomplete*. For example, if only two RNA samples are available for each of the 4 mice, and there are four treatment conditions of interest, then only 2 of the 4 treatments can be applied to each mouse. In this case the design is called an *incomplete block design*. As another example, consider two-color spotted arrays. It is a common practice in each spotted array that two mRNA samples (say a tumor sample and a reference sample) are labelled with green dye and and red dye, respectively. Kerr and Churchill (2001) pointed out that a spotted microarray can be considered as an *experimental block* with a block size of two. When there are more than two experimental conditions of interest, all sample conditions cannot appear on the same array. In these experiments, spotted microarrays can be considered as incomplete blocks. Table 9.3 illustrates an incomplete block design for four treatments.

	Array a_1	Array a_2	Array a_3	Array a_4
Dye d_1	treatment t_1	treatment t_2	treatment t_3	treatment t_4
Dye d_2	treatment t_2	treatment t_3	treatment t_4	treatment t_1

Table 9.3. An incomplete block design for 4 treatments. With only 4 arrays, it is not possible for every pair of treatments to appear together in the same array.

4. Balanced Incomplete Block Design:

One of the most used incomplete block designs is a balanced incomplete block design (BIBD) where all treatments are replicated the same number of times and any pair of treatments appears together equally often within blocks. Because of the fact that treatment comparisons are balanced to blocks (arrays), the design will provide equal precision on treatment comparisons. That is, the same standard error may be used for comparing every pair of treatments.

In the design shown in Table 9.4, every pair of treatments appears together once in an array. Treatment effects are also orthogonal to dyes. We show in Table 9.5 that, when 6 arrays are available, 4 treatments can be compared in a BIBD design; the treatments in this design, however, are not orthogonal to dyes. We demonstrate a BIBD design for 5 treatments in Table 9.6. The 5 treatments in this design are orthogonal to the dyes.

	Array a_1	Array a_2	Array a_3
Dye d_1	treatment t_1	treatment t_2	treatment t_3
Dye d_2	treatment t_2	treatment t_3	treatment t_1

Table 9.4. A balanced incomplete block design for 3 treatments. Every pair of treatments appears together once in an array. Treatments are also orthogonal to dyes.

	a_1	a_2	a_3	a_4	a_5	a_6
Dye d_1	t_1	t_2	t_3	t_4	t_1	t_2
Dye d_2	t_2	t_3	t_4	t_1	t_3	t_4

Table 9.5. A BIBD design for 4 treatments. Treatment comparisons are balanced to arrays. Treatments, however, are not orthogonal to dyes in this design.

	a_1	a_2	a_3	a_4	a_5	a_6	a_7	a_8	a_9	a_{10}
Dye d_1	t_1	t_2	t_3	t_4	t_5	t_1	t_3	t_5	t_2	t_4
Dye d_2	t_2	t_3	t_4	t_5	t_1	t_3	t_5	t_2	t_4	t_1

Table 9.6. A BIBD design for 5 treatments. Treatment comparisons are balanced to arrays. Treatments are also orthogonal to dyes.

5. Split-plot Design:

A split-plot design is a factorial experiment in which a main effect is confounded with blocks (the larger experimental units). In this setting, the blocks are called *whole plots* and the smaller experimental units nested within whole plots are called *subplots*. Let the levels of a factor, say factor A, be randomly assigned to the whole plots and the levels of a second factor, say factor B, be randomly assigned to the subplots within each whole plot. In general, subplots within a whole plot will be more similar than subplots in different whole plots. Consequently, within-whole-plot comparisons will generally be more precise than between-whole-plot comparisons. So the split-plot design is advantageous if the main effects of factor B and the AB interactions are of greater interest than the main effects of factor A alone. Yates[9] (1935) showed that when the number of replications and the experimental conditions are suitable, a split-plot latin square, which eliminates the error variation arising from two types of grouping, may be preferable to randomized blocks.

In studying the contributions of sex, genotype, and age to transcriptional variance in adult fruitflies *Drosophila melanogaster*, Jin *et al.*[10] (2001) conducted an experiment involving two sexes, two genotypes, and two age groups. Six replications including dye swaps were made for each combination of two genotypes and two sexes. A total of 24 two-color cDNA microarrays were used. Using a split-plot design, they directly contrasted the two age groups by always having the 1-week and 6-week adult flies together on the same array block.

We consider a variation of the Jin *et al.* design. The three parts of Table 9.7 show 12 arrays grouped into three sets of four arrays each. Arrays a_1 to a_4 in Table 9.7(a) compare two age groups. Arrays a_5 to a_8 in Table 9.7(b) compare two genotypes. Arrays a_9 to a_{12} in Table 9.7(c) compare two sexes.

(a)

Array	Genotype	Sex	Green-dye Age	Genotype	Sex	Red-dye Age
a_1	type1	M	age1	type1	M	age2
a_2	type1	F	age1	type1	F	age2
a_3	type2	M	age1	type2	M	age2
a_4	type2	F	age1	type2	F	age2

(b)

Array	Sex	Age	Green-dye Genotype	Sex	Age	Red-dye Genotype
a_5	F	age1	type1	F	age1	type2
a_6	F	age2	type1	F	age2	type2
a_7	M	age1	type1	M	age1	type2
a_8	M	age2	type1	M	age2	type2

(c)

Array	Genotype	Age	Green-dye Sex	Genotype	Age	Red-dye Sex
a_9	type1	age1	M	type1	age1	F
a_{10}	type1	age2	M	type1	age2	F
a_{11}	type2	age1	M	type2	age1	F
a_{12}	type2	age2	M	type2	age2	F

Table 9.7. A split-plot design with (a) two age groups, (b) two genotypes and (c) two sexes on each two-color array block.

The design also includes another set of 12 arrays like these that have the dyes reversed. The 12 arrays with dyes reversed are not shown. The experiment thus consists of 24 arrays that allow a split-plot comparison of age groups, genotypes, and sexes. The combined design will provide equally efficient parameter estimates when the effects of sex, genotype, and age group are of equal interest.

6. Designs Involving Nested Factors:

Where each level of one factor can be observed in combination with each level of any other factor in a multi-factor study, the factors are said to be *crossed*. There are many kinds of experiments, however, where factors cannot be crossed. In these experiments the factors may be *nested* and the designs are called *nested designs*. Nesting refers to the condition where all levels of one factor are found within only one level of a second factor. The first factor is then said to be nested within the second factor. In the design structure of an experiment, nested effects occur when the experimental units for one factor are different for each experimental unit of a second factor.

Patient	Sample ID	Tissue Type	Stage
A	A1	leiomyomata	proliferative
A	A2	leiomyomata	proliferative
A	A3	leiomyomata	proliferative
A	A4	leiomyomata	proliferative
A	A5	myometria	proliferative
B	B1	leiomyomata	late secretory
B	B2	leiomyomata	late secretory
B	B3	leiomyomata	late secretory
B	B4	myometria	late secretory

Table 9.8.   Nested design for samples from two patients

For experiments involving human tissues, it might not be possible to obtain RNA materials for a crossed design. For example, in studying the biology of uterine leiomyomata[11], a total of seven uterine tumor tissue samples (leiomyomata) were taken from two patients, A and B, as shown in Table 9.8. Four leiomyomata samples were harvested from patient A, and three leiomyomata samples were harvested from patient B. Patient A is at the proliferative stage, while patient B is at the late secretory stage. The microarray data set contains gene expression readings for thousands of genes. It is of interest to know which genes are expressed in uterine leiomyomata tumor tissue but not in matched normal tissue (myometria) for the same patient. Since the normal and tumor tissues are found within the same stage (proliferative or late secretory), the factors of tissue type and stage are nested in this design. The explicit recognition of nesting here allows for the possibility that differential gene expression for the two tissue types may differ by patient stage.

7. Other Designs:

Although experimental designs often have standard forms or standard structural components, some sophisticated variations have been developed to deal with special circumstances. For example, where the main objective is to compare treatment conditions with the control condition but not to have comparisons among the treatments, Hedayat and Majumdar (1984) consider a class of *balanced test-treatment incomplete block designs*[12] that can be optimal for making treatment-control comparisons.

9.3.   Common Practice in Microarray Studies

Several experimental designs have become standard in microarray studies, and other designs of a more novel nature have been proposed. In this section we look at some of these designs.

9.3.1.   Reference Design

As discussed for spotted arrays in Chapter 3, each array is probed with two samples labelled with different fluorescent dyes. Because there is variation in the amount of RNA from spot to spot, fluorescent intensities are more meaningful when compared in a relative sense. Hence, biologists arrived at the *reference design* in which this kind of variation can be controlled by always having the same reference RNA sample on each spot. A common practice is to compute ratios of the raw signals as estimates of differential expression between the two samples spotted together.

In many studies using spotted microarrays, several treatment conditions are under study and each treatment sample is paired with the same reference sample on each array slide. Kerr and Churchill (2001) refer to this design as a *reference design*. For this design, experimenters usually use one dye, say green (dye d_1), to label the reference sample and another dye, say red (dye d_2), to label the treatment samples of interest. This kind of design is illustrated in Table 9.9. In this illustration, a reference (control) sample (t_0) and three treatments conditions (t_1, t_2, t_3) are under study.

	Array a_1	Array a_2	Array a_3
Dye d_1	reference t_0	reference t_0	reference t_0
Dye d_2	treatment t_1	treatment t_2	treatment t_3

Table 9.9. Reference design for 3 treatments t_1, t_2, and t_3.

For example, to study molecular classification of human lymphomas, Alizadeh, Eisen, Davis, *et al.* (2000) designed the *Lymphochip* by selecting genes that are preferentially expressed in lymphoid cells and genes with suspected roles in processes important in immunology or cancer[13]. A fluorescent cDNA sample, labelled with the Cy5 dye, was prepared from each experimental mRNA sample. A reference cDNA sample, labelled with the Cy3 dye, was prepared from a pool of mRNAs isolated from nine different lymphoma cell lines. Each Cy5-labelled experimental cDNA sample was combined with the Cy3-labelled reference sample and the mixture was hybridized to the microarray. The fluorescence ratio was determined for each gene. Gene expression measurements were obtained for normal and malignant lymphocyte samples using Lymphochip microarrays.

One weakness of the reference design is the fact that half of the intensity readings are derived from the reference sample, which may not be of immediate scientific interest. Specifically, in a reference design with G genes and T treatment samples of interest, there will be a total of $2GT$ intensity readings, with GT of them from the reference sample. If we analyze the data using the analysis of variance (ANOVA) model that will be introduced in Chapter 10, the overall mean and the main effects of arrays, gene, and sample conditions account for $2T + G - 1$ degrees of freedom. The gene-by-condition interaction effects of interest account for $(G-1)T$ degrees of freedom. If the gene-by-array interaction effects are also included in the model, they absorb $(G-1)(T-1)$ degrees of freedom. As noted by Kerr *et al.*[14] (2001c), therefore, no degree of freedom is left to estimate error.

The fact that treatments are not compared directly on the same spot is also viewed as a limitation. To elaborate on the latter point, in the reference design of Table 9.9, for example, if intensity ratios are used to measure fold changes, treatment conditions t_1 and t_2 cannot be compared directly without first being compared with the control sample t_0. Specifically, a comparison of t_2 with t_1 would require, first, a comparison of t_2 with t_0, then t_1 with t_0 and, finally, a comparison of these two comparisons.

Moreover, as we have noted earlier in our discussion of color effects (see subsection 6.2.4), there is an association between gene expression intensity and the dye label. The result is that the dye effect is confounded with the differential expression of treatment and control samples. This fact implies that two scientists working with identical experimental conditions, except for treatment and control samples having opposite dye labels, might possibly arrive at different conclusions.

Simon, Radmacher, and Dobbin[15] (2002) discuss the strengths and weaknesses of the reference design for microarray studies. They provide several arguments in defense of the reference design for microarray studies. These advantages generally depend on the flexibility of the design in constructing arbitrary comparisons across samples. For example, they mention the advantage in cluster analysis because the distances between all pairs of specimens can be evaluated readily. They also mention the advantage for investigating gene expression for specimens from different experiments when a laboratory or several laboratories use a common reference mRNA across experiments. Cross-study comparisons are not as easily made with other designs.

Kerr and Churchill (2001b) considered an *augmented reference design* where, in addition to having every treatment sample compared with the reference sample t_0, a self-comparison array is made that contains the reference sample t_0 labelled with both dyes. Table 9.10 illustrates this kind of design.

	Array a_0	Array a_1	Array a_2	Array a_3
Dye d_1	reference t_0	reference t_0	reference t_0	reference t_0
Dye d_2	reference t_0	treatment t_1	treatment t_2	treatment t_3

Table 9.10. Augumented reference design for 3 treatments t_1, t_2, and t_3.

9.3.2. Time-course Experiment

Time-course experiments are often used in microarray studies because knowing when and where a gene is expressed can provide a strong clue about its biological role. A common set up for the time-course experiment is similar to the reference design as discussed in subsection 9.3.1.

Table 9.11, for example, shows a comparison of expression levels at six time points $t_1, t_2, \ldots, t_6$ with a baseline (time zero) treatment t_0 serving as the reference.

	Array a_1	Array a_2	Array a_3	Array a_4	Array a_5	Array a_6
Dye d_1	time t_0	time t_0	time t_0	time t_0	time t_0	time t_0
Dye d_2	time t_1	time t_2	time t_3	time t_4	time t_5	time t_6

Table 9.11. Time-course design with samples from six time points and a reference sample at time t_0

DeRisi, Iyer, and Brown[16] (1997) conducted a systematic investigation of gene expression of the yeast *Saccharomyces cerevisiae*. In their experiment, cells from yeast culture were inoculated into fresh medium and grown for 21 hours. After an initial 9 hours of growth, samples were harvested at seven successive 2-hour intervals. They labeled cDNA prepared from cells at each successive time point with Cy5, then mixed it with a Cy3-labeled "reference" cDNA sample prepared from cells harvested at the first interval after inoculation. Their dataset consists of more than 43,000 expression-ratio measurements observed from the seven samples.

9.3.3.   Color Reversal

In a reference design, one dye is used to label the reference sample and another dye is used to label other treatment samples. Hence, it can be seen that treatment effects are confounded with dye effects. This confounding of dyes and treatments can be eliminated by repeating the experiment with the dye colors reversed, giving what is generally called a *reversed-color*, *dye-swap*, or *flipped-color* design. An illustrative reference design with reversed color is shown in Table 9.12. Although this design solves the problem of confounding dye and treatment effects, a continuing drawback of this design is that half of the intensity readings are derived from the reference sample.

Dobbins, Shih, Simon[17] (2003) look more carefully at the statistical design of reverse-dye experiments. They examine three types of experiments requiring reverse labelling to control for dye bias, referring to them as 'paired samples, samples from two predefined groups, and reference design data when comparison with the reference is of interest'. They describe efficient designs that control for dye bias in these experimental

	Array a_1	Array a_2	Array a_3
Dye d_1	reference t_0	reference t_0	reference t_0
Dye d_2	treatment t_1	treatment t_2	treatment t_3

	Array a_4	Array a_5	Array a_6
Dye d_2	reference t_0	reference t_0	reference t_0
Dye d_1	treatment t_1	treatment t_2	treatment t_3

Table 9.12. Reference design with color reversal for 3 treatments t_1, t_2, and t_3

contexts and give a sample size formula for these designs. Their general conclusion is that it is usually not efficient to reverse the dyes for every individual sample but rather to increase the number of samples keeping the design balanced with respect to treatment and dye combinations.

9.3.4.   Loop Design

As was described in section 9.3.1, the control sample (t_0) used in the reference design often is of no scientific interest in the study but adds experimental noise to the data. In a study aimed at comparing several treatment conditions, denoted by $t_1, t_2, \ldots, t_k$, Kerr and Churchill (2001b) suggest that a *loop design* be considered. In this design, the k pairs of treatments (t_1, t_2), (t_2, t_3), ..., (t_{k-1}, t_k), (t_k, t_1) are each spotted on two arrays, with the color channel reversed for one array relative to its mate. The unbalanced incomplete block design in Table 9.3 is a loop design for comparing 4 treatments. The balanced incomplete block design in Table 9.4 is a loop design for comparing three treatments. We note also that the design described in Table 9.6 can be considered as a loop design with replication. For comparing four treatments with color reversal, in addition to the four arrays listed in Table 9.3, a loop design would set up four additional arrays for each gene as shown in Table 9.13.

An advantage of the loop design is that it eliminates the need for a reference or control sample because each treatment is compared with another treatment directly. With color reversal, averaging each treatment pair over the two color orders eliminates any additive differential effect from color. A major weakness is that, if there are more than 10 treatment conditions, the loop design can be inefficient. For a discussion regarding the relative efficiencies of loop and reference designs, see Kerr and Churchill (2001b). One practical drawback of the loop design

	a_1	a_2	a_3	a_4	a_5	a_6	a_7	a_8
Dye d_1	t_1	t_2	t_3	t_4	t_1	t_4	t_3	t_2
Dye d_2	t_2	t_3	t_4	t_1	t_4	t_3	t_2	t_1

Table 9.13. A loop design with color reversal for 4 treatments

is that each sample must be labelled with both the red and green dyes, which means doubling the number of labelling reactions. Another drawback is that indirect comparisons may still be required for some pairs of treatments, as is the case for treatments t_2 and t_4 in Table 9.13.

9.3.5.   Example: Time-course Loop Design

This example describes a time-course experiment with a loop design to compare two treatments. The design in Table 9.14 is illustrative of the innovative approaches to microarray study design being taken by scientists in the field[18]. This study design involves a comparison of two treatments A and B. The eight time points covered in the design are 0, 0.5, 1, 2, 4, 6, 12 and 24 hours. The numbers associated with the treatment labels refer to hours of treatment (e.g., $t0$ refers to the specimen taken at time zero without treatment, and $t0.5$ refers to the specimen taken one-half hour after treatment.).

There are two replicates for each treatment at each time point. As a result the design has $8 \times 2 \times 2 = 32$ slides in total. On the first seven slides in each sequence, the treatment condition for each time point is labeled with Cy5 (red) dye and the treatment condition for the preceding time point is labeled with the Cy3 (green) dye. The closing comparison on the eighth slide has $t0$ labeled with Cy5 and $t24$ labeled with Cy3 to complete the loop design. The experimental factors of treatment, dye, and time are orthogonal and balanced in this design. Based upon a loop design, this experiment saves some biological material by not using the reference time zero specimen at each time point, but it is harder to work with and visualize than a reference design. There are no technical complications with the design from an experimental standpoint. The main challenges reside with the analysis, visualization, and interpretation of the data. This design might not suit an experiment where there are tandem effects such as an effect from a treatment and an effect from the solution or medium that comes with the treatment. For instance, in an experiment where the treatment reagent is introduced

		Treatment A:							

Replicate 1

		Array							
		a_1	a_2	a_3	a_4	a_5	a_6	a_7	a_8
	Dye Cy3 green	t0	t0.5	t1	t2	t4	t6	t12	t24
	Dye Cy5 red	t0.5	t1	t2	t4	t6	t12	t24	t0

Replicate 2

		Array							
		a_9	a_{10}	a_{11}	a_{12}	a_{13}	a_{14}	a_{15}	a_{16}
	Dye Cy3 green	t0	t0.5	t1	t2	t4	t6	t12	t24
	Dye Cy5 red	t0.5	t1	t2	t4	t6	t12	t24	t0

		Treatment B:							

Replicate 1

		Array							
		a_{17}	a_{18}	a_{19}	a_{20}	a_{21}	a_{22}	a_{23}	a_{24}
	Dye Cy3 green	t0	t0.5	t1	t2	t4	t6	t12	t24
	Dye Cy5 red	t0.5	t1	t2	t4	t6	t12	t24	t0

Replicate 2

		Array							
		a_{25}	a_{26}	a_{27}	a_{28}	a_{29}	a_{30}	a_{31}	a_{32}
	Dye Cy3 green	t0	t0.5	t1	t2	t4	t6	t12	t24
	Dye Cy5 red	t0.5	t1	t2	t4	t6	t12	t24	t0

Table 9.14. Replicated time-course loop design for two treatments A and B; sample *t0* denotes the baseline control sample

in fresh media, the media alone might induce gene expression changes. To isolate the treatment effect from the media change effect, the media changes should be included only as reference samples for the same time points. The design in Table 9.14 cannot isolate the media change because each sample has its previous time point as reference.

An alternative design might have used the replicate to reverse the dye order for each comparison. If the experiment is one where the time course shows small changes over time (like a slow increase) the design might have a slight disadvantage relative to a reference design if the variability does not allow statistical verification of small changes.

Notes

[1] Fisher, R.A., (1947). *The Design of Experiments*, Oliver and Boyd, Edinburgh, 4th ed.

[2] Comstock, R.E., and Winters, L.M. (1942). *Journal of Agricultural Research*, **64**, 523-532.

[3] Kerr, M.K., and Churchill, G.A. (2001b). *Biostatistics*, **2**, 183-201.

[4] Cochran, W.G., and Cox, G.M., (1992), *Experimental Designs*, Wiley, New York.

[5] Winer, B.J. (1971). *Statistical Principles in Experimental Design*, 2nd ed., New York, McGraw-Hill.

[6] Milliken, G.A. and Johnson, D.E., (1992), *Analysis of Messy Data: Volume 1, Designed Experiments*, Chapman and Hall, Boca Raton.

[7] Neter, J., Kutner, M.H., Nachtscheim, C.J., Wasserman, W. (1996). *Applied Linear Statistical Models*, 4th edition, Richard D. Irwin.

[8] Kerr, M.K., Martin, M., Churchill, G.A. (2000). *Journal of Computational Biology*, **7**, pages 821-824.

[9] Yates, F. (1935). *Journal of Royal Statistical Society, Suppl.* **2**, 181-247.

[10] Jin, W., Riley, R.M., Wolfinger, R.D., White, K.P., Passador-Gurgel, G., and Gibson, G. (2001). *Nature Genetics*, **29**, 389 - 395.

[11] Morton, C.C. (2000). personal communication.

[12] Hedayat, A.S., Majumdar, D. (1984). *Technometrics*, **26**, 363-370.

[13] Alizadeh, A., Eisen, M.B., Davis, R.E., Ma, C., Lossos, I.S., Rosenwald, A., Boldrick, J.C., Sabet, H., Tran,T., Yu, X., Powell, J.I., Yang, L., Marti, G.E., Moore, T., Hudson, J., Lu, L., Lewish, D.B., Tibshirani, R., Sherlock, G., Chan, W.C., Greiner, T.C., Weisenburger D.D., Armitage, J.O., Warnke, R., Levy, R., Wilson, W., Grever, M.R., Byrd, J.C., Botstein, D., Brown, P.O., and Staudt, L.M. (2000). *Nature*, **403**, 503-511.

[14] Kerr, M.K., Martin, M., and Churchill, G.A. (2001c). *Journal of Computational Biology*, **7**, 819-837.

[15] Simon, R., Radmacher, M.D., Dobbin, K. (2002). *Genetic Epidemiology*, **23**, 21-36.

[16] DeRisi, J., Iyer, V.R., and Brown, P.O. (1997). *Science*, **278**: 680-865.

[17] Dobbin, K., Shih, J.H., Simon, R. (2003). *Bioinformatics*, **19**, 803-810.

[18] Björkbacka, H., personal communication, 2003

Chapter 10

ANOVA MODELS
FOR MICROARRAY DATA

Analysis of variance (ANOVA) methods play a major role in statistical analysis in many fields of scientific investigation and now have become an important methodology in microarray studies[1]. In this chapter, the log-linear model for gene expression data is first introduced as a basic model that is a precursor of the ANOVA model. Next, a full ANOVA representation is presented, with explanations about parameter estimation and the connection to normalization and differential gene expression, the latter being the principle quantities of interest in microarray studies. Another section examines the issue of distinguishing differentially expressed genes from unexpressed genes within the ANOVA framework. A subsequent section describes the precise connection between ANOVA methods and the simpler procedure of taking color ratios and examining fold changes in expression. Finally, a last section looks at the important role that mixed models can play in microarray studies.

10.1.   A Basic Log-linear Model

The additive model for background noise assumes that gene intensity w_{gc} is a sum of two independent components. One component is monotonically related to the level of gene concentration in the biological specimen or experimental condition. The second is background noise, unrelated to gene concentration.

An alternative multiplicative model assumes that the logarithm of gene expression, $y_{gc} = \log(w_{gc})$, is a linear model of the following form.

$$y_{gc} = \log(w_{gc}) = \mu + \gamma_g + \tau_c + (\gamma\tau)_{gc} + \epsilon_{gc} \qquad (10.1)$$

Here μ is an overall mean parameter, γ_g and τ_c are main effects that can be considered as normalizing parameters for gene g and experimental condition c, respectively. Kerr and Churchill[2] (2001) use the agricultural word *variety* for the biological specimen, treatment or experimental condition. The term $(\gamma\tau)_{gc}$ is an interaction parameter for gene and experimental condition. This *interaction term* reflects differential expression for gene g in condition c. The last parameter, ϵ_{gc}, is an error term. The overall mean is defined so the error term is centered, i.e., $E(\epsilon_{gc}) = 0$. Thus,

$$E(y_{gc}) = E(\log(w_{gc})) = \mu + \gamma_g + \tau_c + (\gamma\tau)_{gc}. \qquad (10.2)$$

The error term captures all random variability in gene intensity, whether its source is background noise or various sources of variability affecting hybridization. To ensure uniqueness and estimability in the present context, we require the parameters to sum to zero over their respective indices.

Given gene intensity reading w_{gc} and its logarithmic transformation $y_{gc} = \log(w_{gc})$, the equations below show that if the data are normalized by both gene and experimental condition, the resulting normalized data estimate the interaction parameters for gene and condition, i.e., estimate differential expression. With complete data, the estimability constraints imply that the following correspondences exist between the normalizing means for the data and the parameter estimates, denoted by $\hat{\mu}$, $\hat{\gamma}_g$, $\hat{\tau}_c$ and $(\hat{\gamma\tau})_{gc}$.

$$
\begin{aligned}
\hat{\mu} &= \overline{y}_{++} \\
\hat{\gamma}_g &= \overline{y}_{g+} - \overline{y}_{++} \\
\hat{\tau}_c &= \overline{y}_{+c} - \overline{y}_{++} \\
(\hat{\gamma\tau})_{gc} &= y_{gc} - \overline{y}_{g+} - \overline{y}_{+c} + \overline{y}_{++}
\end{aligned}
$$

Here $\overline{y}_{g+}$, $\overline{y}_{+c}$, and $\overline{y}_{++}$ are defined in section 6.2.3. The relationship of model (10.1) to the mRNA concentration ζ_{gc} of gene g in condition c is not explicit. We do know, however, that concentration ζ_{gc} is constant across the experimental conditions for gene g if and only if $(\gamma\tau)_{gc} = 0$ for all c. In particular, if gene g is unexpressed in any of the conditions, so $\zeta_{gc} = 0$ for all c, then the interaction $(\gamma\tau)_{gc} = 0$ for all c.

10.2. ANOVA With Multiple Factors

10.2.1. Main Effects

In addition to the factors *gene* and *specimen (or condition)* as considered in model (10.1), a typical ANOVA model for microarray data often includes other factors such as *array* and, for spotted arrays, *dye*, with each factor having several levels. To mention a few additional factors, there is the *pin tip* used to spot the slide, the *technician* who prepared the array, the *scanner* used, the *grid region* of the slide, and so on. When the main effects of these and other factors are included in the ANOVA model, they serve the role of normalizing the gene expression data.

10.2.2. Interaction Effects

Sets of interaction terms are typically needed in the ANOVA model to account for variability in gene expression not accounted for by main effects. Although all pairwise sets of interaction effects may not be included, those involving gene-by-condition and gene-by-array interactions are usually needed. The gene-by-condition interaction effects are the quantities of scientific interest because these reflect the differential expression of genes across the specimens or experimental conditions.

For example, let $y_{agc} = \log(w_{agc})$ denote the logarithm of the expression measure for *in-situ* oligonucleotide array a, treatment condition c and gene g, where indices $1 \leq a \leq A$, $1 \leq c \leq C$ and $1 \leq g \leq G$. Then one can consider the following ANOVA model.

$$y_{agc} = \mu + \alpha_a + \gamma_g + \tau_c + (\gamma\alpha)_{ga} + (\gamma\tau)_{gc} + \epsilon_{agc} \qquad (10.3)$$

Here symbol μ represents the overall population mean log-expression for all genes. The main effects α_a, τ_c and γ_g for array, treatment condition, and gene, respectively, may be viewed as normalizations of the overall level of gene expression that adjust for study-wide variation contributed by specimen preparation, laboratory procedures, instrument measurement and gene abundance.

Model (10.3) includes interactions of array and experimental condition with gene. The term $(\gamma\alpha)$ represents an interaction of gene and array. The term $(\gamma\tau)$ represents an interaction of gene and treatment condition. The parameters of interest from a genetic point of view are the gene main effects γ_g and the interactions $(\gamma\tau)_{gc}$ between experimental condition and gene. The gene main effects indicate the presence or absence of genes and measure their relative abundance among all biological specimens in the experiment. The interaction terms measure the differentials in gene expression across the different experimental conditions. As both the main effects and the interaction terms are being measured on a logarithmic scale, they reflect relative differences in expression rather than absolute differences. Finally, the interaction terms $(\gamma\alpha)_{ga}$ pick up differential effects on gene expression that are attributable to the array. These might be expected to be insignificant effects, but our experience shows that often this is not the case; gene expression profiles tend to vary by array.

10.3.   A Generic Fixed-Effects ANOVA Model

Assume that the index $l = 1, \ldots, L$ denotes a set of L experimental factors to be considered in the analysis of gene expression data. Let the index $\mathbf{b}$ denote a vector of the form $(b_1, \ldots, b_L)$ where b_l denotes the level of the lth experimental factor. Let $y_{\mathbf{b}}$ denote the intensity measure, possibly transformed and background corrected with respect to the machine reading. Let $E(y_{\mathbf{b}}) = \mu_{\mathbf{b}}$ denote the mean response, and $\epsilon_{\mathbf{b}}$ the error term, centered on zero.

Then, an ANOVA model has the following generic form

$$y_{\mathbf{b}} = E(y_{\mathbf{b}}) + \epsilon_{\mathbf{b}} = \mu_{\mathbf{b}} + \epsilon_{\mathbf{b}}, \tag{10.4}$$

where

$$\mu_{\mathbf{b}} = \beta_0 + \beta_1(b_1) + \beta_2(b_2) + \ldots + \beta_L(b_L) + \sum_{l=1}^{L}\sum_{k>l}^{L} \mathcal{I}_{lk}(b_l, b_k) + \cdots \tag{10.5}$$

Parameter $\beta_l(b_l)$ denotes a main effect for factor l when it has level b_l, for $l = 1, \ldots, L$, respectively. Parameter β_0 is a constant term. Also, if required, an additional index component can be added to label replicated observations at any given factor level combination. To simplify notation, we use parameters

$$\mathcal{I}_{lk}(b_l, b_k)$$

to denote pairwise interaction effects for factors l and k when they have their respective levels b_l and b_k, with $l, k = 1, \ldots, L$. For example, with $L = 3$ factors, the parameter $\mathcal{I}_{13}(b_1, b_3)$, where $(b_1, b_3) = (5, 4)$, signifies the interaction parameter for factors 1 and 3 when these two factors have their levels 5 and 4, respectively. The model can be expanded to include third- and higher-order interaction terms if needed, as indicated by the series of dots in (10.5).

Each set of main-effect parameters and interaction-effect parameters in the ANOVA model has more parameters than can be estimated and, hence, for estimation, each must be subject to one or more estimability constraints. For example, suppose that factor 1 has B_1 levels. Then the main-effect parameters of factor 1 are $\beta_1(b_l)$, $b_l = 1, \ldots, B_1$. Mathematically, only $B_1 - 1$ of these parameters can be estimated because the factor has $B_1 - 1$ degrees of freedom. Hence, this set of main effects is subject to one linear constraint. The estimability constraint can take various forms as may be convenient. For example, the parameters might be constrained to sum to 0 (i.e., $\sum_{b_l} \beta_1(b_l) = 0$) or, alternatively, the last parameter $\beta_1(B_1)$ might be set to zero. These are referred to as the *sum-to-zero constraint* and *set-last-to-zero constraint*, respectively.

The parameters of the ANOVA model may be estimated by various methods. We shall assume that ordinary least squares methods are used here (i.e., the L_2 norm), but the methodology is readily modified for other estimation approaches, such as those based on weighted least squares or least absolute deviations (i.e., the L_1 norm).

We denote the estimated main effects in (10.5) by $\hat{\beta}_l(b_l)$ and the estimated interaction effects by

$$\hat{\mathcal{I}}_{lk}(b_l, b_k).$$

The estimate $\hat{\beta}_l(b_l)$ represents the correction or adjustment of the response to account for the systematic effect on response for all cases where factor l has its b_l level. Thus, for instance, if factor 1 corresponds to *color* and has two levels, red ($b_1 = 1$) and green ($b_1 = 2$), then $\hat{\beta}_1(2)$ represents the systematic effect on $y_\mathbf{b}$ across all readings that have the color green. Subtracting the estimated main effects $\hat{\beta}_l(b_l)$ in ANOVA model (10.5) from $y_\mathbf{b}$ normalizes the intensity data. As a matter of convenience, the estimated constant term $\hat{\beta}_0$ can also be substracted. In other words, the estimated quantities

$$\hat{u}_\mathbf{b} = y_\mathbf{b} - \hat{\beta}_0 - \hat{\beta}_1(b_1) - \hat{\beta}_2(b_2) - \ldots - \hat{\beta}_L(b_L) \qquad (10.6)$$

are the intensity readings that have been normalized for all L factors. We note that if the sets of parameters have been estimated using the sum-to-zero constraint then the $\hat{u}_\mathbf{b}$ will be centered on zero.

10.3.1.   Estimation for Interaction Effects

Henceforth, the set of gene-by-condition interaction effects, in models (10.1) and (10.3), are denoted by

$$\mathcal{I}_{gc} := (\gamma\tau)_{gc},$$

and, for notational convenience, their estimates are denoted by $\hat{\mathcal{I}}_{gc}$. The interaction parameters $\mathcal{I}_{gc}$ and their estimates $\hat{\mathcal{I}}_{gc}$ are subject to estimability constraints that hold simultaneously across conditions $c = 1, \ldots, C$ and genes $g = 1, \ldots, G$. We will adopt the following sum-to-zero constraint form.

$$\sum_c \hat{\mathcal{I}}_{gc} = 0, \quad \text{for each } g \tag{10.7}$$

The constraint implies that $\hat{\mathcal{I}}_{gc}$ is interpreted as the estimated differential expression for gene g under treatment condition c *relative* to the average for all genes and conditions in the study.

To illustrate the kind of quantity that $\hat{\mathcal{I}}_{gc}$ represents, suppose that $\hat{\mathcal{I}}_{gc}$ happens to equal 1.231 with response y being the logarithm of intensity to base 2. Then we know that the absolute expression intensity for gene g in experimental condition c is $2^{1.231} = 2.37$ times the (weighted geometric) mean expression levels for all genes and conditions in the study, other factors in the study being held constant. In other words, $\hat{\mathcal{I}}_{gc} = 1.231$ implies a 2.37 fold over-expression or up-regulation of gene g. As we shall show later, the pattern of the estimates $\hat{\mathcal{I}}_{gc}$ for all conditions c will form the basis of statistical inference about whether a gene g exhibits differential gene expression across the experimental conditions.

10.4.   Two-stage Estimation Procedures

It can be easily seen that parameters for main effects and interaction effects involving the gene factor will be as numerous as the genes themselves and, hence, may number in the thousands. This fact implies that the ANOVA model generally will have a huge number of parameters to estimate.

A *two-stage estimation procedure*[3] for the ANOVA model unbundles the computational problem into a manageable sequence of subproblems. Basically, the first stage estimates those effects that are not indexed by gene and the second stage estimates those that are indexed by gene. Specifically, the normalization of microarray data is carried out in a first stage for all factors except the gene factor. Then, parameter estimates involving genes are derived in a second-stage analysis where the estimation proceeds gene by gene.

We consider a simple design to illustrate the two-stage estimation procedure. We assume for this demonstration that the study contains G readings on gene expression obtained from C treatment conditions and the experiment is replicated R times and use index $r = 1, \ldots, R$ for the replicates. We include a replicate main effect ν_r to absorb scale differences in the replicates. The full ANOVA model is

$$y_{gcr} = \mu + \gamma_g + \tau_c + \nu_r + \mathcal{I}_{gc} + \epsilon_{gcr}. \tag{10.8}$$

The two-stage approach partitions the full model (10.8) into two sub-models, as follows.

$$y_{gcr} = \mu + \tau_c + \nu_r + u_{gcr} \tag{10.9}$$

$$u_{gcr} = \gamma_g + \mathcal{I}_{gc} + \epsilon_{gcr} \tag{10.10}$$

Here the terms u_{gcr} absorb all of the gene-specific effects in the model.

1. The First-stage (The Normalization Step):

   In the first-stage ANOVA model, the main effects of each experimental factor are taken into account, and the analysis is based on the entire study dataset with intensity readings for thousands of genes. The model as described in (10.9) can easily be fitted using a standard ANOVA procedure that is available in most software covering linear statistical analysis. The fitting provides estimates of the overall mean μ and the main effects for condition τ_c, replicate ν_r and the first-stage error terms u_{gcr}. The estimated residuals from the first-stage model, denoted here by $\hat{u}_{gcr}$, are estimates of gene expression that have been normalized (and centered), as described earlier in (10.6).

   Generally speaking, the first stage should be reserved for effects that are not indexed by gene g. It may be quite reasonable, however, to include selected interactions for pairs of factors (excluding gene) that impact response. For instance, dye and array may interact and we can eliminate these interaction effects by using the first-stage ANOVA as a normalization step.

2. The Second-stage (The Differential Expression Step):

Based on the estimated residuals $\hat{u}_{gcr}$ obtained from the first-stage model, the second-stage will be fitted *for each gene g*, $g = 1, \ldots, G$. The main effect for gene g in the study dataset, denoted by γ_g in model (10.10), and interaction effect between gene g and treatment condition c, denoted by $\mathcal{I}_{gc}$, are then estimated by applying a standard ANOVA fitting procedure to the residuals $\hat{u}_{gcr}$. The ANOVA residuals obtained from the second stage are estimates of the error terms ϵ_{gcr}. Although there are typically thousands of genes in a study, second-stage ANOVA is performed for each one. Pertinent estimates are then saved from each analysis.

10.4.1.   Example

To demonstrate ANOVA procedures and the two-stage method for analyzing microarray data, we will consider the Mouse Juvenile Cystic Kidney Data Set, introduced earlier in Section 6.2.2. The study design was presented in Table 6.1. We have identified four sources of variation in this design that we wish to take into account, namely, array, dye (green or red), tissue type (mutant or wild-type), and gene.

In the first-stage ANOVA, we include all effects in the model that are not indexed by gene. We shall include only main effects for array (α_a), dye (δ_d) and tissue type (τ_c), as follows.

$$y_{adgc} = \mu + \alpha_a + \delta_d + \tau_c + u_{adgc} \qquad (10.11)$$

The residuals from this fitted ANOVA model are

$$\hat{u}_{adgc} = y_{adgc} - \hat{y}_{adgc}, \qquad (10.12)$$

where $\hat{y}_{adgc}$ denotes the fitted value. These residuals constitute the normalized microarray data. The normalization is such that the respective sums of the log-intensities across all genes for each array, dye and tissue type are zero. As an illustration, Table 10.1 shows the eight residuals (normalized values) for a particular gene g_k in this study. The consistent pattern of negative residuals suggests that gene g_k has low expression levels across all arrays, dyes, and tissue types, relative to the average for all genes.

The second-stage ANOVA model uses the residuals $\hat{u}_{adgc}$ from the first stage as the response variable. The model we choose to fit in the second stage includes the main effect for gene and gene interactions with array, dye and tissue type. The fitted model is

| Gene g | Main Effects | | | Log-intensity | Fitted value | Residual |
	Array a	Dye d	Tissue type c	y	$\hat{y}$	$\hat{u} = y - \hat{y}$
g_k	1	1	1	6.569	7.854	-1.284
g_k	1	2	1	6.581	7.710	-1.129
g_k	2	1	1	7.143	8.114	-0.971
g_k	2	2	2	6.564	7.607	-1.043
g_k	3	1	2	6.131	7.631	-1.500
g_k	3	2	1	6.420	7.850	-1.430
g_k	4	1	2	6.375	7.571	-1.196
g_k	4	2	2	6.372	7.427	-1.055

Table 10.1. First-stage ANOVA residuals for gene g_k. The residuals represent log-intensity values normalized for arrays a=1,2, dyes d=1,2, and tissue types (conditions c=1,2).

$$u_{adgc} = \gamma_g + (\gamma\alpha)_{ga} + (\gamma\delta)_{gd} + (\gamma\tau)_{gc} + \epsilon_{adgc}, \tag{10.13}$$

The interaction term $(\gamma\tau)_{gc}$, denoted earlier by the simpler notation $\mathcal{I}_{gc}$, captures the differential expression of genes across experimental conditions. Model (10.13) is fitted gene by gene. For each gene, γ_g corresponds to the regression intercept term and the three interaction terms are, in fact, equivalent to main effects. The analysis ignores the slight dependence that is induced in the residuals by the normalization of the first stage analysis which forces the residuals to sum to zero across the factors (array, dye, and tissue type).

Table 10.2 shows the second-stage ANOVA results for the eight residuals of gene g_k. These are produced by a standard regression routine. The sums of squares are, in fact, sequential sums of squares (seq. SS) corresponding to adding the variables in the order shown (array, dye, tissue type).

The estimates of $\mathcal{I}_{gc}$ for gene g_k from the regression output are 0.03545 and -0.03545 for the mutant (type 1) and wild-type (type 2) tissues, respectively. The judgment that remains to be made in this case is whether or not these results suggest that gene g_k is truly differentially expressed in the two tissue types. The investigation of this issue is taken up in the next section. Of course, the kind of results we have presented for gene g_k must also be generated for the remaining 1727 genes in this data set. Thus, this second-stage analysis involves performing 1728 such regression analyses.

Source	Seq SS	df	MS
Array	.225787	3	.075262
Dye	.010822	1	.010822
Tissue type	.005026	1	.005026
Error	.011150	2	.005575
Total	.252784	7	.036112

Table 10.2. Second-stage ANOVA table for gene g_k.

10.5.   Identifying Differentially Expressed Genes

Identifying truly differentially expressed genes in microarray studies is a major statistical challenge that has received the attention of many investigators. A variety of approaches have been proposed. In this section, we consider the issue in the context of classical ANOVA modeling.

If a gene g is not differentially expressed across the experimental conditions, then there is no interaction between gene g and any condition c. Hence, a gene g is differentially expressed or not according to whether the following hypothesis is rejected or not.

$$H_0 : \text{Interaction effect } \mathcal{I}_{gc} = 0, \text{ for all conditions } c \qquad (10.14)$$

We now look at several approaches to testing this hypothesis.

10.5.1.   Standard MSE-based Approach

In a standard application of ANOVA, one would proceed to test whether the sequential sum of squares for experimental condition (tissue type) is large enough to merit rejection of the null hypothesis (10.14). The standard model assumptions require the error terms ϵ to be mutually independent and identical normal random variables. Under these assumptions, an F test can be employed to test the sequential sum of squares for each gene.

To explain the F test for gene g, we denote the mean square for between-treatments by MST_g and the mean square for error by MSE_g. The corresponding degrees of freedom are df_T and df_E, respectively. For expository convenience, we assume these degrees of freedom are the same for all genes. The ratio

$$F_g^* = \frac{\text{MST}_g}{\text{MSE}_g} \qquad (10.15)$$

can be compared to a suitable percentile of the $F(df_T, df_E)$ distribution to decide if hypothesis (10.14) should be rejected or not.

The critical percentile for the test would be selected so as to adjust for multiple comparisons when a large number of genes are being tested simultaneously. The reader is referred to Chapter 11 for a discussion of this aspect of the analysis, including types of error control and algorithms for multiple comparisons. For the demonstration here, we use a Bonferroni adjustment and select the F percentile corresponding to a tail area of

$$\alpha_0 = \frac{\alpha_F}{G_0}, \tag{10.16}$$

where α_F is the specified family type I error probability and G_0 is the anticipated number of undifferentially expressed genes in the test set. The total number of genes G may be used in place of G_0 where the latter is difficult to judge.

For gene g_k in the preceding example, it can be seen from Table 10.2 that the F^* ratio equals $.005026/.005575 = 0.902$. The corresponding degrees of freedom are $df_T = 1$ and $df_E = 2$. Using $\alpha_F = 0.05$ and $G_0 = G = 1728$, it follows that $\alpha_0 = 0.05/1728 = 0.0000290$ and the critical percentile of $F(1, 2)$ is $34,591$, a huge value. Comparison of F^* with this percentile leads to the conclusion that gene g_k is not differentially expressed. Other genes could be similarly tested.

For a microarray study in which only two conditions are being compared (say, a control condition and a treatment condition), a t-test may be used in lieu of the F-test. In this case, the t^* statistic takes the form

$$t_g^* = \frac{\hat{d}_g}{s(\hat{d}_g)}, \tag{10.17}$$

where $\hat{d}_g$ and $s(\hat{d}_g)$ are the estimated log-expression difference between treatment and control and its estimated standard error, respectively. The estimated standard error depends on the value of MSE_g. In our earlier interaction notation,

$$\hat{d}_g = \hat{I}_{g2} - \hat{I}_{g1}. \tag{10.18}$$

For gene g_k, regression output gives $t^* = -0.0708956/0.0746653 = -0.9495$. The critical percentile of the t distribution in this case corresponds to a tail area of $\alpha_0/2 = 0.0000145$ and equals 186, again a huge value. Except for computational rounding, it can be seen that $F^* = (t^*)^2$ for the case of two experimental conditions.

10.5.2.   Other Approaches

The preceding test methodology uses the observed mean square error (MSE) for the t or F statistics at the level of the individual gene. (Recall that MSE enters the denominator of both t and F statistics.) Experience with microarray data sets has made investigators cautious about assuming a normal error term for the ANOVA model. Data anomalies and non-normal features of the error term are frequently encountered in microarray data and make MSE values susceptible to distortion. Moreover, the problem is aggravated by the fact that many microarray experimental designs provide few degrees of freedom for the error term at the individual gene level (as occurs in the preceding case illustration).

Some investigators, such as Dudoit *et al.*[4] (2002), are successful in using tests based on t and F statistics. Other investigators have chosen alternative strategies for dealing with the problems. Some use permutation tests to avoid assumptions about the error distribution, although this approach also may depend on having reasonably large degrees of freedom at the individual gene level. The permutation test approach is taken up in detail in Chapter 12.

Another strategy is adopted in Efron *et al.*[5] (2001) and in the SAM software, Tusher *et al.*[6] (2001) and Chu *et al.*[7] (2002), where a variance-offset is used to improve reliability of the test statistics. They compute statistics of the form

$$\frac{d_g}{s_0 + s_g} \tag{10.19}$$

for each gene g, where d_g is a score statistic (such as a difference) for differential expression, s_g is a standard deviation, and s_0 is an offset parameter. Parameter s_0 is chosen as a particular percentile of all s_g values and referred to as a 'fudge factor' in the SAM documentation. The offset parameter s_0 helps to stabilize the ratios. Efron *et al.* (2001) state that setting $s_0 = 0$ (i.e., omitting the constant) is a 'disasterous choice' in their application (page 1156). We note that setting $s_0 = 0$ would give either a standard t^* statistic or the square root of a standard F^* statistic, as defined in the preceding section.

10.5.3.   Modified MSE-based Approach

Drawing a parallel with the SAM approach in the case of an F statistic, a ratio of the following form can be computed for each gene g.

$$F_g^*(\text{adjusted}) = \frac{\text{MST}_g}{\text{MSE}_g(\text{adjusted})}, \tag{10.20}$$

where

$$\text{MSE}_g(\text{adjusted}) = w\text{MSE}_g + (1 - w)\text{MSE}_0. \tag{10.21}$$

Here w is a fractional weight and $\text{MSE}_0 > 0$ is a selected offset parameter. In essence, the denominator is a shrinkage estimator with MSE_g being shrunk toward MSE_0. Notice that if $w = 1$, the adjusted F statistic reduces to the standard F statistic. At the other extreme, $w = 0$ gives the constant MSE_0 as the denominator.

Choosing appropriate values of MSE_0 and weight w allows the ratio $F_g^*(\text{adjusted})$ to be optimized. This approach is in the same spirit as that of Efron *et al.* (2001) and the SAM software system by Tusher et al (2001) and Chu *et al.* (2001), but differs in that the latter involves a weighting of a *standard deviation* and an offset parameter whereas the former involves a weighting of a *mean square error* (MSE value) and an offset parameter.

The approach of using a fixed divisor[8] for the ratio $F_g^*(\text{adjusted})$ corresponds to the case with $w = 0$ in (10.20), and (10.21). To understand this approach, let σ^2 denote the variance of the error term ϵ in the standard ANOVA model. Then, assuming that df_T is the same for all genes, it follows that $df_T \text{MST}_g/\sigma^2$ follows a $\chi^2_{df_T}$ distribution under the null hypothesis (10.14). Furthermore, if at least 50 percent of genes are not differentially expressed (generally a reasonable assumption), then

$$\frac{df_T}{\chi^2_{df_T}(0.5)} \text{Median}(\text{MST}) \tag{10.22}$$

estimates σ^2, where $\text{Median}(\text{MST})$ and $\chi^2_{df_T}(0.5)$ denote the medians of the MST_g values and the $\chi^2_{df_T}$ distribution, respectively. Thus, the test statistic

$$C_g = \chi^2_{df_T}(0.5) \frac{\text{MST}_g}{\text{Median}(\text{MST})} \tag{10.23}$$

should be distributed approximately as $\chi^2_{df_T}$ if H_0 is true for gene g, i.e., if gene g is not differentially expressed. A comparison of C_g with an appropriate critical percentile of the $\chi^2_{df_T}$ distribution can test whether gene g is differentially expressed. Statistic C_g in (10.23) corresponds to $df_T F_g^*(\text{adjusted})$ with the offset parameter in (10.20) receiving full weight $(w = 0)$ and being set equal to

$$\text{MSE}_0 = \frac{df_T}{\chi^2_{df_T}(0.5)} \text{Median}(\text{MST}). \tag{10.24}$$

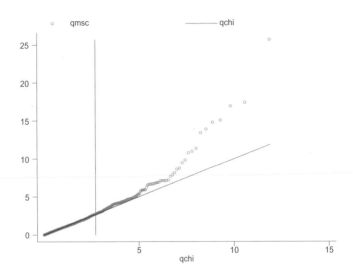

Figure 10.1. χ_1^2 plot of statistics C_g for all genes, together with a line of identity. Departures of points from the line of identity at large values of C_g are indicative of differentially expressed genes. The vertical line marks the 90th percentile of the χ_1^2 distribution and the approximate end of the linear segment of the plot.

For the example, Figure 10.1 shows a χ_1^2 plot of the C_g statistics for the 1728 genes. (Recall that $df_T = 1$.) A line of identity is also plotted in the figure. A vertical line appears at the 90th percentile point of the horizontal scale.

It is evident that the plot is quite linear for at least 90 percent of the genes. Genes that have large values of C_g that lie well above the line of identity can be concluded to be differentially expressed. The list of the dozen genes with the largest values of C_g is identical to that given in Lee *et al.* (2002b) as genes that differentiate between mutant and wild-type tissues. Among the dozen genes, three genes are up-regulated in wild-type tissue and the remainder are up-regulated in mutant tissue. For this data set, we have Median(MST) = 0.027630. As $\chi_1^2(0.5) = 0.454936$ and $df_T = 1$, it follows from (10.24) that the offset parameter is

$$\text{MSE}_0 = (df_T)\,\text{Median(MST)}/\chi_1^2(0.5) = 0.060733.$$

Gene $g = g_k$, which featured in our earlier demonstrations, has $\text{MST}_g = 0.005026$ and, hence, a C_g value of $0.005026/0.060733 = 0.0828$, which is very small and leads to the conclusion that gene g_k is not differentially expressed.

10.6.   Mixed-effects Models

The ANOVA model (10.3) assumes that the main effects and interaction effects of the model are fixed, not random. In some studies, however, it may be quite reasonable to treat some of these effects as random and, more specifically, to assume they are normally distributed. For example, the main effects $\alpha_a, a = 1, \ldots, A$, for arrays may very well be a random sample from a normal population of array effects. A mixed-effects model approach to the analysis of microarray data is discussed in this section. The mixed model approach provides different parameter estimates and, hence, possibly different substantive results. The implications of random effects remain to be investigated in depth.

Wolfinger et al.[9] (2001) use mixed models to assess gene significance for cDNA microarray data from the Saccharomyces cerevisiae swi/snf mutation study conducted by Sudarsanam et al.[10] (2000). The study investigates mutants deleted for a gene encoding one conserved (Snf2) or one unconserved (Swi1) component, each in either rich or minimal media. The four experimental conditions are arrayed in triplicate. The same wild-type strain is used as a reference sample in all twelve arrays and is labeled with Cy5 in channel 2, while the experimental strains are labeled with Cy3 in channel 1.

Let y_{gca} be the logarithm to base 2 of the background-corrected measurement from gene g, $g = 1, \ldots, 6917$, treatment condition c, $c = 1, \ldots, 5$, and array a, $a = 1, \ldots, 12$. Treatment condition here refers to the type of cDNA sample, i.e., the four treatment samples snf2-rich, snf2-minimal, swi1-rich, swi1-minimal, plus the reference sample from a wild-type strain. Instead of forming ratios, they let the two observations for each gene on each array be indexed by treatment condition.

The authors apply the following first-stage ANOVA normalization model.

$$y_{gca} = \mu + \tau_c + \alpha_a + \mathcal{I}_{ac} + u_{gca} \tag{10.25}$$

where μ represents an overall mean, τ_c is the main effect for treatment condition c, α_a is the main effect for array a, $\mathcal{I}_{ac}$ is the interaction effect of array a and treatment condition c, and u_{gca} is an error term. Let $\hat{u}_{gca} = y_{gca} - \hat{y}_{gca}$ denote the residuals from this model, computed by subtracting the fitted values for the effects from the y_{gca} values. Because the experiment does not employ a reverse color design, the dye effect is completely confounded with treatment differences, and hence the dye effect is not included in the model.

The authors next consider the following second-stage ANOVA model for each gene.

$$u_{gca} = \gamma_g + \mathcal{I}_{gc} + \mathcal{I}_{ga} + \epsilon_{gca}. \qquad (10.26)$$

In model (10.26) all effects are indexed by gene g and are assumed to serve similar roles to those from the normalization model in (10.25), but at the gene level. The gene-by-array interaction term $\mathcal{I}_{ga}$ models the effects for each spot. In addition to standard stochastic assumptions, Wolfinger *et al.* (2001) assume that the effects α_a, $\mathcal{I}_{ac}$, u_{gca}, $\mathcal{I}_{ga}$, and ϵ_{gca} are all normally distributed random variables with zero means and variance components $\sigma^2{}_{\alpha}$, $\sigma^2{}_{\alpha\tau}$, $\sigma^2{}_{u}$, $\sigma^2{}_{\alpha\gamma_g}$, $\sigma^2{}_{\epsilon_g}$, respectively. These random effects are assumed to be independent both across their indices and with each other. The remaining terms in the models are assumed to be fixed effects, and thus both models (10.25) and (10.26) are mixed models. Variance components were estimated by the method of restricted maximum likelihood (REML). Note that the original notation has been modified to be consistent with usage in this chapter.

The estimates of primary interest are the gene-by-condition interaction effects $\mathcal{I}_{gc}$ which measure the effects of treatment conditions for each gene. Differences between these effects can be tested by using mixed-model t-tests of all possible pairwise comparisons within a gene. The degrees of freedom (df) for the t-test can be set equal to the df for error from the second-stage ANOVA model. In their article, Wolfinger *et al.* also demonstrate how increasing the number of replications can increase the statistical power of the analysis.

Wernisch, Kendall, Soneji, *et al.*[11] (2003) also apply mixed ANOVA models to quantify the various sources of error in microarray replicates. Their model differs from the model of Wolfinger *et al.* in that Wernisch *et al.* introduce common variance components for all genes. Significance values for differential expression are obtained by a hierarchical bootstrapping scheme applied to scaled residuals.

10.7.   ANOVA for Split-plot Design

As pointed out in the preceding chapter, some microarray studies involve nested or split-plot designs. In these designs, the error term has distinct components corresponding to main plots and subplots. The split-plot feature bears on the analysis when deciding what error variance is appropriate for judging significant differences.

To illustrate the split-plot feature, consider ANOVA model (10.3). The random error term ϵ_{agc} can be decomposed into two components.

$$\epsilon_{agc} = \epsilon_{ac}^{(1)} + \epsilon_{agc}^{(2)} \tag{10.27}$$

The two components, having superscripts (1) and (2), denote the array and sub-array error components, respectively. The components of the error term in (10.27) have zero means by definition. They may be assumed to be independent and possibly also to be normally distributed. To show the significance of recognizing the separate error components, note that in comparing gene expression for genes on the same array, only the variance of error component $\epsilon_{agc}^{(2)}$ applies. The error component $\epsilon_{ac}^{(1)}$ is common to both genes because they lie on the same array and, hence, is canceled in considering the difference.

The error components in (10.27) assume two levels of nesting. In general, nesting can involve more than two levels and can be more intricate in structure, depending on the specifics of the design. For example, in model (10.3), arrays may be nested within each treatment condition, which may be a reason to break error component $\epsilon_{ac}^{(1)}$ into a sum of two further components, say, $\epsilon_{ac}^{(1)} = \epsilon_c^{(1a)} + \epsilon_{ac}^{(1b)}$.

The experimental setting of model (10.27) is of the *split-plot design* type with the arrays being one experimental unit and the spots on arrays being a nested experimental sub-unit.

As described in section 9.2, Jin *et al.*[12] (2001) investigated the effects of sex, genotype, and age on transcriptional variance in adult fruit flies. Having six replications for each combination of two genotypes and two sexes, their experiment consisted of 24 cDNA arrays with 48 separate labeling reactions. Their experiment involved a split-plot design such that both sex and genotype are evaluated at the whole-plot level, while age and dye are evaluated at the sub-plot level. Because of the split-plot nature of the design, the error structure of the ANOVA mixed model corresponds to the array mean square for sex and genotype terms, and the error mean square for age and dye terms.

Treatment of microarray experiments as split-plot designs was also considered by Emptage *et al.*[13] (2003). Using the array as the larger size experimental unit (the whole plot) and the spot on the array as the smaller size experimental unit (the subplot), they discussed model equations appropriate to different designs.

10.8.   Log Intensity Versus Log Ratio

It is common in the analysis of spotted cDNA microarray data to take the ratio of expression intensities of the experimental and reference specimens on each spot of the array (Chen *et al.*[14] (1997); Eisen *et al.*[15] (1998)). Kerr and Churchill[16] (2001c) point out that, although it is convenient to speak of fold change in expression, it can also be misleading because ratios expressing fold change in fluorescence do not necessarily correspond to fold change in actual expression. Ratios of fluorescent intensities do not necessarily account for differential behavior of dyes or variations between samples or arrays. These effects must be accounted for to obtain unbiased estimates of expression ratios. Direct approaches to normalization require preprocessing steps, and ratios can be very sensitive to how these steps are carried out.

With regard to the use of the log-ratio of red and green intensities in spotted cDNA microarray analyses, we point out that the correlation of these intensities at the same spot which results from the varying amount of deposited DNA is accounted for in ANOVA model (10.5) by the inclusion of appropriate interaction terms. With a single array slide, for instance, inclusion of the gene-by-dye interaction term suffices.

Where there are multiple arrays, the third-order effects of interactions among genes, dyes, and arrays might be included to capture this source of variability. The microarray design must include a dye-color reversal feature to allow these important interactions to be estimated. For spotted cDNA microarrays, these intensities are output from the two color channels of the scanner. With reversed-color designs, there will be two such ratios, with the two colors reversed between the experimental and reference specimens. An ANOVA approach to the analysis of microarray data can work with the ratio data directly (usually in the form of logarithms of the ratios) or with the absolute intensity data themselves. The purpose of this section is to show the analytical correspondence of these two approaches.

A stylized context for a microarray study can be used to show the desired correspondence. In this stylized context, we assume that the microarray data have been normalized and that we are focusing on a single gene. Typical experimental factors that must be accounted for at the level of an individual gene are the array, dye (channel color) and treatment condition. In this case, the ANOVA model has the form

$$y_{adc} = E(y_{adc}) + \epsilon_{adc} = \gamma + \alpha_a + \delta_d + \tau_c + \epsilon_{adc} \qquad (10.28)$$

The subscripts denote array a, dye d, and condition c, $a = 1, \ldots, A$, $d = 1, \ldots, D$ and $c = 1, \ldots, C$. The response variable y_{adc} denotes the observed normalized log-intensity, $E(y_{adc})$ denotes the mean of y_{adc}, γ is the mean level of expression of the gene across all readings, α_a is the effect of array a, δ_d is the effect of dye d, τ_c is the effect of treatment condition c and ϵ_{adc} is the error term.

We draw several matters to your attention. First, as the model is being considered at the level of the individual gene, α_a, δ_d, and τ_c correspond to interaction terms in a full ANOVA model for the entire gene set while γ corresponds to the main effect for the gene. Second, the reference specimen is taken to be one level of the treatment factor; for convenience we take this to be the last level C. Third, as there are two dye colors, the dye factor has two levels, i.e., $D = 2$. Fourth, we assume that a complete factorial design is employed. Fifth, we assume no replication for this demonstration.

Consider the typical ratio analysis for one treatment condition spotted on one array; without loss of generality, say, treatment condition c_1 spotted on array a_1. Assuming that the dye color for condition c_1 is d_1 and that the dye color for reference control condition c_0 is d_2, the log-ratio of the two intensities for the gene is

$$\log(w_{111}/w_{120}) = y_{111} - y_{120} \qquad (10.29)$$
$$= \delta_1 - \delta_2 + \tau_1 - \tau_0 + \epsilon_{111} - \epsilon_{120}$$

As a reversed-color design is employed, the reverse situation is found on a second array, say, array a_2. The log-ratio for the reversed-color array is

$$\log(w_{221}/w_{210}) = y_{221} - y_{210} \qquad (10.30)$$
$$= \delta_2 - \delta_1 + \tau_1 - \tau_0 + \epsilon_{221} - \epsilon_{210}$$

Finally, if we average these two differences (the equivalent of taking the logarithm of the geometric mean of the two ratios), we obtain

$$\tfrac{1}{2}\{\log(w_{111}/w_{120}) + \log(w_{221}/w_{210})\} \qquad (10.31)$$
$$= \tfrac{1}{2}\{(y_{111} - y_{120}) + (y_{221} - y_{210})\}$$
$$= (\tau_1 - \tau_0) + \tfrac{1}{2}(\epsilon_{111} - \epsilon_{120}) + \tfrac{1}{2}(\epsilon_{221} - \epsilon_{210})$$

Thus, the geometric mean of the log ratios estimates the difference in treatment effects for the treatment (c_1) and control (c_0) conditions in the study, i.e., the difference $\tau_1 - \tau_0$. It is in this precise sense that the color ratio and ANOVA analyses correspond.

Given this correspondence, one may ask why the ANOVA method with fluorescent intensity readings as the outcome variable may be favored over the use of color ratios. The reason is that the ANOVA statistical machinery is so standard, familiar, and flexible that it is easy to incorporate many of the other relevant design elements of microarray studies into the analysis, including other experimental factors, and to deal with statistical issues, such as missing values and model diagnostics.

Notes

[1] Kerr, M.K., Martin, M., and Churchill, G.A. (2001c). *Journal of Computational Biology*, **7**, 819-837.

[2] Kerr, M.K., Churchill, G.A. (2001b). *Biostatistics*, **2**, 183-201.

[3] Lee, M.-L.T., Lu, W., Whitmore, G.A., Beier, D. (2002b). *Journal of Biopharmaceutical Statistics*, **12(1)**, 1-19.

[4] Dudoit, S., Fridlyand, J., and Speed, T.P. (2002). *Journal of the American Statistical Association*, **97**, 77-87.

[5] Efron, B., Tibshirani, R., Storey, J.D., Tusher, V. (2001). *Journal of American Statistical Association*, **96**, 1151-1160.

[6] Tusher, V.G., Tibshirani, R., Chu, G. (2001). *Proceedings of the National Academy of Sciences, USA*, **98**, 5116-5121.

[7] Chu, G., Narasimhan, B., Tibshirani, R., Tusher, V. (2002). http://www-stat.stanford.edu/~ tibs/clickwrap/sam

[8] Lee, M.-L.T., Bulyk, M.L., Whitmore, G.A., Church, G.M. (2002a). *Biometrics*, **58**, 129-136.

[9] Wolfinger, R.D., Gibson,G., Wolfinger, E.D., Bennett, L., Hamadeh, H., Bushel, P., Afshari, C., Paules, R. (2001). *Journal of Computational Biology*, **8**, 625-637.

[10] Sudarsanam, P., Vishwanath, R.T., Brown, P.O., and Winston, F. (2000). *Proceedings of the National Academy of Sciences, USA*, **97**, 3364-3369.

[11] Wernisch, L., Kendall, S.L., Soneji, S., Wietzorrek, A., Parish, T., Hinds, J., Butcher, P.D., and Stoker, N.G. (2003). *Bioinformatics*, **19**, 53-61.

[12] Jin, W., Riley, R.M., Wolfinger, R.D., White, K.P., Passador-Gurgel, G., and Gibson, G. (2001). *Nature Genetics*, **29**, 389 - 395.

[13] Emptage, M.R., Hudson-Curtis, B., Sen, K. (2003). *Journal of Biopharmaceutical Statistics*, **13**, 159-178.

[14] Chen, Y., Dougherty, E.R., and Bittner, M.L. (1997). *Journal of Biomedical Optics*, **2**, 364-374.

[15] Eisen, M., Spellman, P.T., Brown, P.O. and Botstein, D. (1998). *Proceedings of the National Academy of Sciences, USA*, **95**, 14863-14868.

[16] Kerr, M.K., Martin, M., and Churchill, G.A. (2001c). *Journal of Computational Biology*, **7**, 819-837.

Chapter 11

MULTIPLE TESTING
IN MICROARRAY STUDIES

11.1.   Hypothesis Testing for Any Individual Gene

Deciding whether a particular gene is differentially expressed across experimental conditions frequently employs classical hypothesis testing. The test alternatives for gene g may be stated generically as follows:

The null hypothesis H_0: Gene g is not differentially expressed

The alternative hypothesis H_1: Gene g is differentially expressed

A statistical test of hypotheses exposes an investigator to two types of error. A *type I error* (false positive) occurs if the alternative hypothesis H_1 is concluded when, in fact, the null hypothesis H_0 is true. A *type II error* (false negative) occurs if H_0 is concluded when, in fact, the alternative H_1 is true. The *power* of an hypothesis test is defined as the probability of concluding the alternative hypothesis H_1 when, in fact, H_1 is true.

A principal aim of a microarray study is to have a high probability of declaring a gene to be differentially expressed if it is truly differentially expressed, while keeping the probability of making a false declaration of differential expression acceptably low. As microarray studies typically involve the simultaneous testing of hundreds or thousands of genes for differential expression, the probabilities of producing incorrect test conclusions (false positives and false negatives) must be controlled for the whole gene set. In this chapter, we consider the important problem

	Test Declaration: Not Called as Diff. Expressed	Test Declaration: Called as Diff. Expressed	Number of Genes in Tests
The Null Hypothesis H_0 is True: Gene g in the Test is Not Differentially Expressed	A_0	R_0	G_0
The Alternative H_1 is True: Gene g in the Test is Differentially Expressed	A_1	R_1	G_1
Total Number of Genes in Tests	A	R	G

Table 11.1. Multiple testing framework. G: total number of genes being tested, G_0: the unknown number of truly unexpressed genes, G_1: the unknown number of truly differentially expressed genes, R_0: the number of false positives, A_1: the number of false negatives.

encountered in hypothesis testing in microarray studies, namely, the problem of *simultaneous inference* or *multiple testing* when investigating data for the whole set of genes.

11.2.   Multiple Testing for the Entire Gene Set

Benjamini and Hochberg[1] (B&H) (1995) point out that, when pursuing multiple inferences, researchers tend to select the (statistically) significant ones for emphasis, discussion, and support of conclusions. An unguarded use of single-inference procedures results in a greatly increased false positive (significance) rate. Hence, the need for multiple testing methods. For a good review of earlier development in multiple hypothesis testing, see Hochberg and Tamhane[2] (1987), and Shaffer[3] (1995). The challenge of multiple testing in microarray studies has been discussed by many authors from various perspectives, including Wolfinger *et al.*[4] (2001), Efron *et al.*[5] (2001), Storey and Tibshirani[6] (2001), Dudoit *et al.*[7] (2002), Lee and Whitmore[8] (2002c), and Dudoit *et al.*[9] (2003), among others.

11.2.1.   Framework for Multiple Testing

In Table 11.1, we consider the framework adapted from Benjamini and Hochberg (1995). It is useful for understanding the problem of multiple testing and the control of inferential errors in microarray studies.

This framework postulates that for any given gene, there are, in fact, only two possible situations. Either the gene is not differentially expressed (null hypothesis H_0 is true) or it is differentially expressed (alternative hypothesis H_1 is true). The *test declaration* or decision is either that the gene is differentially expressed (H_0 rejected) or that it is not differentially expressed (H_0 accepted). Thus, there are four possible test outcomes for each gene corresponding to the four combinations of true hypothesis and test declaration.

The total number of genes being tested is G with G_1 and G_0 being the unknown numbers that are truly differentially expressed and not differentially expressed, respectively. Usually, G_0 will be much larger than G_1 and, indeed, in some studies it may be uncertain if any gene is actually differentially expressed (i.e., it may be uncertain if $G_1 > 0$).

The counts of the four test outcomes are shown by the entries A_0, A_1, R_0, and R_1 in the multiple testing framework. These counts are random variables in advance of the analysis of the study data. The counts A_0 and A_1 are the numbers of true and false negatives (i.e., true and false declarations that genes are not differentially expressed). The counts R_1 and R_0 are the numbers of true and false positives (i.e., true and false declarations of genes being differentially expressed). The totals A and R are the numbers of genes that the study declares are not differentially expressed (H_0 accepted) and are differentially expressed (H_0 rejected), respectively.

We index the genes for which H_0 and H_1 hold by the sets $\mathcal{G}_0$ and $\mathcal{G}_1$, respectively. We must remember, of course, that the memberships of these index sets are unknown because we do not know in advance if any given gene is differentially expressed or not. The central problem of multiple testing is to classify the genes into two sets that match $\mathcal{G}_0$ and $\mathcal{G}_1$ as closely as possible. The classification should be done in a manner that minimizes the scientific cost of misclassification, with costs being appropriately defined.

11.2.2. Test Statistic for Each Gene

The test decision for any gene g is taken on the basis of a summary statistic which we will denote here by v_g. In different applications discussed in Chapter 14, the summary statistic may be a standard normal statistic z, t statistic, F statistic, χ^2 statistic or other test statistic. Under the null hypothesis H_0 that gene g is not differentially expressed, v_g is an outcome from a null probability density function $f_0(v)$.

The P-value of the statistic v_g is defined as the level of significance of the test for which H_0 would just be rejected. We denote this P-value for gene g by p_g. Thus, p_g is a measure of how discordant the evidence is with H_0, with smaller values being evidence against H_0 in favor of H_1. We will limit our discussion of the multiple testing issue to the P-values for the G genes. In some heuristic sense, if p_g is small then we should decide that gene g belongs to the set of differentially expressed genes. The challenge is to know how 'small' the value of p_g must be before this decision is taken.

The test statistics v_g and their corresponding p_g are derived from the same microarray data set and may show varying degrees of statistical dependence from gene to gene. This implies that test outcomes for different genes may be probabilistically dependent. For example, a subarray of spots on a microarray slide may share an excess of fluorescence because of contamination of the slide and the genes corresponding to the affected spots will then have test statistics that share this common influence. Multiple testing must allow for dependence if it is assumed or known to be present. There are practical cases where dependence is a major concern and others where it is not and independent p_g may be assumed.

We wish to emphasize in our discussion of dependence that we are not discussing biological dependencies of differential expression levels among genes (i.e., co-regulation). These kinds of dependencies are certainly going to be present in every microarray study. For example, H_1 may be true for a group of genes because they are differentially expressed together under given experimental conditions. The focus of our concern is whether the random variation in the v_g is intercorrelated among genes.

11.2.3. Two Error Control Criteria in Multiple Testing

Multiple testing requires the choice of an error control criterion. Here we present two criteria that have received wide acceptance.

1. *Familywise Error Rate* (FWER) α_F:

   The number of false positives in the multiple testing framework is given by R_0 in Table 11.1. The familywise type I error probability, denoted by α_F here, is the probability that one or more false positives occur, i.e.,

   $$\alpha_F = P(R_0 \geq 1). \tag{11.1}$$

   The familywise type I error probability α_F is sometimes referred to as the familywise error rate (FWER)[10]. The error control criterion could call for α_F to be at a specified level. The subscript F is used to remind us that the error probability applies to the whole *family* of G tests.

2. *False discovery rate* (FDR) ϕ_F:

The proportion of null hypotheses that are falsely rejected in the multiple testing framework is given by R_0/R. The proportion is defined as zero if there are no rejected null hypotheses (i.e., if $R = 0$). The expected value of this ratio, namely,

$$E(\frac{R_0}{R}) \tag{11.2}$$

is defined as the *false discovery rate* or FDR for short, and was proposed by Benjamini and Hochberg (1995) as an error control criterion. The criterion could call for the FDR to have the specified level ϕ_F. Again, the subscript F is used to remind us that the error probability applies to the whole family of G tests.

11.2.4.   Implementation Algorithms

The actual implementation of multiple tests may employ various algorithms in conjunction with a chosen error control criterion. We now discuss several of these algorithms.

1. *Šidàk method:*

If the G multiple tests being performed in Table 11.1 have independent test statistics, then their P-values, $p_g, g = 1, \ldots, G$, are also independent. The Šidàk method assumes such independence. It also assumes that the null hypothesis H_0 holds for all genes G, i.e., that $G = G_0$. These assumptions lead to the following decision rule:

Classify gene g as differentially expressed (i.e., reject H_0) if

$$p_g < \alpha_0, \tag{11.3}$$

where the critical value α_0 is chosen so that

$$\alpha_0 = 1 - (1 - \alpha_F)^{1/G}. \tag{11.4}$$

For example, if the familywise type I error probability $\alpha_F = 0.10$ is specified and there are a total of $G = 2000$ genes being tested, then

(11.4) gives $\alpha_0 = 0.0000527$. Thus, a gene would be declared to be differentially expressed only if it has an extremely small P-value.

The explanation for the form of (11.4) is that if the null hypothesis H_0 is true for every gene and there are to be no false positives, then H_0 must be accepted for every gene. Under independence, the probability of this event is $(1-\alpha_0)^G = 1-\alpha_F$. Appropriate rearrangement gives the expression in (11.4).

Under the assumptions of the Šidàk method, the number of false positives will be binomially distributed with mean value

$$E(R_0) = G\alpha_0. \qquad (11.5)$$

Thus, α_0 might be specified in terms of a target value for the expected number of false positives rather than a specification for α_F as called for in (11.4). When testing a large number of genes, it may be quite reasonable for the expected number of false positives to be moderately large. For example, if $G = 2000$, an investigator may feel that $E(R_0) = 4.0$ is tolerable. The formula (11.5) then implies that $4.0 = 2000\alpha_0$ or $\alpha_0 = 0.002$. Notice that $\alpha_0 = 0.002$ implies from (11.4) that $1 - \alpha_F = (1 - 0.002)^{2000} = 0.0182$ or $\alpha_F = 0.982$. Thus, a focus on tolerable levels for the expected false positive rate shows that large values of the familywise type I error risk α_F may be reasonable in microarray studies.

The Šidàk method is judged by some analysts to be conservative in the sense of leading to somewhat fewer genes identified as differentially expressed than by some competing methods (described below). This feature follows from the assumption that the gene set has no differentially expressed genes. The large number of genes in many microarray studies, combined with the fact that a high proportion are usually not differentially expressed, give the Šidàk method only a mild conservatism, as the case illustration that follows in the next section will demonstrate.

2. *Bonferroni method:*

This method is classical and depends on a probability inequality known as the Bonferroni inequality. It is the analog of the Šidàk method for the case where the P-values for the tests are possibly dependent. The method is valid for any degree or form of dependence. Like the Šidàk method, the Bonferroni method assumes that the null hypothesis H_0 holds for all genes G, i.e., that $G = G_0$. Hence, the two methods share the same conservatism. The Bonferroni decision rule is the following:

Classify gene g as differentially expressed (i.e., reject H_0) if

$$p_g < \alpha_0, \tag{11.6}$$

where the critical value α_0 is chosen so that

$$\alpha_0 = \frac{\alpha_F}{G}. \tag{11.7}$$

For example, if $\alpha_F = 0.10$ is specified and $G = 2000$ then (11.7) gives $\alpha_0 = 0.10/2000 = 0.0000500$. Notice that this value of α_0 is slightly smaller than that provided by the Šidàk method in the same illustration (0.0000527).

The explanation for the form of (11.7) is that if H_0 is true for every gene and there is to be no false positive then H_0 must be accepted for every gene. Under dependence, the Bonferroni inequality states that the probability of this event will be no smaller than $1 - G\alpha_0$. Thus, setting $G\alpha_0$ equal to α_F guarantees that the actual familywise type I error risk will not exceed α_F. In future discussion, we will treat α_F as the actual risk rather than an upper bound.

The assumption of dependent tests in the Bonferroni method implies that the number of false positives will not have a definite distribution form. Nevertheless, the expected number of false positives is still given by the following formula.

$$E(R_0) = G\alpha_0 = \alpha_F \tag{11.8}$$

The righthand equality here follows from (11.7). The fact that the expected number of false positives $E(R_0)$ and familywise type I error risk α_F are equal implies that $E(R_0)$ must be less than one with the Bonferroni method.

As already noted, the Bonferroni method is conservative in the sense of leading to somewhat fewer genes identified as differentially expressed than some competing methods. The conservatism is slightly greater than the Šidàk method because it is also accommodating arbitrary dependence among the test P-values. For most microarray studies, the conservatism has only a small influence, as the case illustration that follows in the next section will demonstrate.

3. *Holm method:*

The Holm method (Holm, 1979) involves a sequential application of the Bonferroni method to the ordered P-values, $p_{(1)} \leq \ldots \leq p_{(G)}$. Like the Bonferroni method, the Holm method is appropriate for P-values that are possibly dependent. Unlike the Šidàk and Bonferroni methods, however, it does anticipate that some genes may be differentially expressed. The Holm decision rule is the following:

Let J be the largest index for which

$$p_{(j)} \leq \frac{\alpha_F}{G - j + 1} \quad \text{for } j = 1, \ldots, J. \tag{11.9}$$

Then classify genes corresponding to ranks $(1), \ldots, (J)$ as differentially expressed (i.e., reject H_0 for these genes).

Observe that this method is a *step-up method* in which each successively larger P-value is tested until the requisite condition is not satisfied.

The rationale of the method is the following. The method starts by comparing the smallest P-value to α_F/G. If $p_{(1)} \leq \alpha_F/G$ then the gene with rank (1) is classified as differentially expressed. If this inequality doesn't hold, then all genes are classified as not differentially expressed. This step is none other than an application of the Bonferroni method to the full set of G genes. If the rank (1) gene is classified as differentially expressed then it is set aside and the Bonferroni method is applied to the remaining $G - 1$ genes. The P-value of the second rank gene, namely $p_{(2)}$, is compared to $\alpha_F/(G-1)$, and so on, for the remaining genes. At each stage, where genes ranked $(1), \ldots, (J)$ are being considered, the familywise type I error risk for the remaining unclassified genes is bounded above by α_F.

When the Holm method stops, the J genes classified as differentially expressed are assumed to approximate the index set $\mathcal{G}_1$ and J is an estimate of G_1 in Table 11.1.

Because the total number of genes being tested, G, is typically quite large compared to the number of differentially expressed genes G_1, the comparison value $\alpha_F/(G - j + 1)$ will typically vary over a small dynamic range; specifically, from α_F/G to α_F/G_0 as index j ranges over $1, \ldots, G_1 + 1$. For example, if $\alpha_F = 0.10$, $G = 2000$, $G_1 = 200$ and $G_0 = 1800$ then $\alpha_F/G = 0.10/2000 = .0000500$ while $\alpha_F/G_0 = 0.10/1800 = 0.0000556$. As the former value is the one given by the Bonferroni method, it can be seen that the Holm method may classify a few more genes as differentially expressed, namely, those genes having values of p_g in the interval between 0.0000500 and 0.0000556.

4. *Benjamini and Hochberg FDR method:*

The Benjamini and Hochberg (B&H) method, like the Holm method, accommodates situations where some genes may be differentially expressed. The method has a hybrid requirement with respect to dependence of the P-values. Specifically, it requires P-values for tests where H_0 is true to be independent but places no requirement for independence on tests where H_0 is false. The B & H decision rule involves a sequential consideration of the ordered P-values, $p_{(1)} \leq \ldots \leq p_{(G)}$ as follows:

Let J be the largest index j for which

$$p_{(j)} \leq \frac{j}{G} \phi_F \qquad (11.10)$$

Then classify genes corresponding to ranks $(1), \ldots, (J)$ as differentially expressed (i.e., reject H_0 for these genes).

Observe that this method is a *step-down method* in which each successively smaller P-value is tested until the requisite condition is satisfied.

Benjamini and Hochberg prove that their method, under the stated assumptions, ensures that the false discovery rate is no larger than $(G_0/G)\phi_F$. Thus, it controls the FDR for any number of true null hypotheses (i.e., any number of genes that are not differentially expressed). Where no genes are differentially expressed (so $G_0 = G$), the upper bound becomes ϕ_F, the specified FDR requirement.

When the B & H method stops, the J genes classified as differentially expressed are assumed to approximate the index set $\mathcal{G}_1$ and J is an estimate of G_1 in Table 11.1.

The comparison value in the B & H method will typically vary from ϕ_F/G to $(G_1/G)\phi_F$ as index j ranges over $1, \ldots, G_1$, which will be a

Gene g	Differential Expression $\hat{d}_g$	Standard-ized Value z_g	Two-sided P-value p_g	Test Algorithm			
				Šidàk	Bonferroni	Holm	B&H
		Up-regulated in Mutant Tissue					
g_{1560}	1.249	5.395	.0000001	DE	DE	DE	DE
g_{1038}	1.028	4.442	.0000089	DE	DE	DE	DE
g_{1347}	1.014	4.379	.0000119	DE	DE	DE	DE
g_{401}	0.948	4.094	.0000424	DE	DE	DE	DE
g_{293}	0.918	3.968	.0000725				DE
g_{1198}	0.902	3.898	.0000969				DE
g_{408}	0.830	3.587	.0003344				DE
g_{1691}	0.773	3.338	.0008435				
g_{1238}	0.758	3.277	.0010499				
		Up-regulated in Wild-Type Tissue					
g_{1584}	-0.809	-3.496	.0004718				DE
g_{1224}	-0.816	-3.527	.0004197				DE
g_{1229}	-0.956	-4.131	.0000362	DE	DE	DE	DE

Table 11.2.   Distinguished genes that show substantial differential expression between mutant and wild-type tissues, according to four multiple testing algorithms.

large dynamic range if many genes are differentially expressed. For example, if $\phi_F = 0.10$, $G = 2000$, $G_1 = 200$ and $G_0 = 1800$ then $\phi_F/G = 0.10/2000 = .00005$ while $(G_1/G)\phi_F = (200/2000)0.10 = 0.01$. Note that the FDR specification here, as reflected by ϕ_F, states that at most 10 of every 100 rejections of H_0 are to be false discoveries of differential expression.

11.2.5.   Example of Multiple Testing Algorithms

To illustrate the preceding algorithms, we return to the Mouse Juvenile Cystic Kidney Data Set in section 6.2.2. Recall that mutant and wild-type kidney tissues were being compared with reference to $G = 1728$ genes. See the study design in Table 6.1. In one analysis, reported in Lee *et al.*[11], a standard normal statistic z_g was computed for each gene to identify up- and down-regulated genes in the mutant tissue relative to wild-type tissue. The statistic measured the standardized difference in gene expression between mutant and wild-type tissue. Table 11.2 is compiled from this article and gives the largest twelve of these statistics (in absolute value). The statistic z_g corresponds to test statistic v_g in our notation. We have now appended the two-sided P-values p_g for these statistics to the table.

To compare and contrast the test results produced by the algorithms, we adopt values of $\alpha_F = \phi_F = 0.10$. The table shows the genes declared as differentially expressed (DE) according to each of the four algorithms. In this case illustration, the Šidàk, Bonferroni, and Holm methods select identical sets of genes as differentially expressed. The B & H method chooses a larger set. It must be remembered that the B & H method is employing a different error control criterion than the other three methods (FDR versus FWER), so equating ϕ_F with α_F does not make the two criteria equivalent in terms of their error control. In the final analysis, the superiority of a particular algorithm depends on the validity of the inferences produced, i.e., on the actual error rates in classifying genes as differentially expressed or not. The preceding comparison does not settle this question. What is found generally is that these kinds of algorithms tend to produce similar lists with differences appearing at the margin where the evidence in support of one hypothesis or the other is more mixed.

Some references on multiple testing (e.g., Westfall and Young[12], 1993), propose the calculation of adjusted *P*-values which can be compared directly with the specified level of familywise type I error control. For example, for the Holm method, the adjusted *P*-values would have the following form

$$p^a_{(j)} = \max\left[p^a_{(j-1)}, (G - j + 1)p_{(j)} \right] \quad \text{for } j = 1, \ldots, G, \qquad (11.11)$$

where the $p^a_{(j)}$ denote the adjusted *P*-values with $p^a_{(0)}$ defined as zero. The adjusted *P*-values are compared to α_F and hypothesis H_0 is rejected for the gene with rank (j) if $p^a_{(j)} \le \alpha_F$. The adjusted and unadjusted sets of *P*-values select the same genes as differentially expressed.

The multiple testing algorithms presented here are widely used and easily applied. Many variations on these algorithms, having different rationales and refinements, are available. The reader is referred to the cited references for more information about some of these alternative methods.

11.2.6. Concluding Remarks

The standard implementations of the Šidàk and Bonferroni methods assume that all G genes are not differentially expressed because the number G was used in linking the familywise type I error risk α_F with the critical value α_0. In contrast, the Holm method and B&H method, like other sequential testing algorithms, adjust the cut-off criterion as consecutive genes are 'classified' as being differentially expressed in the algorithm. The number of genes so classified is an estimate of count G_1. Holm's method, for example, counts down from G to $G - J + 1$ as expressed genes are detected. In this case, J estimates G_1.

Later, in Chapter 14 on power and sample size, we use the anticipated count G_0 of undifferentially expressed genes in implementing the Šidàk and Bonferroni methods and, thus, remove the conservative nature of their usual implementation. Note that both the Šidàk and Bonferroni methods could be implemented using a 'guess' about the value of G_0. The Bonferroni implementation, for example, would then use a cutoff of $\alpha_0 = \alpha_F/G_0$ rather than $\alpha_0 = \alpha_F/G$. If G_0 is well chosen then the Bonferroni method would not be more conservative than Holm's method.

Notes

[1] Benjamini, Y., Hochberg, Y. (1995). *Journal of Royal Statistical Society,* **B 57**, 289-300.

[2] Hochberg, Y. and Tamhane A.C. (1987). *Multiple Comparison Procedures.* John Wiley and Sons, New York.

[3] Shaffer, J.P. (1995). *Annual Review of Psychology,* **46**, 561-584.

[4] Wolfinger, R.D., Gibson, G., Wolfinger, E.D., Bennett,L., Hamadeh, H., Bushel, P., Afshari, C., Paules, R. (2001). *Journal of Computational Biology,* **8**, 625-637.

[5] Efron, B., Tibshirani, R., Storey, J.D., Tusher, V. (2001). *Journal of the American Statistical Association,* **96**, 1151-1160.

[6] Storey, J.D., Tibshirani, R. (2001). Technical Report, Stanford University, 2001.

[7] Dudoit, S., Fridlyand, J., and Speed, T.P. (2002). *Journal of the American Statistical Association,* **97**, 77-87.

[8] Lee, M.-L.T., Whitmore, G.A. (2002c). *Statistics in Medicine,* **21**, 3543-3570.

[9] Dudoit, S., Shaffer, J.P., and Boldrick, J.C. (2003). *Statistical Sciences,* **18**, 71-103.

[10] Hochberg, Y. and Tamhane A.C. (1987). *Multiple Comparison Procedures.* John Wiley and Sons, New York.

[11] Lee, M.-L.T., Lu, W., Whitmore, G.A., Beier, D. (2002b). *Journal of Biopharmaceutical Statistics,* **12(1)**, 1-19.

[12] Westfall, P.H. and Young, S.S. (1993). *Re-sampling Based Multiple Testing: Examples and Methods for P-value Adjustment,* John Wiley and Sons, New York.

Chapter 12

PERMUTATION TESTS
IN MICROARRAY DATA

It was noted in Chapter 10 that some standard tests for differential expression at the gene level, such as t- and F-tests based on normal-error theory, may be unreliable because the distributional assumptions do not hold. A family of tests called *permutation tests* or *randomization tests* offer an alternative testing approach which relies on relatively weak assumptions and yet are quite powerful and simple to apply with available software. Ludbrook and Dudley[1] (1998), for example, argue that permutation tests are, in fact, superior to traditional testing approaches in many biomedical contexts. Good[2] (2000) provides a comprehensive treatment of the subject and demonstrates the great practical value of this testing methodology.

The basic concepts underlying permutation tests are set out in the first subsection. Applications of the methodology to microarrays is discussed next. Finally, a case illustration is presented in the last section.

12.1.   Basic Concepts

To convey the basic concepts, consider a conventional testing situation in which n experimental units are randomly divided into two groups of n_1 units and n_2 units, respectively, where $n = n_1 + n_2$. The group of n_1 units is subjected to a control condition, and the group of n_2 units is subjected to a treatment condition. An appropriate response measure y_{ij} is recorded for unit $j, j = 1, \ldots, n_i$, in group $i = 1, 2$. The null hypothesis H_0 of interest postulates that there is no difference in the response pattern for units subjected to the treatment and control conditions. If this null hypothesis is true, the random assignment of experimental units to treatment and control conditions implies that all possible arrangements of the n observations into two groups of n_1 and n_2 cases are *equally probable*. Each arrangement may be viewed as a permutation of the n response values with the first n_1 values assigned to group 1 and the remainder assigned to group 2. We denote a particular permutation by index π. The theory of permutations tells us that there are A such permutations or arrangements in this case, where

$$A = \frac{(n_1 + n_2)!}{n_1! n_2!}. \tag{12.1}$$

For example, if $n_1 = 5$ and $n_2 = 6$, then $A = 11!/(5!6!) = 462$.

In this situation, it is desired to test if the response patterns for the treatment and control conditions share a common feature (such as the same mean or the same variance) or if they differ on that feature. The mean response is a common choice for the feature of interest and we will use it here to illustrate the basic concepts. Thus, we choose the test statistic of interest to be the *observed difference*

$$d^* = \frac{\sum_{j=1}^{n_2} y_{2j}}{n_2} - \frac{\sum_{j=1}^{n_1} y_{1j}}{n_1} = \bar{y}_2 - \bar{y}_1, \tag{12.2}$$

between the two groups, where $\bar{y}_i$ denotes the mean response for group i. The asterisk (*) reminds us that d^* is the observed difference in group means for the actual data gathered in the study. We are inclined to accept the null hypothesis H_0 (postulating identical response patterns) if $|d^*|$ is small and to reject H_0 if $|d^*|$ is large.

If H_0 is true, so the response patterns under treatment and control conditions are identical, then each permutation π of the n response values can be imagined to be a possible realization of the experimental study and can be analyzed accordingly. This analysis would yield a calculated difference in mean response for the two artificial groups created by the permutation. Denote this *calculated difference* corresponding to permutation π by d_π.

The permutation procedure yields a total of A such differences d_π, where A is given by equation (12.1). One of these permutations corresponds to the actual pattern of response for the study. If π^* happens to be that special permutation then $d_{\pi^*} = d^*$, i.e., the calculated difference for that particular permutation will match the observed difference in the study.

A *P*-value for a test of hypotheses gauges the consistency of the statement in the null hypothesis with the statistical evidence. Specifically, the P-value is the probability under the null hypothesis that the test statistic would match or be less consistent with H_0 than the actual statistic observed in the study. P-values are referred to as one- or two-sided according to whether the null hypothesis is one- or two-sided. In a permutation test, therefore, if the null hypothesis were true then the P-value is the fraction of the A calculated differences d_π that are greater or equal to the observed difference d^* in absolute value, i.e.,

$$P\text{-value} = \frac{\text{count}_\pi\left(|d_\pi| \geq |d^*|\right)}{A} \qquad (12.3)$$

Table 12.1 shows a simple illustration of a permutation test involving $n_2 = 3$ treatment units having the responses 8.6, 7.2, and 5.6 and $n_1 = 2$ control units having responses 6.0 and 4.8, as shown at the top of the table. The test involves $A = 5!/(2!3!) = 10$ permutations. The first in the list corresponds to the actual pattern of response and, hence, is marked by an asterisk (*). The list of permutations includes all possible arrangements of the two control and three treatment units, designated by labels 1 and 2, respectively. The second last column shows the calculated difference in means d_π between treatment and control units for each permutation. The actual difference in the study is $d^* = 1.733$, as shown for permutation 1. The last column notes whether the absolute difference for a permutation equals or exceeds the observed absolute difference of 1.733 (yes or no). Three of the 10 permutations satisfy the inequality so the two-sided P-value for the permutation test is $3/10 = 0.30$. For comparison, a standard two-sided t-test, assuming normality and equal treatment and control variances, gives a P-value of 0.246.

The smallest possible P-value occurs when $|d^*|$ takes the most extreme value among all possible eligible permutations. In this situation, the P-value equals $1/A$ (or a larger value if two or more permutations match this most extreme outcome). Thus, $1/A$ is the smallest P-value that can be given by a single permutation test. In the case where $n_1 = 5$, $n_2 = 6$ and $A = 462$, for instance, the smallest possible P-value would be $1/462 = 0.0022$. This technical observation is important because it shows that the permutation test cannot signal a significant difference, i.e., a small P-value, unless a reasonably large number of permutations are possible. This implies that n_1 and n_2 must be large enough to make A large. For example, in a case like that in Table 12.1 with $n_1 = 2$ and $n_2 = 3$, we have $A = 10$ and the smallest possible P value would be $1/10 = 0.10$, which would not constitute strong evidence against a null hypothesis.

| Permutation | Observation | | | | | Mean | $\|d_\pi\| \geq 1.733$ |
π	8.6	7.2	5.6	6.0	4.8	Difference d_π	Yes (Y) or no (N)
1*	2	2	2	1	1	1.733*	Y*
2	2	2	1	2	1	2.067	Y
3	2	2	1	1	2	1.067	N
4	2	1	2	2	1	0.733	N
5	2	1	2	1	2	-0.267	N
6	1	2	2	2	1	-0.433	N
7	1	2	2	1	2	-1.433	N
8	1	2	1	2	2	-1.100	N
9	1	1	2	2	2	-2.433	Y
10	2	1	1	2	2	0.067	N

Table 12.1. Simple demonstration of a permutation test for a difference of means between a treatment group (label 2) and a control group (label 1) of sizes $n_1 = 2$ and $n_2 = 3$, respectively. The calculated P-value is 3/10=0.30. The asterisk * denotes the observed difference.

The number of permutations grows quickly with the magnitudes of n_1 and n_2. For example, if $n_1 = n_2 = 10$, $A = 20!/(10!10!) = 184,756$. Computing all possible permutations in this situation is not practical, especially if many such tests must be performed, as is the case in microarray studies. A common strategy is to estimate the P-value using a random sample of all permutations. Conventional sampling theory can guide the choice of the number of permutations needed to estimate the P-value within a given margin of error. The random sample of permutations can be selected with or without replacement as may be computationally convenient.

12.2.   Permutation Tests in Microarray Studies

Permutation tests are attractive for microarray studies because they rely on few assumptions and can be applied in a straightforward manner, albeit a computationally intensive one.

12.2.1.   Exchangeability in Microarray Designs

Permutation tests can take into account many experimental design features such as multiple treatments, matched observations, and blocking factors. The concept of exchangeable random variables is useful in understanding permutation tests applied to the complex experimental settings found in microarray studies. If $\mathbf{y} = (y_1, \ldots, y_k)$ is a vector of k random variables with joint density function $g(\mathbf{y})$ and $\mathbf{y}_\pi = \pi(\mathbf{y})$ is any permutation of the components of $\mathbf{y}$ then the k random variables are *exchangeable* if $g(\mathbf{y}) = g(\mathbf{y}_\pi)$ for all vectors $\mathbf{y}$ and all permutations π. In essence, realizations of exchangeable variables are statistically indistinguishable.

The implication of exchangeability for permutation tests is the following. With experimental randomization, the postulated truth of the null hypothesis H_0 implies that some observations are exchangeable that would not be exchangeable if H_0 did not hold. In permutation tests, one need only identify those *additional* sets of observations that would be exchangeable *if the null hypothesis were true* and calculate the test statistic for each of the additional permutations that this expanded exchangeability offers. The emphasis in the last sentence is important. Some observations in an experiment may be exchangeable whether the null hypothesis holds or not. For example, experimental replicates are exchangeable observations. Permutations of such exchangeable observations would not alter the experimental findings. We let A_0 denote the baseline number of such permutations. On the other hand, if H_0 holds then additional sets of observations become exchangeable. The number of possible permutations of exchangeable observations when H_0 is true might equal, say, A_1, where $A_1 > A_0$. Hence, the additional permutations offered by assuming the truth of H_0 number $A = A_1/A_0$.

To illustrate the idea of exchangeability, consider the example in Table 12.1. Whether H_0 is true or not, the two observations within the control group are exchangeable, as are the three observations within the treatment group. Thus, the baseline permutations for this design number $A_0 = 2!3! = 12$. Under the null hypothesis that control and treatment have identical response patterns, all five observations become exchangeable, giving $A_1 = 5! = 120$ possible permutations. Thus, H_0 contributes an additional $A = A_1/A_0 = 120/12 = 10$ permutations. These 10 permutations are precisely those listed in Table 12.1.

As a second example, consider an experiment involving a randomized complete block design. Suppose the design has 16 blocks and, within each block, six experimental units are randomly assigned in pairs to three treatment conditions. Consider each block for the moment. Each of the three sets of random pairs in a block are exchangeable whether the three treatments have a differential impact on response or not. Thus, there are $2!2!2! = 8$ possible permutations of the paired observations for the block that would not alter the experimental findings. Under the null hypothesis of no treatment differences for the three treatment conditions, the six observations within each block become exchangeable. These six observations can be arranged in $6! = 720$ different permutations if H_0

holds. Thus, assuming H_0 holds gives us $6!/(2!2!2!) = 720/8 = 90$ additional permutations for the permutation test, namely, all possible arrangements of the six observations into three sets of pairs. The sixteen blocks are not exchangeable, however, because block effects are anticipated. Thus, each block provides 90 extra permutations under H_0 and, as the permutations are independently applied to each block, the total number of additional permutations for all 16 blocks taken together equals $A = 90^{16} = 1.853 \times 10^{31}$, a huge number. In this case, a random sample of all possible permutations would be selected to estimate the P-value for a test of no differences among the three treatments.

12.2.2.   Limitation of Having Few Permutations

In analyzing microarray data, tests for differential expression are conducted at the gene level. The permutation test can be done for each gene separately and a P-value generated for each of the G genes in the microarray study. The P-values must then be subjected to the same multiple comparison procedures as other testing methods. The multiple comparison procedures attempt to control the false positive rate or the false discovery rate for the whole family of tests. Both of these types of error control require that the number of permutations A for each gene not be too small if the permutation test is applied independently to each gene. Recall that $1/A$ is the smallest P-value for a permutation test in this setting.

For example, suppose a design offers $A = 70$ permutations for testing each gene and the microarray study involves 8400 genes, most of which are not differentially expressed. If $1/A$ is used as the P-value for classifying a gene as differentially expressed, then the list of differentially expressed genes would be expected to have as many as $G(1/A) = 8400/70 = 120$ false positives, which may be a very inconvenient number for the investigator and may swamp the number of true positives.

Next we discuss how some investigators have circumvented this limitation of having too few permutations for individual tests.

12.2.3.   Pooling Test Results Across Genes

The permutation test for any single gene, say gene g, yields an observed test statistic as well as a set of test statistics for the permutations. We now use symbol v for the generic form of our test statistic and B for the number of permutations. We denote the observed test statistic by v_g^* and the B permutation test statistics by $v_{bg}, b = 1, \ldots, B$. The number

B is less than or equal to A, depending on whether the complete set of A permutations is included or only a random sample.

The *family null hypothesis* H_0 for a microarray study states that none of the G genes is differentially expressed. Under this H_0, it is plausible to assume that the permutation test statistics v_{bg}, for all b and all g, are drawn independently from a common null distribution of such test statistics with some probability density function, say, $f_0(v)$. In other words, not only are all permutations for any single gene equally probable but the permutation test statistics from different genes can be pooled. The pooling of permutation test results across genes offers a refined basis for testing for differential expression that is free of the limitation imposed by applying the test to each gene in isolation.

To illustrate how the pooled results might be employed, we refer to the SAM[3] software package. In that package, the observed test statistics v_g^* are ordered from smallest to largest across all G genes to form the order statistics $v_{(1)}^* \leq \cdots \leq v_{(G)}^*$. Here a subscript of type (j) refers to the jth ranked statistic. The same order statistics are formed from the test statistics for each permutation b, giving $v_{b(1)} \leq \cdots \leq v_{b(G)}$ for $b = 1, \ldots, B$. The permutation order statistics are averaged for each order as follows.

$$\overline{v}_{(j)} = \frac{1}{B} \sum_{b=1}^{B} v_{b(j)} \quad \text{for } j = 1, \ldots, G \tag{12.4}$$

The average $\overline{v}_{(j)}$ estimates the expected value of the jth order statistic of a sample of size G from the null p.d.f. $f_0(v)$. To test for differential expression, the SAM software compares $\overline{v}_{(j)}$ with $v_{(j)}^*$. If the difference

$$v_{(j)}^* - \overline{v}_{(j)} \tag{12.5}$$

exceeds a threshold parameter Δ then the gene corresponding to $v_{(j)}^*$ is judged to be differentially expressed. The gene is called *significant negative* or *significant positive* according to whether the difference is negative or positive.

12.3.   Lipopolysaccharide-*E.coli* **Data Set**

We introduce a new microarray data set to illustrate the use of permutation tests.

Macrophages were removed from six mice. The cells from each mouse were divided onto six plates. Three of the six plates served as the control reference RNAs (denoted by $t = 0$). The remaining three plates received a mock treatment (buffer change incubated for two hours), an *E.coli* lipopolysaccharide (LPS) endotoxin treatment (10 ηg/ml for two hours) and an *E.coli* bacteria treatment (approximately 1 *E.coli* strain bacterium per macrophage for two hours). These three treatments are labelled as $t = 1, 2, 3$, respectively. Thus, the microarray data are derived from six slides (a=1,...,6) for each of six mice, giving a total of 36 slides. The six slides for each mouse compare the control (t=0) with the three treatments (t=1,2,3) in a reversed-color arrangement. Each slide has 13,028 genes (spots) that are investigated in this analysis, including one positive control and fifteen negative controls. The design layout for each mouse is shown in Table 12.2.

	Array a					
Treatment t	a=1	a=2	a=3	a=4	a=5	a=6
Color Cy3 $k = 3$	t=0	t=1	t=0	t=2	t=0	t=3
Color Cy5 $k = 5$	t=1	t=0	t=2	t=0	t=3	t=0

Table 12.2.   Experimental design for each mouse for the LPS-*E.coli* microarray study

The raw response measure provided for statistical analysis is a background-corrected intensity reading, denoted later by w. The readings are derived using *masliner*[4] from scans at several power levels that provide an accurate resolution of gene expression without undue influence from detection and saturation threshold effects.

12.3.1.   Statistical Model

The microarray data were analyzed separately for each mouse, using the following two-stage ANOVA model.

Stage 1:

$$y_{tkag} = \mu + \tau_t + \kappa_k + \alpha_a + u_{tkag} \qquad (12.6)$$

Stage 2:

$$u_{tkag} = \gamma_g + (\tau\gamma)_{tg} + (\kappa\gamma)_{kg} + (\alpha\gamma)_{ag} + \epsilon_{tkag} \qquad (12.7)$$

The background-corrected intensity reading was subjected to an affine (natural) logarithmic transformation to yield the response variable $y_{tkag} = \log(a+w_{tkag})$. Here $a > 0$ is a constant chosen separately for each mouse to ensure positive readings. Symbol μ represents the overall population mean log-expression for all genes. Symbols τ, κ, α, and γ are the main effects corresponding to the factors: treatment/control, color, array/slide, and gene, respectively. Terms of form $(\tau\gamma)$ represent interactions of the treatment effect τ and gene effect γ. The first-stage ANOVA normalizes the response data for main effects contributed by treatment, color, and array. Model (12.6) yields the residuals $\hat{u}_{tkag}$ as normalized log-intensity readings. These residuals are then used in the second-stage analysis. Model (12.7) includes a main effect for gene as well as sets of pair-wise interaction terms for treatment/control, color and array with gene. The term ϵ is a random error. The error term has mean zero by definition. We assume that the error terms are mutually independent but make no assumption about their variance or distribution form. The indexes have the following ranges, as illustrated in the design layout: $t = 0, 1, 2, 3$ for treatment/control, with 0 denoting the control; $k = 3, 5$ for dyes Cy3 and Cy5; $a = 1, \ldots, 6$ for the six arrays (slides) and $g = 1, \ldots, 13028$ for genes.

The potential for differential expression across treatments 1, 2, and 3 is of scientific interest here. Two of the six slides for each mouse compare the control (0) with one of the three treatments (1,2,3), first in one color order and then in reverse order. Thus, the residuals from the first-stage model can be rearranged to yield an estimate of $(\tau\gamma)_{tg} - (\tau\gamma)_{0g}$, the difference in differential gene expression between treatment t and the control (0). For notational convenience, we denote the $(\tau\gamma)_{tg} - (\tau\gamma)_{0g}$ terms by $\mathcal{I}_{tg}$.

The estimate of $\mathcal{I}_{1g}$ for treatment $t = 1$, denoted by $\hat{\mathcal{I}}_{1g}$, is calculated as follows:

$$\hat{\mathcal{I}}_{1g} = \frac{(u_{151g} - u_{031g}) + (u_{132g} - u_{052g})}{2}, \tag{12.8}$$

Corresponding calculations yield $\hat{\mathcal{I}}_{2g}$ and $\hat{\mathcal{I}}_{3g}$ for treatments $t = 2$ and $t = 3$, respectively.

The quantity in (12.8) equals the log-geometric mean of the two (reversed) color ratios. The averaging in (12.8) cancels the gene main effect and the array-gene and color-gene interaction terms that appear in (12.7). The $\hat{\mathcal{I}}_{tg}$ are the quantities of scientific interest here, as the differences $(\tau\gamma)_{tg} - (\tau\gamma)_{0g}$ capture differential expression among the treatments 1, 2, and 3, relative to control 0, for each gene g.

The data set here is massive, and care is needed to do the analysis without inordinate computation. The first-stage ANOVA is straightforward and is done using a standard software routine. The residuals w_{tkag} are saved from that first-stage analysis. Because of the balanced nature of the design, the second-stage analysis can be done by subtracting estimated effects from the first-stage residuals, gene by gene. This happens to be computationally simpler than running the ANOVA routine a second time for each gene.

12.3.2. Permutation Testing and Results

The scientists were primarily interested in genes that are differentially expressed in a pair-wise comparison of the LPS treatment (treatment 2) and the *E. coli* treatment (treatment 3). Thus, we focus on the differences $\hat{\mathcal{I}}_{3g} - \hat{\mathcal{I}}_{2g}$. We now demonstrate the use of the SAM software (2002) to carry out permutation tests to identify differentially expressed genes with respect to these two treatments. The normalized gene expression data that enter the analysis for each gene are the parameter estimates $\hat{\mathcal{I}}_{2g}$ and $\hat{\mathcal{I}}_{3g}$, arranged in six blocks (each mouse is considered as a block with two treatments) as shown in Table 12.3. We now add the index m to these estimates, $m = 1, \ldots, 6$, to represent the six mice.

	Mouse (Block) m					
Treatment t	1	2	3	4	5	6
LPS $t = 2$	$\hat{\mathcal{I}}_{2g1}$	$\hat{\mathcal{I}}_{2g2}$	$\hat{\mathcal{I}}_{2g3}$	$\hat{\mathcal{I}}_{2g4}$	$\hat{\mathcal{I}}_{2g5}$	$\hat{\mathcal{I}}_{2g6}$
E.coli $t = 3$	$\hat{\mathcal{I}}_{3g1}$	$\hat{\mathcal{I}}_{3g2}$	$\hat{\mathcal{I}}_{3g3}$	$\hat{\mathcal{I}}_{3g4}$	$\hat{\mathcal{I}}_{3g5}$	$\hat{\mathcal{I}}_{3g6}$

Table 12.3.   Data layout for each gene in a SAM analysis comparing treatments 2 and 3 in the LPS-*E.coli* microarray study. The data are treated as two-class data, blocked by mouse.

Figure 12.1 and Figure 12.2 show the SAM plot and SAM output for this illustration. Several adjustments and specifications were required to prepare the data for analysis in SAM. First, the data were treated as two-class data, blocked by mouse. Second, the logarithms of the expression intensity data were transformed to base 2. Third, all settings were given default values. The SAM Δ parameter, which controls the rejection region for the test of each gene, was chosen so that 35 genes were identified as significant (29 positive, 6 negative with $\Delta = 0.51056$). This specification for Δ was somewhat subjective, and other values could be chosen to produce either a longer or shorter list of significant genes.

Input Parameters

Imputation Engine	Row Average Imputer
Data Type	Two Class, unpaired data
Data in log scale?	TRUE
Number of Permutations	64
Blocked Permutation?	TRUE
RNG Seed	1234567
(Delta, Fold Change)	(0.51056,)
(Upper Cutoff, Lower Cutoff)	(1.93370, -2.25565)

Computed Quantities

Computed Exchangeability Factor S0	0.073734246
S0 percentile	0.4
False Significant Number (Median, 90 percentile)	(0.80104, 11.85545)
False Discovery Rate (Median, 90 percentile)	(2.28870, 33.87271)
Pi0Hat	0.80104

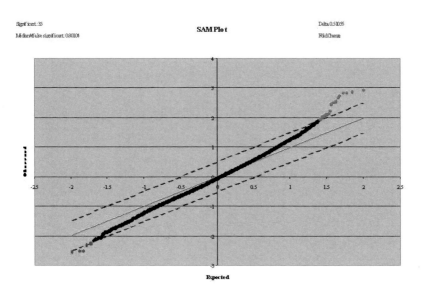

Figure 12.1. Significance Analysis of Microarrays (SAM)

29 Positive Significant Genes

Gene Name	Gene ID	Score(d)	Numerator(r)	Denominator(s+s0)	Fold Change	q-value (%)
M008726_01	M008726_01	2.9183343	0.526715	0.180484809	1.43393	1.63478348
M011396_01	M011396_01	2.8481222	0.39960985	0.140306427	1.32160	1.63478348
M012241_01	M012241_01	2.8280116	0.33279095	0.117676658	1.25921	1.63478348
M007393_01	M007393_01	2.8275862	0.793987383	0.28080042	1.76417	1.63478348
M000262_01	M000262_01	2.735945	0.485186517	0.177337819	1.39947	1.63478348
M007250_01	M007250_01	2.6750644	0.851184883	0.318192295	1.85681	1.63478348
M009967_01	M009967_01	2.5336556	0.351530133	0.138744244	1.27248	1.63478348
M011450_01	M011450_01	2.5003538	0.458737467	0.183469026	1.37719	1.63478348
M000297_01	M000297_01	2.4930723	0.577557983	0.231665159	1.50285	1.63478348
M000327_01	M000327_01	2.4390895	0.357849733	0.14671447	1.28806	1.63478348
M010363_01	M010363_01	2.4283745	0.3886732	0.160054886	1.31136	1.63478348
M001947_01	M001947_01	2.2218025	0.304497917	0.137049951	1.23574	1.63478348
M000707_01	M000707_01	2.121964	0.588872567	0.277512992	1.53727	1.63478348
M005695_01	M005695_01	2.1196553	0.34736665	0.163878842	1.26804	1.63478348
M003069_01	M003069_01	2.0865237	0.581398517	0.278644577	1.52364	1.63478348
M006453_01	M006453_01	2.0784973	0.342421817	0.16474489	1.27569	1.63478348
M011054_01	M011054_01	2.0763981	0.6784251	0.326731711	1.64327	1.63478348
M000711_01	M000711_01	2.0596922	0.347606133	0.168766056	1.27652	1.63478348
M003519_01	M003519_01	2.0551741	0.502825283	0.244663114	1.43128	1.63478348
M010442_01	M010442_01	2.0194313	0.3785933	0.187475211	1.29509	1.63478348
M007817_01	M007817_01	2.0193947	0.24910585	0.12335669	1.18697	1.63478348
M004654_01	M004654_01	2.0135277	0.70317515	0.34922547	1.65963	1.63478348
M008784_01	M008784_01	2.0118947	0.32961905	0.163835143	1.25916	1.63478348
M009144_01	M009144_01	2.0108957	0.336430617	0.167303863	1.26702	1.63478348
M006950_01	M006950_01	1.9854465	0.3224464	0.162404977	1.25438	1.63478348
M000610_01	M000610_01	1.9413727	0.537318867	0.276772655	1.47493	1.63478348
M008012_01	M008012_01	1.9378335	0.378348067	0.195242816	1.31146	1.63478348
M011215_01	M011215_01	1.934338	0.447350717	0.231268116	1.38234	1.63478348
M007711_01	M007711_01	1.933695	0.440462317	0.227782722	1.37422	1.63478348

6 Negative Significant Genes

Gene Name	Gene ID	Score(d)	Numerator(r)	Denominator(s+s0)	Fold Change	q-value (%)
M012664_01	M012664_01	-2.5597954	-0.351276233	0.137228246	0.78448	1.63478348
M001068_01	M001068_01	-2.5351718	-0.372147	0.146793601	0.77421	1.63478348
M000917_01	M000917_01	-2.5339346	-0.312651717	0.123385869	0.80483	1.63478348
M001152_01	M001152_01	-2.3271432	-0.379553767	0.163098588	0.77231	1.63478348
M012591_01	M012591_01	-2.2607922	-0.301769633	0.133479595	0.81363	1.63478348
M007425_01	M007425_01	-2.255653	-0.287646433	0.127522468	0.81811	1.63478348

Figure 12.2. Tabulated output from SAM

The median number of false positives among the 35 significant genes is reported as 0.80104, giving a false discovery rate of 2.29%. Under the null hypothesis of no differential expression between treatments 2 and 3, we note that the pair of observations for each mouse are exchangeable, giving 2! permutations. As there are six mice (blocks), the full set of permutations numbers $(2!)^6 = 64$.

In section 10.5.3, we mentioned that the SAM software uses a variance-offset to improve reliability of the test statistics. The software computes statistics (in our notation) of the form $v_g^* = r_g/(s_0 + s_g)$ for each gene g, where r_g is a score statistic for differential expression, s_g is a standard deviation and s_0 is an offset parameter. Parameter s_0 is chosen as a particular percentile of all s_g values (see section 14.1 of the SAM Users Guide and Technical Document). The offset parameter s_0 helps to stabilize the ratios. We also showed in section 10.5.3 that the approach of using a fixed divisor corresponds to that in SAM if the s_0 offset parameter were made large enough to dominate the ratio (swamping the effect of s_g for each gene). In this case, significant genes would be chosen strictly on the basis of the score statistic r_g in the numerator of the SAM ratio. The SAM output shows the numerator values. As the test statistic will tend to be larger if r_g is larger (in absolute value), one would expect that lists of significant genes would be similar with the two approaches.

If the fixed divisor approach as discussed in section 10.5.3 is used in the computations, the gene with ID M007250-01 has the largest score statistic among all genes. If the SAM approach is used, however, this gene appears in the 6th position in the table in Figure 12.2 with $r_g = 0.85118488$, which equals the value of the difference $\hat{\mathcal{I}}_{3g} - \hat{\mathcal{I}}_{2g}$ in our notation here. A comparison of the two lists of significantly differentially expressed genes identified by the two procedures (with a fixed divisor versus $s_g + s_0$ in the test statistic) shows a moderate degree of overlap. There are 12 genes common to both lists of the top 35 significant genes. It is often the case that different lists of significant genes are obtained with different statistical assumptions and methods.

Notes

[1] Ludbrook, J. and Dudley, H. (1998). *The American Statistician*, **52**, 127-132.

[2] Good, P. (2000). *Permutation Tests: A Practical Guide to Resampling Methods for Testing Hypotheses*, 2nd edition, Springer, New York.

[3] Chu, G., Narasimhan, B., Tibshirani, R., Tusher, V. (2002). Significance Analysis of Microarrays (SAM): Users Guide and Technical Document, version 1.21., http://www-stat.stanford.edu/~tibs/clickwrap/sam

[4] Dudley, A.M., Aach, J., Steffen, M.A., and Church, G.M. (2002). *Proceedings of the National Academy of Sciences, USA*, **99**, 7554-7559.

Chapter 13

BAYESIAN METHODS
FOR MICROARRAY DATA

Bayesian models and methods play an important role in statistical analysis generally and in the analysis of microarray data in particular. Some of the mathematically most sophisticated methods for microarrays involve Bayesian principles and techniques. This chapter provides an overview of some ideas underlying the Bayesian approach to microarray data analysis.

Many models have been proposed to relate gene expression readings w_{gc} to true concentrations ζ_{gc}. In the next section, we present mixture models for gene expression. In a subsequent section, we consider such models for differential gene expression. Finally, later sections look at the Bayesian aspect of these models in an explicit way and show how this aspect can be very useful for both interpretation and analysis.

13.1. Mixture Model for Gene Expression

The simple observation that a gene is either expressed or not expressed in a given experimental setting implies that the observed intensity for any gene can be represented by a mixture model, which we now describe. For the present, the discussion focuses on a given experimental condition and notation for the condition (usually index c) is suppressed.

First, let p_1 denote the probability that any given gene g is expressed under the experimental condition prior to observing any experimental evidence. As this probability is not indexed by g, it is assumed that all genes are equally likely to be expressed. (This assumption can be modified if the experimenter holds different prior beliefs about different genes being expressed.)

Second, let $f(w|\zeta)$ denote the probability density function (p.d.f.) of observing expression intensity w when the concentration of the gene in the experimental condition equals ζ. As the gene is unexpressed if ζ is zero, it follows that $f(w|0)$ represents the p.d.f. of the background noise random variable. Observe the implicit assumption that the p.d.f. $f(w|\zeta)$ depends on the gene only through the value of ζ and, specifically, that the background noise component has the same p.d.f. for all genes under the given experimental condition.

Third, let $q(\zeta)$ denote the p.d.f. of gene concentration for an expressed gene under the experimental condition. The value of ζ_g for an expressed gene g is a random draw from this p.d.f.. Putting these three specifications together, the marginal p.d.f. $f(w)$ of the observed gene intensity under the experimental condition has the following mixture form.

$$f(w) = p_1 f_1(w) + p_0 f_0(w), \qquad (13.1)$$

where $p_0 = 1 - p_1$, $f_0(w) = f(w|0)$ is the p.d.f. of the background noise, and

$$f_1(w) = \int_0^\infty f(w|\zeta) q(\zeta) d\zeta$$

is the marginal density function of intensity for an expressed gene.

Mixture models are central to a Bayesian perspective on microarray data analysis. The model in equation (13.1) has been used in the analysis of microarray data – see, for example, Lee *et al.*[1] (2000) and Efron *et al.*[2] (2001). The first term on the righthand side of (13.1) represents the contribution to the marginal p.d.f. of observed intensity if a gene is expressed with probability p_1 in the experimental condition and the second term is the contribution if the gene is unexpressed in the condition. Although we are not in a testing framework here, test terminology is convenient so subscripts '0' and '1' are used for the null and alternative hypotheses that the gene is either unexpressed or expressed according to whether $\zeta = 0$ or $\zeta > 0$.

To visualize mixture model (13.1), one may consider the histogram of intensities w_{gc} for all genes $g = 1, \ldots, G$ under any given experimental condition c. The histogram of the w_{gc} values is a graphical estimate of the p.d.f. $f(w)$. The challenge of microarray analysis is to estimate the components of the mixture model (13.1) from the intensity data used to construct the histogram.

The components of model (13.1) are not equally accessible in terms of estimating their form from microarray data. The background noise p.d.f.

$f(w|0)$ is easiest to estimate because generally the designated gene set $\mathcal{G}$ contains many genes that are not expressed in any of the experimental conditions. The microarray may also contain control spots that are intentially included by design so they would not be expressed. Finally, replicates contain information about the background noise p.d.f., $f_0(w)$.

The p.d.f. $f(w|\zeta)$ for the observed intensity of an expressed gene having true concentration ζ is more difficult to assess. Parameter ζ is an unobserved property of the gene in the biological population defined by the experimental condition, whereas w is the observed intensity reading for the gene under that condition. If the gene were replicated on r spots under the same experimental condition, giving replicated intensity readings $w_1, \ldots, w_r$, then these r observations constitute a sample from p.d.f. $f(w|\zeta)$ and could be used to infer its form. The number of replicates r, however, is not generally large and the replicates are subject to random variation, so precise knowledge of $f(w|\zeta)$ is difficult to gather. True replicates are also very difficult to construct.

Finally, the most difficult p.d.f. to estimate is $q(\zeta)$, the density function of true gene concentrations for expressed genes in the designated gene set $\mathcal{G}$. This density function will be unique to the biological population defined by the experimental condition. P.d.f. $q(\zeta)$ is the object of scientific interest in a microarray study, but the study provides no direct observation of this distribution. As the designated set $\mathcal{G}$ is finite, the microarray study involves a finite number of ζ_g values drawn from $q(\zeta)$. Estimates of the ζ_g for all expressed genes provide a rough profile of $q(\zeta)$. As yet, we have not explained how such estimates are obtained from the microarray data.

13.1.1. Variations on the Mixture Model

The additive background noise model (5.1) in Chapter 5 states that $w_{gc} = x_{gc} + B_{gc}$ where x_{gc} represents latent gene expression and B_{gc} is background noise for gene g and condition c. This additive specification for background noise allows further development of the mixture model in (13.1) and, specifically, the elaboration of the p.d.f. $f(w|\zeta)$. If noise is the only source of randomness in microarray readings, then x is a fixed quantity. Then, for any gene with a given concentration ζ, one might assume that x is proportional to ζ, say, $x = \kappa\zeta$. Parameter κ is a multiplier, possibly different from one experimental condition to another, that determines the contribution of the true gene concentration to the quantity of light, radiation, or other physical medium that is read from the array. With this assumption, it follows that the difference $w - \kappa\zeta$ is simply background noise and, therefore, that

$$f(w|\zeta) = f(w - \kappa\zeta|0) = f_0(w - \kappa\zeta). \tag{13.2}$$

This model entails a translation of the scale of expression measurement from w to $w - \kappa\zeta$. The model hypothesizes that subtracting the true quantity of gene expression from the reading leaves pure background noise. Several variations of this model have been proposed. All of these variations have the mathematical implication that the form of the p.d.f. $f(w|\zeta)$, or its equivalent in other models, can be inferred directly from that of the noise distribution.

Different forms of mixture model (13.1) arise when the raw expression data are monotonically transformed. If the transformation is $y = t(w)$ then w in (13.1) is replaced by y. For example, y may denote the logarithm of the raw intensity measurement, i.e., $y = \log(w)$. In mathematical terms, if $f^*(y)$ is the p.d.f. of y then $f(w)$ and $f^*(y)$ are related by

$$f^*(y)dy = f[t^{-1}(y)]dt^{-1}(y),$$

where t^{-1} denotes the inverse transformation. The other p.d.f.s in the mixture model are similarly related. The mixture model (13.1) might then be rewritten for y as

$$f^*(y) = p_1 f_1^*(y) + p_0 f_0^*(y).$$

Changing notation for the p.d.f.s with each change of variable is potentially confusing so, for expository convenience, the notation $f(\cdot)$, $f_1(\cdot)$, and $f_0(\cdot)$ is used as generic notation, whichever variable may be under consideration. The context will make the functional form of the p.d.f.s clear.

Different families of distributions have been proposed for the component p.d.f.s $f(w|\zeta)$, $q(\zeta)$, and $f_0(w)$ of the mixture model, although the settings in which they have been presented differ somewhat from the one used here. The idea of the mixture model is sometimes explicitly developed in these articles and at other times is only implicit in the presentation. Lee *et al.* (2000), for example, present an illustration of mixture model (13.1) for log-intensities in a simple microarray experiment. They model the component distributions $f_1(y)$ and $f_0(y)$ as normal densities with $y = \log(w)$. They compute empirical Bayes estimates of the model parameters. Baldi and Long[3] (2001) also model log-expression data using the normal distribution family. Newton *et al.*[4] (2001) have used the gamma distribution family as a model for gene expression data and explicitly develop the mixture model. Ibrahim *et al.*[5] (2002) use the lognormal distribution family as a model for expression data. Rocke and Durbin[6] (2001) use a hybrid model in which the components of the additive noise model, x_{gc} and B_{gc}, are assumed to be independent lognormal and normal random variables, respectively. Efron *et al.* (2001) propose a non-parametric approach to modeling the components of the mixture model.

It was noted already that the background noise p.d.f. is the most accessible. The next subsection gives an illustration of estimating the background noise p.d.f. $f_0(w)$ as a gamma distribution. In later sections, we include other case illustrations to show how the components of mixture models are estimated from microarray data.

13.1.2.   Example of Gamma Models

Newton *et al.* (2001) and Kendziorski *et al.*[7] (2003), among others, have studied the family of gamma distributions as a suitable model for gene expression data. The following formulation is motivated by their work but departs from it in some respects.

A gamma distribution for a general variable r with shape parameter α and scale parameter β has p.d.f.

$$p(r|\alpha, \beta) = \frac{1}{\beta^\alpha \Gamma(\alpha)} r^{\alpha-1} \exp(-r/\beta). \qquad (13.3)$$

The mean and variance of the distribution are given by $\alpha\beta$ and $\alpha\beta^2$, respectively. Notation $\Gamma(a)$ denotes the gamma function.

To illustrate use of the gamma model for modeling background noise, we examine the background noise p.d.f. $f_0(w) = f(w|0)$, using data from the Mouse Juvenile Cystic Kidney Data Set. We use the data from Array 1 for mutant tissue on the green channel (see Table 6.1 in subsection 6.2.2). Using a similar strategy to that adopted for the study of background noise in Chapter 5, it is assumed that observed intensities w_{gc} below a certain level constitute pure noise (so $w_{gc} = B_{gc}$ because x_{gc} is zero). We choose the median value of CH1I (value 2875) for this cutoff on the assumption that at least 50 percent of genes in the set of 1728 genes are not expressed. We have used maximum likelihood estimation to fit a truncated gamma distribution to this sample, obtaining parameter estimates $\hat{\alpha} = 1.492$ and $\hat{\beta} = 2770$. The estimated mean is therefore $1.492(2770) = 4133$.

Figure 13.1 shows a plot of the fitted percentiles of the gamma background noise distribution against the observed expression intensities. The line of identity is also plotted. The alignment of the line and quantile plot at the lower end of the scale indicates the adequacy of fit of the gamma model to background noise. Observe that the alignment continues until around 6000 and then the observed intensities diverge upward, suggesting the presence of the latent real expression component x_{gc} in these genes. We caution that a look at the lower end of the distribution at higher resolution suggests some systematic departures from the gamma distribution so the model is not ideal (we do not show that detail here).

The appropriateness of the gamma distribution to describe the conditional p.d.f. $f(w|\zeta)$ for expressed genes (i.e., where $\zeta > 0$) is not explored here and is an open question. Even more uncertain is the nature of the p.d.f. $q(\zeta)$ for expressed genes in mutant tissue.

13.2.   Mixture Model for Differential Expression

The mixture model (13.1) applies to a single intensity measurement w or, possibly, to a transformation of a single intensity measurement, such as $y = \log(w)$. Yet, as explained in previous chapters, microarray studies are mainly concerned with the identification of genes or sets of genes that are differentially expressed across experimental conditions. Thus, interest centers not on single intensity readings but on comparisons of expression levels. Importantly, the mixture model can be extended to differential gene expression across two or more experimental conditions. The form of the mixture model remains essentially the same in this case but the components require a new interpretation. The ideas are developed in the next two subsections. In these subsections, we use notation v_g to represent a summary statistic of differential expression for gene g; a summary form that differs from one application to another. We let θ_g denote the corresponding parameter that is being estimated by v_g. Parameter θ_g measures the true extent of differential expression.

13.2.1.   Mixture Model for Color Ratio Data

Consider a two-color cDNA array comparing a treatment condition (c=1) with a control condition (c=0). Assume that a reversed-color design is used so two spots are devoted to each gene g. Let $w_{g1}^{(R)}$ and

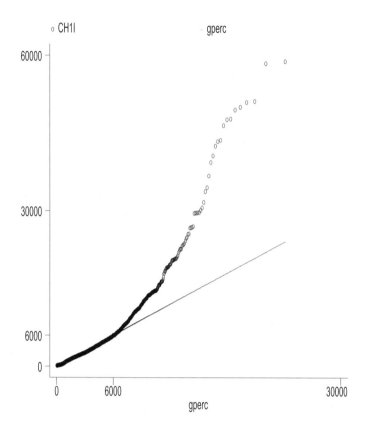

Figure 13.1. Plot of fitted percentiles of the gamma background noise distribution to observed expression intensities for data from the green channel of Array 1 (mutant tissue) of the Mouse Juvenile Cystic Kidney Data Set. The line of identity is also plotted. The alignment of the line and quantile plot at the lower end of the distribution indicates the adequacy of fit of the gamma model for background noise.

$w_{g0}^{(G)}$ be the normalized intensity readings of one spot and $w_{g1}^{(G)}$ and $w_{g0}^{(R)}$ be the readings of the other, where superscripts (R) and (G) denote red and green intensities, the subscripts $g0$ denote readings for gene g from the control condition, and the subscripts $g1$ denote readings for gene g from the treatment condition. The following geometric mean measures the differential expression for gene g between the treatment and control conditions and, hence, is our summary measure.

$$v_g = \left(\frac{w_{g1}^{(R)} w_{g1}^{(G)}}{w_{g0}^{(G)} w_{g0}^{(R)}} \right)^{1/2} \tag{13.4}$$

Observe that multiplicative color and spot effects cancel out in measure v_g.

Denoting the ratio of true concentrations of gene g in the two conditions by $\theta_g = \zeta_{g1}/\zeta_{g0}$, it is clear that v_g will estimate θ_g. Statistic v_g will tend to be close to one if $\theta_g = 1$ and will tend to differ from one otherwise. A value of v_g much less than one would indicate gene g is down-regulated in the treatment condition relative to the control condition (i.e., $\theta_g < 1$) while v_g much greater than one would indicate gene g is up-regulated in the treatment condition relative to the control condition (i.e., $\theta_g > 1$). In this context, we shall refer to θ_g as the *differential expression ratio* for gene g. We set $\theta_g = \infty$ if $\zeta_{g0} = 0$.

Dropping the subscript g, the mixture model (13.1) takes the following form for the geometric-mean statistic v.

$$f(v) = p_1 f_1(v) + p_0 f_0(v), \tag{13.5}$$

where $p_0 = 1 - p_1$, $f_0(v) = f(v|1)$ and

$$f_1(v) = \int_0^\infty f(v|\theta)q(\theta)d\theta.$$

The p.d.f. $f(v|\theta)$ represents the distribution of the geometric-mean statistic when the differential expression ratio is θ. Thus, the p.d.f. $f(v|1)$ represents the null distribution of the geometric-mean statistic when there is no differential expression (i.e., when $\theta = 1$). The p.d.f. $q(\theta)$ describes the distribution of differential expression ratios for genes in the designated gene set $\mathcal{G}$ that are truly differentially expressed in the biological populations being compared. Note that $q(\theta)$ may have a probability mass at $q(0)$ and $q(\infty)$ corresponding to genes for which either $\zeta_{g1} = 0$ or $\zeta_{g0} = 0$, respectively, in the treatment and control conditions.

Many researchers have looked at probability models for color ratios along the lines described here. One tractable representation that also is realistic assumes that the logarithms of the four w observations entering expression (13.4) have a joint multivariate normal distribution. In this case, v_g has a lognormal distribution. In the next section, this representation is linked to the ANOVA model for microarray data based on logarithms of expression measurements.

Our preliminary development in this section has been motivated in part by the work reported in Newton *et al.* (2001). These authors assume that the color ratio for a spot is a ratio of two independent gamma distributions with identical shape parameters and derive a number of

interesting results in the context of hierarchical Bayes modeling.   A
later section will give more specifics on hierarchical Bayes modeling and
the results of these and other authors. Newton *et al.* did not explicitly consider the reverse-color design but the following variation of their
setup preserves the mathematical features of their approach. Consider
the following statistic for each gene.

$$v_g = \frac{w_{g1}^{(R)} + w_{g1}^{(G)}}{w_{g0}^{(G)} + w_{g0}^{(R)}} = \frac{t_{g1}}{t_{g0}} \tag{13.6}$$

Observe that statistic v_g is the ratio of the total expression for treatment
1 to the total expression for treatment 0 (the control) for the two array
spots. The totals are denoted by t_{g1} and t_{g0}, respectively, in (13.6).

Under appropriate assumptions, the ratio of the two totals in (13.6)
can be modeled as a ratio of two independent gamma random variables
with a common shape parameter. The requisite assumptions are: (1)
the four readings are independent gamma random variables, (2) the numerator readings share a common scale parameter, say β_1, as do the
two denominator readings, say β_0, and (3) the color and spot effects do
not alter the scale parameter but are additive in determining the shape
parameter. With these assumptions, the numerator and denominator
totals t_{g1} and t_{g0} are independent gamma random variables with a common shape parameter α and scale parameters β_1 and β_0, respectively.
The differential expression parameter is

$$\theta = \beta_1/\beta_0$$

in this context. Under the preceding conditions, the Newton *et al.* hierarchical Bayes development will continue to apply. This development is
taken up in the next section.

The gamma model introduces a technical element that ties in with
much earlier discussion related to data normalization. In the differential
expression setting, the statistic v_g estimates the true differential expression parameter θ_g. It has been assumed implicitly in the modeling that
statistic v_g encapsulates all information in the data about the unknown
parameter θ_g, i.e., that v_g is a sufficient statistic for θ_g. In the log-normal
model for the color-ratio statistic (13.4), statistic v_g is sufficient. In the
gamma model for the color-ratio statistic (13.6), statistic $v_g = t_{g1}/t_{g0}$ is
not sufficient. It is found that the product statistic

$$u_g = \sqrt{t_{g1}t_{g0}}$$

plays a role. Statistic u_g is a measure of the total quantity of expression for gene g and, thus, is closely related to the intensity-dependent normalization discussed in Chapter 6. A little mathematics shows that the joint density function for the pair (t_{g1}, t_{g0}) can be transformed into a joint p.d.f. for pair (u_g, v_g) with the following form.

$$f(u, v | \lambda, \theta) = \frac{\lambda^{-2\alpha} u^{2\alpha-1}}{\Gamma(\alpha)^2 v} \exp\left[-\frac{u}{\lambda} \left(\sqrt{\theta/v} + \sqrt{v/\theta} \right) \right] \qquad (13.7)$$

Here $\lambda = \sqrt{\beta_0 \beta_1}$. In looking at the mathematical form of this p.d.f., it can be seen that v will tend to be a more precise estimator of θ for larger values of u. This tendency is consistent with observed values of v_g having less dispersion with larger values of u_g. This phenomenon has been found in many applications. Thus, the gamma model has a realistic feature in this respect.

The marginal distribution of v is obtained from (13.7) by integrating out variable u, giving

$$f(v | \theta) = \frac{\Gamma(2\alpha)}{\Gamma(\alpha)^2} \frac{\theta^\alpha v^{\alpha-1}}{(\theta + v)^{2\alpha}}. \qquad (13.8)$$

This formula, in different notation, appears as equation (2) in Newton *et al.* (2001).

13.2.2.   Relation of Mixture Model to ANOVA Model

In the ANOVA models described in Chapter 10, differential expression effects are captured by the gene-condition interaction terms, which were denoted there by $\mathcal{I}_{gc}$, with estimates denoted by $\hat{\mathcal{I}}_{gc}$. It was noted that concentration ζ_{gc} is constant across experimental conditions for gene g if and only if $\mathcal{I}_{gc} = 0$ for all c. In particular, if gene g is unexpressed in any of the conditions, so $\zeta_{gc} = 0$ for all c, then $\mathcal{I}_{gc} = 0$ for all c.

In Chapters 10 and 14, we present several summary statistics of differential gene expression of the general form $v_g = h(\hat{\mathbf{I}}_g)$, where $\hat{\mathbf{I}}_g = (\hat{\mathcal{I}}_{g1}, \ldots, \hat{\mathcal{I}}_{gC})'$ denotes the estimated differential expression vector for gene g across all C experimental conditions and h is any function specified by the investigator that captures the particular differential expression features that are of scientific interest. We previously studied two particular types of functions in detail; one in which $h(\hat{\mathbf{I}}_g)$ is a linear function of $\hat{\mathbf{I}}_g$ and another in which it is a quadratic function of $\hat{\mathbf{I}}_g$.

Letting $\boldsymbol{\mathcal{I}}_g = (\mathcal{I}_{g1}, \ldots, \mathcal{I}_{gC})'$, the parameter

$$\theta_g = h(\boldsymbol{\mathcal{I}}_g)$$

captures the true extent of differential gene expression. In this context, we note that $\theta_g = 0$ if and only if gene g is not differentially expressed across the experimental conditions, i.e., $h(\mathbf{0}) = 0$. Dropping the index g, the mixture model then takes the following general form in this situation:

$$f(v) = p_1 f_1(v) + p_0 f_0(v), \tag{13.9}$$

where $p_0 = 1 - p_1$, $f_0(v) = f(v|0)$ and

$$f_1(v) = \int f(v|\theta) q(\theta) d\theta.$$

The integral defining $f_1(v)$ is a definite integral over the domain of θ.

As one demonstration of the mixture model (13.9), consider a design in which treatment (1) and control (0) conditions are compared in a reversed-color cDNA experiment, as in the previous subsection. Fitting the ANOVA model to the normalized log-intensity data yields interaction estimates $\hat{\mathcal{I}}_{g1}$ and $\hat{\mathcal{I}}_{g0}$. The estimated expression differential for the two conditions is given by

$$v_g = \hat{\mathcal{I}}_{g1} - \hat{\mathcal{I}}_{g0}.$$

This difference is none other than the logarithm of the geometric mean color ratio (13.4) in the previous subsection. The corresponding true expression differential is $\theta_g = \mathcal{I}_{g1} - \mathcal{I}_{g0}$. The parameter θ_g here equals the logarithm of the differential expression ratio introduced in the previous subsection. Thus, the mixture models for the geometric mean color-ratio and the ANOVA interaction difference are mathematically equivalent.

In the previous demonstration and in many other applications where statistic v involves a comparison, contrast or other linear combination of differential expression effects, the statistic v can assume both positive and negative values reflecting (in some general sense) down-regulation and up-regulation of genes, respectively. In these situations, it is useful to divide the mixture component $p_1 f_1(v)$ into two subcomponents as follows[8]

$$p_1 f_1(v) = p_- f_-(v) + p_+ f_+(v), \tag{13.10}$$

where $p_1 = p_- + p_+$. The elements in (13.10) with labels '$-$' and '$+$' correspond to the down-regulated and up-regulated components, respectively. A later section will show the usefulness of this partitioning of the $p_1 f_1(v)$ component.

13.2.3.   Bayes Interpretation of Mixture Model

We now take up the Bayesian approach to the analysis of microarray data in an explicit fashion. Mixture model (13.9) can be given an immediate Bayesian interpretation. The probabilities p_1 and p_0 are *prior probabilities* for a gene being differentially expressed or not, in advance of having the study data. The p.d.f. $q(\theta)$ is a *prior distribution* for the true extent of differential expression for the gene. The parameter θ_{gc} for a gene g in experimental condition c will be drawn from this density function. The relative magnitudes of the mixture components, given the summary statistic v, are the posterior probabilities that the gene is differentially expressed or not. In notation, we have

$$P(E_1|v) = \frac{p_1 f_1(v)}{f(v)}, \quad P(E_0|v) = \frac{p_0 f_0(v)}{f(v)}, \tag{13.11}$$

where E_0 and E_1 denote the events that the gene is not differentially expressed and differentially expressed, respectively.

The posterior probabilities give the relative evidence in favor of a gene being differentially expressed or not. The probabilities remind us that establishing the true situation for a gene is not a certainty. A posterior probability of $P(E_1|v) = 0.5$ states that the odds are even that the gene in question is differentially expressed while a posterior probability of $P(E_1|v) = 0.999$ represents odds of 999 to 1 in favor of differential expression.

The posterior distribution of θ, given v, may be written down directly from the mixture model, as follows.

$$P(\theta|v) = \begin{cases} p_0 f_0(v)/f(v) & \text{for } \theta = 0 \\ \\ p_1 f_1(v|\theta)q(\theta)/f(v) & \text{for } \theta \neq 0 \end{cases} \tag{13.12}$$

The precise form of this posterior distribution requires knowledge of the prior probability p_1 and the prior p.d.f. $q(\theta)$. We discuss later how this probability and p.d.f. might be assessed. Observe that distribution (13.12) can be used to make probability statements about the true extent of differential expression, as measured by θ, given the observed statistic

v. First, the posterior probability $P(E_1|v)$ gives the probability that the gene is differentially expressed. Second, assuming that the gene is differentially expressed, the most probable value of θ, denoted here by $\hat{\theta}$, is the point of maximum density of $P(\theta|v)$, i.e., the solution to

$$\max_{\theta \neq 0} f_1(v|\theta)q(\theta) \tag{13.13}$$

Under quite general conditions, $\hat{\theta}$ will tend to be a value that lies between v and the modal value of $q(\theta)$ and, hence, is a shrinkage estimator.

The discussion in the previous section pointed out a case where the statistic v is not a sufficient statistic for θ. Where that is the case, a refinement can be made in the posterior probability distribution for θ in (13.12). We will not pursue that refinement here.

The component probabilities p_0 and p_1 and density functions $f_0(v)$ and $f_1(v)$ in (13.11) may depend on unspecified parameters, in which case the probabilities $P(E_0|v)$ and $P(E_1|v)$ are, in fact, conditional posterior probabilities because they are conditional on the unspecified parameters. A full Bayesian approach would require a specification of a joint prior distribution for these unknown parameters in order to eliminate their conditionality. The elicitation of appropriate prior distribution forms is a challenging aspect of the Bayesian approach. Ibrahim *et al* (2002), for example, discuss this issue in a microarray context. Another approach to dealing with any unspecified parameters is the empirical Bayes approach that is described in the next section.

13.3. Empirical Bayes Methods

Microarray studies often involve thousands of genes but, for any single gene, expression readings may only be gathered from a small number of sample specimens. This context is suitable for empirical Bayes methods because these allow inferences for any given gene to draw on information from other genes. Early development of empirical Bayes methods can be found in Robbins[9] (1951), Robbins and Hannan[10] (1955), Robbins[11] (1964), Efron and Morris[12] (1973), Efron and Morris[13] (1975), and Carlin and Louis[14] (1996), among others. An interesting brief review of the historical development of empirical Bayes methods can be found in Efron[15] (2001). In this article, Efron discusses the connection between empirical Bayes methods and applications in microarray studies.

The principle of empirical Bayes is useful for estimating the parameters of the mixture models described in previous sections. We take mixture model (13.9) as a case in point. To fit mixture model (13.9), we must decide on the form of the distributions $f_1(v)$ and $f_0(v)$ and estimate any unknown parameters of these distributions, as well as the prior probabilities p_1 and $p_0 = 1 - p_1$. In the case where v may assume both positive and negative values, it is necessary to make assumptions about the forms of both the negative and positive subcomponents as defined in (13.10).

Investigators may proceed in a variety of ways. Some researchers have adopted a non-parametric approach for estimating the density functions $f_0(v)$ and $f_1(v)$ and prior probabilities p_0 and p_1. Efron *et al.* (2001), for example, estimate a ratio corresponding to $f_0(v)/f(v)$ using a non-parametric approach that combines randomization and permutation, logistic regression fitting and numerical smoothing. Observe from (13.11) that this ratio, together with an estimate of p_0, gives an estimate of the posterior probability $P(E_0|v)$ that a gene is not differentially expressed. The statistic v in our notation corresponds to their statistic z.

Lee *et al.* (2002b) adopt a completely parametric approach. In their application, $f_0(v)$ is taken to be a normal distribution and $f_-(v)$ and $f_+(v)$ are taken to be reverse-Weibull and Weibull distributions, respectively. They then use empirical Bayes methods to estimate the unknown parameters. Newton and Kendziorski[16] (2003) provided an overview of parametric empirical Bayes methods for microarrays. They considered the gamma-gamma and lognormal-lognormal models and presented the software they developed for the empirical Bayes hierarchical modeling approach. The reader is referred to these articles for more details.

13.3.1.   Example of Empirical Bayes Fitting

To demonstrate the essence of empirical Bayes fitting, we consider a case example.

In Chapter 12, a microarray data set referred to as the LPS-*E. coli* data set was introduced. Primary interest centered on whether genes are differentially expressed in a pair-wise comparison of the LPS treatment (treatment 2) and the *E.coli* treatment (treatment 3). The comparison focuses on the difference $v_g = \hat{\mathcal{I}}_{3g} - \hat{\mathcal{I}}_{2g}$. The corresponding parameter is $\theta_g = \mathcal{I}_{3g} - \mathcal{I}_{2g}$. For a parametric representation, let $f_0(v)$ be a normal density function with mean μ_0 and standard deviation σ_0. Furthermore, let $f_-(-v)$ and $f_+(v)$ be Weibull p.d.f.s. The Weibull density function for a general variable r, with parameters α and β, has the following form.

$$p(r) = \frac{\beta r^{\beta-1}}{\alpha^\beta} \exp\left[-\left(\frac{r}{\alpha}\right)^\beta\right] \quad \text{for } r > 0 \qquad (13.14)$$

Parameter α is a scale parameter and, in fact, corresponds to the 63rd percentile of the distribution. Parameter β is a shape parameter. Larger values of β correspond to distributions that are less variable, less right-skewed and shaped more like the normal distribution. It is assumed that $f_-(-v)$ follows a Weibull p.d.f. with parameters (α_-, β_-) and that $f_+(v)$ follows a Weibull p.d.f. with parameters (α_+, β_+). As the scale for the former p.d.f. is reversed (note the p.d.f. argument is $-v$), we refer to it as a reverse-Weibull distribution.

The Weibull distribution family is used in this illustration because it is a flexible distribution for fitting. The true shape of distribution $f_1(v)$ is unknown, being an unknown convolution of distributions for background noise and true differential expression for genes in the designated set $\mathcal{G}$. Other candidate distributions for $f_1(v)$ have been proposed and also seem to work well (e.g., the gamma and lognormal). Additional research is required to have a better understanding of the form of $f_1(v)$. An advantage of the non-parametric approach, such as that described in Efron et al. (2001), is that it lets the microarray data themselves give an empirical estimate of $f_1(v)$ without the constraints of a given distribution family.

To implement the empirical Bayes fitting of the mixture model, we set up the sample log-likelihood function as a function of the unknown parameters. The function to be maximized in this application is

$$\log L(\text{parameters}) = \sum_{g=1}^{G} \log f(v_g) \qquad (13.15)$$

Table 13.1 shows the parameters and their estimates. One unusual feature arises in this application. A single gene (gene g_{7069}) is outlying in the negative domain, having a difference statistic of $v_g = -0.3584$. In maximizing the log-likelihood function, it is found that the routine attempts to fit a degenerate density function to this single observation. We therefore take $f_-(v)$ to be degenerate and give it parameter estimates consistent with this degeneracy (specifically, zero variance). To estimate the remaining parameters for the model, the p.d.f. $f_-(v_g)$ is dropped as a component of $f(v_g)$ in the likelihood function (13.15). The outlying observation (gene g_{7069}) is also dropped from the likelihood calculation.

The parameter estimates suggest that about 18 percent of the 13,028 genes in the study are differentially expressed and virtually all of these

Parameter	Estimate
p_0	0.8249
p_-	0.0001
p_+	0.1750
μ_0	-0.0238
σ_0	0.0589
α_-	0.3584
α_+	0.1248
β_-	∞
β_+	1.5488

Table 13.1.   Parameter estimates for the Weibull-normal mixture model.
The estimated model component for down-regulated genes is degenerate because only
a single case is outlying in the negative direction of v_g.

are up-regulated in *E.coli*. For up-regulated genes, the estimate 0.1248 for α_+ suggests that 63 percent of the differentially expressed genes have log-differences in expression below this level. In terms of fold change, this 63rd percentile corresponds to a fold change of $\exp(0.1248) = 1.13$. Thus, most differentially expressed genes are only modestly up-regulated. The distribution of differences for differentially expressed genes is unimodal and somewhat right skewed because the parameter estimate of β_+ is 1.5488, which is larger than one but only moderately large. The one gene that is down-regulated in the *E.coli* treatment corresponds to a fold change of $\exp(.3584) = 1.43$.

Substitution of the parameter estimates in Table 13.1 into the mixture model (13.9), with positive and negative subcomponents defined by (13.11), gives estimates of the densities $f(v_g)$, $f_-(v_g)$ and $f_+(v_g)$ for every gene g. The posterior probabilities $P(E_1|v_g)$ and $P(E_0|v_g)$ can then be estimated from (13.11).   In total, 1689 genes have estimated posterior probabilities of differential expression exceeding 0.50 and 270 genes have estimates exceeding 0.99. All but one of these genes (gene g_{7069} being the exception) are up-regulated in the *E.coli.* treatment. Table 13.2 shows estimates of $P(E_1|v_g)$ for a fragment of the 270 genes for which this estimate exceeds 0.99, including gene g_{7069}. Probability values of 1.0000 in the table are rounded and do not represent certainty.

How well does the fitted normal-Weibull mixture model fit the observed differential expression data v_g? Figure 13.2 shows an overlay plot of the empirical cumulative distribution function (c.d.f.) of the v_g and the c.d.f. of the fitted mixture model. The two curves are barely distin-

Gene g	Differential Expression v_g	Est. Posterior Prob. of Being Diff. Expressed $\hat{P}(E_1|v_g)$
Down-regulated in *E.coli*.		
g_{7069}	-0.3584	1.0000
Up-regulated in *E.coli*.		
g_{4485}	0.2035	0.9904
g_{3627}	0.2035	0.9904
g_{7707}	0.2037	0.9905
g_{8955}	0.2038	0.9906
g_{1274}	0.2044	0.9908
...	...	...
g_{4654}	0.4874	1.0000
g_{7393}	0.5504	1.0000
g_{6551}	0.5636	1.0000
g_{7250}	0.5900	1.0000

Table 13.2.   A fragment of the 270 genes having estimated posterior probabilities of being differentially expressed exceeding 0.99.

guishable. Closer examination shows that they do not depart vertically from each other by more than 0.0052 at any point.

13.4.   Hierarchical Bayes Models

The mixture models for gene expression, see (13.1), and for differential expression, see (13.5) and (13.9), have already been given a Bayes interpretation. We now look at this Bayesian interpretation more carefully in order to see the potential for the use of hierarchical Bayes modeling of microarray data.

Consider the mixture model for differential expression (13.5) as a specific case. The sampling distribution in this microarray context is $f(v|\theta)$, which represents the probability density of differential expression statistic v given the true differential expression parameter θ. Sampling distribution $f(v|\theta)$ has one distinguished case corresponding to the situation where the gene is not differentially expressed in any experimental condition. In this special case, $\theta = 1$ and the sampling distribution $f(v|\theta)$ takes the form $f(v|1)$ which corresponds to pure background noise. Hierarchical Bayes modeling attempts to model the uncertainty that the investigator has about the unknown parameters of this sampling distribution. The reference to 'hierarchy' comes from the fact that the

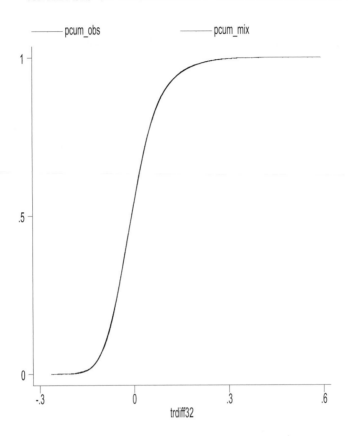

Figure 13.2. An overlay of the empirical c.d.f. of the observed differential expression data v_g and the c.d.f. of the fitted mixture model. The two curves are barely distinguishable.

uncertainly may be modeled with several tiers of prior information, as follows.

The density function $f(v|\theta)$ has θ as the primary *first-tier parameter* and possibly has other additional parameters (called nuisance parameters) that have not yet been explicitly set out. (The shape parameter α in the gamma model of differential expression is one such nuisance parameter.) We shall view these nuisance parameters as fixed unknowns in this exposition of ideas. The hierarchical Bayes approach will initially give the first-tier parameters a joint prior probability distribution to reflect uncertainty about their values in advance of observing data. In mixture model (13.5), this prior distribution has already been specified. It states that θ equals 1 with probability p_0 and takes a specific non-unitary value with probability density $p_1 q(\theta)$.

We now move to the second tier. The prior distribution for θ has its own parameters, so-called *second-tier parameters*, namely, p_0 and $p_1 = 1 - p_0$ and the parameters of the p.d.f. $q(\theta)$ which have not yet been given an explicit form in the notation. Our empirical Bayes approach to fitting the mixture model has estimated these second-tier parameters of the prior distribution; see Table 13.1 for the estimates obtained in an earlier case illustration. A hierarchical Bayes approach may wish to model the uncertainty in these second-tier parameters by setting additional prior distributions on (p_0, p_1) and on the parameters of density function $q(\theta)$. The hierarchical approach may even pursue this development to third or fourth tiers. We will pause, however, at the second tier and study what is gained by hierarchical modeling of several tiers of parameters.

13.4.1. Example of Hierarchical Modeling

We return to the adaptation of the Newton *et al.* (2001) hierarchical Bayes model as an example. The development picks up where it left off in subsection *Mixture Model for Color Ratio Data* (subsection 13.2.1). The subscript for gene g is dropped for the time being. The notation continues to be that of the book rather than of Newton *et al.*

The expression totals for treatment 1 and treatment 0 (the control), denoted by t_1 and t_0, are modeled as independent gamma random variables with a common shape parameter α and respective scale parameters β_1 and β_0. This specification gives the sampling distribution for the summary statistics (t_0, t_1). A transformed version of this sampling distribution appears in (13.7). The transformations involve both the parameters, which are transformed from (β_1, β_0) to (θ, λ), and the statistics, which are transformed from (t_0, t_1) to (u, v). The parameters $(\alpha, \beta_0, \beta_1)$ are the first-tier parameters.

Newton *et al.* give prior distributions to the parameters β_0 and β_1. Under the assumption that the gene is differentially expressed, the reciprocals of these scale parameters are assumed to be independent draws from a common gamma distribution having shape parameter a and scale parameter b (this book's notation). If the gene is not differentially expressed, then the scale parameters β_0 and β_1 are equal to a common value, say β, and the reciprocal of this common value is assumed to be drawn from the same gamma distribution with shape parameter a and scale parameter b. First-tier parameter α is treated as a nuisance parameter. Finally, the authors give the second-tier parameter pair (p_0, p_1) a beta distribution with parameters $(2, 2)$ in order to 'stabilize the computations' and enable 'a nice interpretation of the output' (see page 45 of their article). This distribution has p.d.f. $h(p) = 6p(1 - p)$. The authors take the hierarchical development no further. They obtain empirical Bayes estimates of the first-tier parameter α and second-tier parameters a and b and then proceed to investigate the implications of their model. Table 13.3 summarizes the model.

Variables	Hierarchical Feature: Distribution Form	Parameters
1. Sampling Distribution for Summary Statistics		
t_0, t_1	Independent gamma	α, β_0, β_1
2. First-tier Parameters		
α	No distribution assumed	Estimated
β_0, β_1	Independent reciprocal-gamma if gene is diff. expressed	a, b
$\beta_0 = \beta_1 = \beta$	Reciprocal-gamma if gene is not diff. expressed	a, b
3. Second-tier Parameters		
a, b	No distribution assumed	Estimated
$p_0, p_1 = 1 - p_0$	Beta	Fixed at $(2, 2)$

Table 13.3. An adaptation of the Newton et al. (2001) hierarchical Bayes model for microarray data in an experiment with treatment and control conditions. The hierarchy involves two tiers and ends by estimating prior parameters α, a and b and fixing the parameters of the beta distribution for (p_0, p_1).

Newton *et al.* (2001) investigate numerous implications for their hierarchical Bayes model. One of these is illustrated here. The posterior distribution of the differential expression parameter θ for a differentially expressed gene is given by (in the book notation)

$$P(\theta|t_0, t_1, \alpha, a, b) \propto \theta^{-(\alpha+a+1)} \left(\frac{1}{\theta} + \frac{t_0 + b}{t_1 + b} \right)^{-2(\alpha+a)}. \qquad (13.16)$$

As pointed out by the authors, the ratio $(t_1 + b)/(t_0 + b)$ acts as a shrinkage estimator of the differential expression parameter θ and will lie closer to unity than the ratio of totals in (13.6).

As a further development of hierarchical modelling, reference can be made to Newton *et al.*[17] (2003), which considers a semiparametric hierarchical mixture method for modeling microarray data.

Notes

[1] Lee, M.-L.T., Kuo, F.C., Whitmore, G.A., Sklar, J. (2000). *Proceedings of the National Academy of Sciences, USA*, **97**, 9834-9839.

[2] Efron, B., Tibshirani, R., Storey, J.D., Tusher, V. (2001). *Journal of American Statistical Association*, **96**, 1151-1160.

[3] Baldi, P. and Long, A.D. (2001). *Bioinformatics*, **17**, 509-519.

[4] Newton, M.A., Kendziorski, C.M., Richmond, C.S., Blattner, F.R., Tsui, K.W. (2001). *Journal of Computational Biology*, **8**, 37-52.

[5] Ibrahim, J.G., Chen, M.-H., Gray, R.J. (2002). *Journal of the American Statistical Association*, **97**, 88-99.

[6] Rocke, D.M., Durbin, B. (2001). *Journal of Computational Biology*, **8**, 557-569.

[7] Kendziorski, C.M., Newton, M.A., Lan, H. and Gould, M.N. (2003). *Statistics in Medicine*. (In press).

[8] Lee, M.-L.T., Lu, W., Whitmore, G.A., Beier, D. (2002b). *Journal of Biopharmaceutical Statistics*, **12(1)**, 1-19.

[9] Robbins, H. (1951). *Proc. Second Berkeley Symposium*, **1**, 131-148. Univ. Calif. Press.

[10] Robbins, H., Hannan, J. (1955). *Annals of Mathematical Statistics*, **26**, 37-51.

[11] Robbins, H. (1964). *Annals of Mathematical Statistics*, **35**, 1-20.

[12] Efron, B., Morris, C. (1973). *Journal of American Statistical Association*, **68**, 117-130.

[13] Efron, B., Morris, C. (1975). *Journal of American Statistical Association*, **70**, 311-319.

[14] Carlin, B.P., Louis, T.A. (1996). *Bayes and Empirical Bayes Methods for Data Analysis*, Chapman and Hall, New York.

[15] Efron, B. (2001). Technical Report No.2001-30B/219, Department of Statistics, Stanford University, Stanford, California.

[16] Newton, M.A., Kendziorski, C.M. (2003). In: *The Analysis of Gene Expression Data: Methods and Software*, Parmigiani, G., Garrett, E.S., Irizarray, R.A., and Zeger, S.L., eds. 255-271. Springer, New York.

[17] Newton, M.A., Noueiry, A., Sarkar, D., Ahlquist, P. (2003). Technical Report No. 1074, Department of Statistics, University of Wisconsin, Madison.

Chapter 14

POWER AND SAMPLE SIZE CONSIDERATIONS AT THE PLANNING STAGE

In this chapter, methods are presented for calculating requisite sample sizes and power at the planning stage of general microarray studies. As discussed in Chapter 9, experimental designs for microarray studies vary widely and it is important to determine what statistical power a particular design may have to uncover a specified level of differential expression. Selected interaction parameters in the ANOVA model measure differential expression of genes across experimental conditions. The methodology takes account of the multiple testing that is part of all such studies. The link to implementation algorithms for multiple testing is described. A Bayesian perspective on power and sample size determination is also presented. The discussion encompasses choices of experimental design and replication for a study. Practical examples and case studies are used to demonstrate the methods. The examples show forcefully that replication of a microarray experiment can yield large increases in statistical power. This chapter refers to spotted arrays in the discussion and illustrations but the proposed methodology is equally applicable to expression data from *in-situ* oligonucleotide arrays. The chapter discusses interpretation and use of the sample size and power tables for microarray studies that appear in Appendices A and B.

Many researchers have recognized the importance of planning microarray studies in terms of their design and size. Simon *et al*[1] (2002), for example, give an incisive overview of issues related to planning including study objectives, sample selection, replication, and analysis strategies.

14.1.   Test Hypotheses in Microarray Studies

The key statistical quantity in a microarray study is the differential expression of a gene in a given experimental condition. Our study of statistical power centers on an analysis of variance (ANOVA) model that incorporates a set of interaction parameters reflecting differential gene expression across experimental conditions[2] (2002). We take the response variable for the ANOVA model as the logarithm (to base 2) of the machine reading of intensity and refer to it simply as the *log-intensity*. Thus, if w is the intensity measurement, the response variable in the ANOVA model is taken as $y = \log_2(w)$. We assume that w is positive so the logarithm is defined. If the readings are background corrected then we assume that only corrected readings with positive values are used in the analysis.

As in Chapter 10, the gene-by-condition interaction effects $(\gamma\tau)_{gc}$ in equation (10.3) capture the differential expression of genes across experimental conditions. To simplify notation, we use the symbol

$$\mathcal{I}_{gc} = (\gamma\tau)_{gc} \qquad (14.1)$$

to denote the interaction term, with their estimates denoted by $\hat{\mathcal{I}}_{gc}$. Here, indices g and c refer to gene g and condition c, with ranges $g = 1, \ldots, G$ and $c = 1, \ldots, C$, respectively. We will adopt the constraint form where their (weighted) sums equal zero. That is,

$$\sum_c \hat{\mathcal{I}}_{gc} = 0, \quad \text{for each } g \qquad (14.2)$$

The sum-to-zero constraint implies that $\hat{\mathcal{I}}_{gc}$ is interpreted as the estimated differential expression intensity for gene g under condition c *relative* to the average for all genes and conditions in the study. For illustrations of the kind of quantity that $\hat{\mathcal{I}}_{gc}$ represents, see section 10.3.1.

Let $\boldsymbol{\mathcal{I}}_g = (\mathcal{I}_{g1}, \mathcal{I}_{g2}, \ldots, \mathcal{I}_{gC})'$ denote the column vector of interaction parameters for gene g across experimental conditions $c = 1, \ldots, C$, where the prime denotes transposition. With respect to differential gene expression, the null hypothesis H_0 and alternative (research) hypothesis H_1 of interest for any given gene g can be stated in terms of the interaction vector $\boldsymbol{\mathcal{I}}_g$ as follows:

$H_0 : \boldsymbol{\mathcal{I}}_g = \mathbf{0}$, the zero vector, indicating that gene g is not differentially expressed across the experimental conditions

$H_1 : \boldsymbol{\mathcal{I}}_g = \boldsymbol{\Delta}$, where $\boldsymbol{\Delta} = (\Delta_1, \Delta_2, \ldots, \Delta_C)'$ is a specified nonzero vector denoting the hypothesized alternative values for interaction terms $\mathcal{I}_{gc}$, $c = 1, \ldots, C$, indicating that gene g is differentially expressed across the experimental conditions. Quantity Δ_c is the hypothesized log-2 difference for the cth experimental condition.

The non-zero vector $\boldsymbol{\Delta}$ specified in the alternative hypothesis H_1 is a vector of hypothesized values for differential expression levels on the log-2 scale that the investigator desires to detect. For instance, a study may include four experimental conditions such that conditions $c = 1$ and $c = 2$ replicate a treatment condition and conditions $c = 3$ and $c = 4$ replicate a control condition. In this illustrative study, it may be desired to detect any gene g that has a differential expression pattern of form $\boldsymbol{\Delta} = (1.5, 1.5, -1.5, -1.5)'$. This pattern is equivalent to testing for an 8-fold up-regulation under treatment relative to control, i.e., $2^{1.5-(-1.5)} = 2^3 = 8$. The specified vector $\boldsymbol{\Delta}$ in H_1 is shown to be common for all genes g. This specification simplifies the following development but can be relaxed if it is felt that the hypothesized interaction effects should be different for different gene groups.

As discussed in Chapter 11, a test of hypotheses exposes an investigator to two types of error. A *type I error* occurs if the research hypothesis H_1 is concluded when in fact the null hypothesis H_0 is true. A *type II error* occurs if H_0 is concluded when H_1 is true. The *power* of any hypothesis test is defined as the probability of declaring the research hypothsis H_1 when, in fact, H_1 is true. A principal aim of a microarray study is to have a high probability of declaring a gene to be differentially expressed if it is truly differentially expressed, while keeping the probability of making a false declaration of differential expression acceptably low. Thus, in formal terms, a key design objective of a microarray study is to achieve high test power while controlling type I error risk.

In an actual microarray study, genes that are truly differentially expressed will generally do so to different degrees, some weakly some strongly. Therefore, the components $\mathcal{I}_{gc}$ of the interaction parameter vectors $\boldsymbol{\mathcal{I}}_g$ will have values that vary over a continuum as g varies. It is important for us to stress here, however, that this distribution of true expression levels does not directly enter into the power calculation in the planning stage. Instead, the alternative hypothesis H_1 refers only to a single non-zero vector $\boldsymbol{\mathcal{I}}_g$, specifically, the investigator-specified vector $\boldsymbol{\Delta}$. It is this nonzero vector that is to be used as the reference differential expression pattern for a power calculation. The assumption is that the alternative vector $\boldsymbol{\Delta}$ (or any vector that lies an equivalent 'distance' from zero) represents a pattern of differential expression that the investigator wishes to detect with high probability (i.e., with high power).

14.2.   Distributions of Estimated Differential Expression

In many applications, it is reasonable to assume that the estimated differential expression vector $\hat{\mathcal{I}}_g$, where $\hat{\mathcal{I}}_g = (\hat{\mathcal{I}}_{gc}, c = 1, \ldots, C)'$, has an approximate multivariate normal distribution with a mean zero and covariance matrix Σ under the null hypothesis H_0. The claim to a normal approximation is especially strong where the microarray study involves repeated observations of gene expression across conditions so that the interaction estimates $\hat{\mathcal{I}}_{gc}$ are averages of independent log-intensity readings. An appeal to the central limit theorem then supports the assumption of approximate normality. Likewise, under the alternative hypothesis H_1, $\hat{\mathcal{I}}_g$ also has an approximate multivariate normal distribution with the same covariance matrix but now with non-zero mean Δ. We note that the covariance matrix Σ will have rank $C - 1$ because of the interaction sum constraint.

14.3.   Summary Measures of Estimated Differential Expression

On the basis of the ANOVA modeling approach, different statistics may be used to summarize differential expression for single genes in microarray studies. We shall calculate power for some summary measure

$$V_g = h(\hat{\mathcal{I}}_g) \tag{14.3}$$

of the estimated differential expression vector $\hat{\mathcal{I}}_g$ for gene g, where h is any function specified by the investigator that captures the particular differential expression features that are of scientific interest in the statistical test. Later, we shall study two particular types of functions in detail; one in which $h(\hat{\mathcal{I}}_g)$ is a linear function of $\hat{\mathcal{I}}_g$ and another in which it is a quadratic function of $\hat{\mathcal{I}}_g$.

The variable V_g is a random variable for gene g that will have some realization v_g in the microarray study. Under null hypothesis H_0, summary measure V_g has a probability density function (p.d.f.) that we denote by $f_0(v)$. Similarly, under the alternative hypothesis H_1, summary measure V_g has a p.d.f. that we denote by $f_1(v)$. We shall show that it is the statistical distance between these two density functions, in a precise sense, that defines the level of power for a microarray study.

14.4. Multiple Testing Framework

As microarray studies typically involve the simultaneous study of thousands of genes, the probabilities of producing incorrect test conclusions (false positives and false negatives) must be controlled for the whole gene set. As discussed in Chapter 11, the challenge of multiple testing in microarray studies has been discussed by many authors from various perspectives. In this chapter, we are considering power and sample size calculations at the planning stage of a study before data are available. The perspective on multiple testing is therefore different than when applying a multiple testing procedure to study results that are already in hand.

Because the development of this chapter is closely related to Chapter 11, we again present the Benjamini and Hochberg[3] (1995) framework in Table 14.1. This table appeared earlier as Table 11.1 in Chapter 11. It is useful for understanding the problem of multiple testing and the control of inferential errors in microarray studies. The framework in the table postulates that there are, in fact, only two possible situations for any gene. Either the gene is not differentially expressed (hypothesis H_0 is true) or it is differentially expressed at the level described by the alternative hypothesis H_1. Thus, as discussed earlier, the hypothesis testing framework abstracts from the reality of genes having varying degrees of differential expression. The test declaration (decision) is either that the gene is differentially expressed (H_0 rejected) or that it is not differentially expressed (H_0 accepted). Thus, there are four possible test outcomes for each gene corresponding to the four combinations of true hypothesis and test declaration.

The total number of genes being tested is $G = G_0 + G_1$, where G_1 and G_0 are the true numbers of genes that are differentially expressed and not differentially unexpressed, respectively. At the planning stage of any microarray study, the counts G_1 and G_0 are unknown but fixed numbers. The counts of the four test outcomes are shown by the entries A_0, A_1, R_0 and R_1 in the multiple testing framework. The counts A_0 and A_1 are the numbers of true and false negatives (i.e., true and false declarations that genes are not differentially expressed). The counts R_1 and R_0 are the numbers of true and false positives (i.e., true and false declarations that genes are differentially expressed). The numbers $A = A_0 + A_1$ and $R = R_0 + R_1$ are the total numbers of genes that the microarray study declares are not differentially expressed (i.e. H_0 accepted), and are differentially expressed (i.e. H_0 rejected), respectively. These counts of test outcomes, however, are random variables in advance of the analysis of the microarray gene expression data.

	Test Declaration: Not Called as Diff. Expressed	Test Declaration: Called as Diff. Expressed	Number of Genes in Tests
The Null Hypothesis H_0 is True: Gene g in the Test is Not Differentially Expressed	A_0	R_0	G_0
The Alternative H_1 is True: Gene g in the Test is Differentially Expressed	A_1	R_1	G_1
Total Number of Genes in Tests	A	R	G

Table 14.1. Multiple testing framework. G: total number of genes being tested, G_0: true number of undifferentially expressed genes, G_1: true number of differentially expressed genes, R_0: number of false positives, A_1: number of false negatives.

The respective proportions of genes that are undifferentially and differentially expressed are given by:

$$p_0 = G_0/G \quad \text{and} \quad p_1 = G_1/G = 1 - p_0$$

As the counts G_0 and G_1 are generally unknown, so are the proportions p_0 and p_1. We show later that, for power and sample size considerations, the values of these counts and proportions *must be anticipated* at the planning stage of a study. Usually, the anticipated number (G_0) of undifferentially expressed genes will be much larger than the number (G_1) of differentially expressed genes. Indeed, in some studies it may be uncertain if any gene is actually differentially expressed (i.e., it may be uncertain if $G_1 > 0$). We index the genes for which H_0 and H_1 hold by the sets $\mathcal{G}_0$ and $\mathcal{G}_1$, respectively. We must remember, of course, that the memberships of these index sets are unknown because we do not know in advance if any given gene is differentially expressed or not.

The test outcomes counted by R_0 are false positives reflecting type I errors. We use α_0 to denote the probability of a type I error for any single gene g in the index set $\mathcal{G}_0$ under the selected decision rule. Thus,

$$
\begin{aligned}
\alpha_0 &= \text{Prob}\{\text{Gene } g \text{ falsely declared as differentially expressed}\} \\
&= \text{Prob}\{\text{Type I error for gene } g \text{ in } \mathcal{G}_0\} \\
&= E(R_0)/G_0.
\end{aligned}
\tag{14.4}
$$

Likewise, the test outcomes counted by A_1 are false negatives reflecting type II errors. We use β_1 to denote the probability of a type II error for any single gene in the index set $\mathcal{G}_1$ under the decision rule. Thus,

$$
\begin{aligned}
\beta_1 &= \text{Prob}\{\text{Gene } g \text{ falsely declared as not differentially expressed}\} \\
&= \text{Prob}\{\text{Type II error for gene } g \text{ in } \mathcal{G}_1\} \\
&= E(A_1)/G_1.
\end{aligned}
\tag{14.5}
$$

The *power* of any hypothesis test is defined as the probability of concluding H_1 when, in fact, H_1 is true. In the context of multiple testing, power is defined as the expected proportion of differentially expressed genes that are correctly declared to be differentially expressed. Thus,

$$
\begin{aligned}
\text{Power} &= 1 - \beta_1 \\
&= \text{Prob}\{\text{Gene } g \text{ correctly declared as differentially expressed}\} \\
&= \frac{\text{Expected number correctly called as differentially expressed}}{\text{Actual number differentially expressed}} \\
&= \frac{E(R_1)}{G_1}.
\end{aligned}
\tag{14.6}
$$

Another related performance measure in multiple testing is the *false discovery rate* or FDR for short, proposed by Benjamini and Hochberg (1995). The FDR measure looks at error controls from a different angle in Table 14.1. Instead of conditioning on the true but unknown state of whether genes are differentially expressed or not, the FDR measure takes the number of genes that are *declared* as differentially expressed as its denominator. FDR is defined as the expected ratio of the number of false positives to the number of genes declared as differentially expressed. With reference to the notation in the multiple testing framework, the FDR is defined as the following expected value.

$$
\text{FDR} = E\left(\frac{R_0}{R}\right)
\tag{14.7}
$$

The ratio R_0/R in (14.7) is taken as 0 if $R = 0$. FDR is considered again in the discussion of a Bayesian perspective on power in section 14.10.

14.5.　Dependencies of Estimation Errors

As explained in Chapter 10, the vector estimates $\hat{\mathcal{I}}_g$ may be probabilistically dependent for different genes in the same microarray study. This implies that test outcomes for different genes may be probabilistically dependent. We again emphasize in our discussion of dependence that we are not discussing biological dependencies of differential expression levels among genes (i.e., co-regulation). These kinds of dependencies are certainly going to be present in every microarray study. For example, H_1 may be true for a group of genes because they are differentially expressed together under given experimental conditions. The focus of our concern is whether *estimation errors* in the components of $\hat{\mathbf{I}}_g$, representing departures between observed and true values, are intercorrelated among genes. The summary measures V_g for the affected genes and, hence, their test outcomes (H_0 or H_1) will then reflect this dependence in estimation errors. We can envisage practical cases where dependence may be a major concern and others where it may be minor.

Given the potential for some dependence of the errors in the vector estimates $\hat{\mathbf{I}}_g$, even after careful modeling of effects, we consider two different ways to proceed. If the dependency is judged to be substantial or we wish to be conservative in the control of false positives, we may adopt a Bonferroni approach, which we describe shortly. On the other hand, if the dependency is judged to be insignificant, we may wish to calculate power or sample size on the assumption that the vector estimates are mutually independent.

14.6.   Familywise Type I Error Control

There are several ways of specifying the desired control over type I errors in the planning context of multiple testing. We consider two ways:

(1) A specification of the *familywise type I error probability* α_F

This specification refers to the probability of producing one or more false positives for genes in index set $\mathcal{G}_0$, which we denote by α_F. As mentioned earlier in Chapter 11, the familywise type I error probability α_F is sometimes referred to as the familywise error rate (FWER)[4]. Thus, in the notation of the preceding multiple testing framework, we have

$$\begin{aligned} \alpha_F &= \text{Prob}\{\text{One or more false positives}\} \\ &= \text{Familywise type I error probability} \\ &= P(R_0 \geq 1). \end{aligned} \tag{14.8}$$

(2) A specification of the *expected number of false positives* $E(R_0)$.

The expected number of false positives is oftened called the per-family error rate (PFER). This specification refers to the expected number of genes in index set $\mathcal{G}_0$ for which H_0 is incorrectly rejected, i.e., the quantity $E(R_0)$.

In the following development, we show the connection between the type I error risk for an individual test, denoted earlier by α_0, and the multiple testing control quantities α_F and $E(R_0)$.

We first define an acceptance interval $\mathcal{A}$ for the summary statistic V_g that gives the desired α_0 risk for a test on a single gene. Specifically, we wish to use the following decision rule to make a judgement call on whether any gene g is differentially expressed or not:

If $v_g \in \mathcal{A}$, declare gene g is not differentially expressed (conclude H_0).

If $v_g \notin \mathcal{A}$, declare gene g is differentially expressed (conclude H_1).

$$(14.9)$$

As the null hypothesis H_0 is true for any single gene $g \in \mathcal{G}_0$, it follows that the observed summary measure V_g derived from the microarray data will fall in acceptance interval $\mathcal{A}$ with the following probability.

$$P(V_g \in \mathcal{A} \mid g \in \mathcal{G}_0) = \int_{\mathcal{A}} f_0(v)dv = 1 - \alpha_0 \quad \text{for each gene } g \in \mathcal{G}_0$$

$$(14.10)$$

As we demonstrate later, we can use (14.10) to calculate $\mathcal{A}$ from knowledge of the form of the null p.d.f. $f_0(v)$. Interval $\mathcal{A}$ is chosen to be the shortest among those intervals satisfying (14.10).

We now describe two testing approaches, depending on whether the estimation errors in $\hat{\mathcal{I}}_g$ are independent or not. We refer to these as the Šidák and Bonferroni approaches, respectively.

14.6.1. Type I Error Control: the Šidák Approach

If the estimation errors of interaction effects $\hat{\mathcal{I}}_g$ can be considered as independent among genes, the familywise type I error probability α_F and the type I error probability α_0 for an individual test for any gene $g \in \mathcal{G}_0$ are connected as follows for the gene index set $\mathcal{G}_0$.

$$
\begin{aligned}
P(R_0 = 0) &= \text{Prob\{No false positive\}} \\
&= \prod_{g=1}^{G_0} P(\text{declaring } H_0 \text{ for gene } g \mid g \in \mathcal{G}_0) \\
&= (1 - \alpha_0)^{G_0} \\
&= 1 - \alpha_F
\end{aligned}
$$

$$(14.11)$$

In most microarray studies, G_0 is large and, hence, even a small specification for α_0 will translate into a large value for the familywise type I
error probability α_F. In addition, in most studies, it is uncertain what
number of genes are differentially expressed. In this situation, an investigator may wish to assume that all genes are not differentially expressed
(so $G_0 = G$) and change the exponent in (14.11) from G_0 to G.

With the assumption of independence among estimation errors, the
random variable R_0 follows a binomial distribution with parameters G_0
and α_0. Thus, the expectation $E(R_0)$ equals $\alpha_0 G_0$. When G_0 is reasonably large and α_0 is small, the *number of false positives* R_0 that will
arise under the assumption of independence will follow an approximate
Poisson distribution with mean parameter

$$E(R_0) = \alpha_0 G_0 \approx -\ln(1 - \alpha_F). \qquad (14.12)$$

For example, if the familywise type I error α_F is 0.20 and G_0 is large, the
Poisson mean is $E(R_0) = -\ln(0.80) = 0.223$. In this case, the probability of experiencing no false positive is $\exp(-0.223) = 0.80$. The probability of exactly one false positive is $0.223\exp(-0.223) = 0.223(0.80) =$
0.18. The probability of experiencing two or more false positives is therefore 0.02. Because of the direct connection between α_F and the mean
$E(R_0)$ in this case, either value may be used to specify the desired control
over the familywise type I error risk.

As another example, if an investigator feels that expecting 2.5 false
positives is tolerable, then this specification implies that $E(R_0) = -\ln(1-
\alpha_F) = 2.5$ and, hence, a familywise type I error probability of $\alpha_F =
1 - \exp(-2.5) = 0.918$. This α_F value may appear very high. The illus-
tration reminds us, however, that a large value of α_F may be reasonable
in microarray studies where a few false positives among thousands of
genes must be tolerated in order to avoid many false negatives (i.e., to
avoid missing many differentially expressed genes). The design of a microarray study involves a careful balancing of costs of false positives and
false negatives. The connection between α_F and α_0 in this last example
is

$$\alpha_0 = \frac{E(R_0)}{G_0} \approx \frac{-\ln(1 - \alpha_F)}{G_0}. \qquad (14.13)$$

For instance, if G_0 happens to equal 5000 then, $\alpha_0 = 2.5/5000 = -[\ln(1-
0.918)]/5000 = 0.00050$.

Under the approach of independent estimation errors represented by rule (14.11), we may wish to focus more directly on the number of false positives by using the following property of order statistics for simple random samples: The k_1th lowest and k_2th highest order statistics of the summary measures v_g for genes $g \in \mathcal{G}_0$ span an expected combined tail area of $k/(G_0 + 1)$ where $k = k_1 + k_2$. This property may be used to set the acceptance interval $\mathcal{A}$ based on the anticipated values of extreme order statistics under the null p.d.f. $f_0(v)$. Specifically, the acceptance interval in (14.10) might be defined by the following specification.

$$\alpha_0 = \frac{k}{G_0 + 1} \qquad (14.14)$$

Substitution of (14.14) into (14.11) gives the following implied value for the familywise type I error probability for this rule.

$$\alpha_F = 1 - \left(1 - \frac{k}{G_0 + 1}\right)^{G_0} \approx 1 - \exp(-k) \quad \text{for large } G_0 \qquad (14.15)$$

The mean number of false positives for this rule is approximately k, i.e., $E(R_0) \approx k$. Although the form of (14.14) is motivated by the theory of order statistics in which k is a whole number, (14.14) and (14.15) can be used with fractional values of k.

14.6.2. Type I Error Control: the Bonferroni Approach

The Bonferroni procedure is widely used in statistics for error control where simultaneous inferences are being made. The procedure makes use of the following well-known Bonferroni probability inequality to control the familywise type I error probability.

$$
\begin{aligned}
P(V_g \notin \mathcal{A} \text{ for one or more genes } g \in \mathcal{G}_0) &\leq \sum_{g \in \mathcal{G}_0} P(V_g \notin \mathcal{A} \mid g \in \mathcal{G}_0) \\
&= G_0 P(V_g \notin \mathcal{A} \mid g \in \mathcal{G}_0) \\
&= G_0 \alpha_0
\end{aligned}
$$

$$(14.16)$$

This inequality holds regardless of the dependency among the estimated differential expression vectors $\hat{\mathcal{I}}_g$ of the gene set.

Again, letting α_F denote the desired level of the familywise type I error probability for the microarray study, the acceptance interval in

(14.10) may be defined by specifying the type I error probability for any gene g to be

$$\alpha_0 = P(V_g \notin \mathcal{A} \mid g \in \mathcal{G}_0) = \frac{\alpha_F}{G_0} \qquad (14.17)$$

That is, for the Bonferroni procedure, the individual error rate α_0 is defined as the desired familywise error rate α_F divided by the total number of genes having no differential expression. This definition of the acceptance interval $\mathcal{A}$ guarantees that the following inequality holds for the familywise type I error probability.

$$P(\text{one or more false positives}) \leq \alpha_F \qquad (14.18)$$

Thus, the inequality in (14.18) assures us that the Bonferroni procedure keeps the familywise type I error probability at the desired level α_F or lower. In subsequent discussion in the Bonferroni context, we refer to α_F as the familywise type I error probability although the inequality (14.18) implies that the true error probability may be somewhat lower. We note, for given G_0 and α_F, that the Bonferroni rule (14.17) will always choose a wider acceptance interval $\mathcal{A}$ than the rule based on the Šidák approach in (14.11).

With respect to the expected number of false positives, using the Bonferroni procedure in (14.17) provides the following result.

$$E(R_0) = \alpha_0 G_0 = \alpha_F \qquad (14.19)$$

It can be seen that the expected number of false positives equals the familywise type I error probability in this case. Thus, necessarily, the expected number $E(R_0)$ cannot exceed one (although the actual number R_0 is not so constrained).

Unlike the independence approach discussed in the preceding section, there is no direct link between the probability distribution for the number of false positives R_0 and the familywise type I error probability α_F under the Bonferroni approach. The Bonferroni procedure controls the chance of incurring one or more false positives but provides no probability statement about how many false positives may be present if some do occur (i.e., the approximate Poisson distribution does not apply).

14.7. Familywise Type II Error Control

As with type I errors, we can quantify type II error control in several ways in the context of multiple testing. We focus on two ways.

(1) The *familywise type II error probability* β_F (or, equivalently, one minus the *familywise power level*).

This measure refers to the probability of producing one or more false negatives for genes in index set $\mathcal{G}_1$, which we denote by β_F. Thus, in the notation of the preceding multiple testing framework, we have

$$
\begin{aligned}
\beta_F &= P(\text{one or more false negatives}) \\
&= \text{Familywise type II error probability} \\
&= P(A_1 \geq 1).
\end{aligned}
\tag{14.20}
$$

The corresponding familywise power level is then $1 - \beta_F$.

(2) The *expected number of true positives* $E(R_1)$.

This measure refers to the expected number of genes in index set $\mathcal{G}_1$ that are correctly declared as differentially expressed, i.e., the quantity $E(R_1)$.

In the following development, we show the connection between the type II error risk for an individual test, denoted earlier by β_1, and the values of the multiple testing quantities β_F and $E(R_1)$.

The power of the test for any *single* gene that is differentially expressed at the level defined in H_1 equals $1 - \beta_1$. This declaration is equivalent to having the summary measure $V_g = h(\hat{\boldsymbol{\mathcal{I}}}_g)$ for the gene in question fall outside the acceptance interval $\mathcal{A}$. We denote this rejection interval by the complement $\mathcal{A}^c$. The power for a single differentially expressed gene is therefore given by

$$
P(V_g \in \mathcal{A}^c \mid g \in \mathcal{G}_1) = \int_{\mathcal{A}^c} f_1(v)dv = 1 - \beta_1 \quad \text{for any gene } g \in \mathcal{G}_1.
\tag{14.21}
$$

In essence, therefore, $1 - \beta_1$ is fixed by the rejection interval which, in turn, is fixed by the specified control on the familywise type I error risk and the specification for the alternative hypothesis H_1. Our use of p.d.f. $f_1(v)$ for this power calculation means that we are examining the power for any and all hypothesized alternative differential gene expression vectors $\boldsymbol{\Delta}$ whose estimates map into the same random variable $V_g = h(\hat{\boldsymbol{\mathcal{I}}}_g)$ having the p.d.f. $f_1(v)$.

As with type I errors, we encounter the Šidák and Bonferroni formulas for power, depending on whether estimation errors in $\hat{\boldsymbol{\mathcal{I}}}_g$ are independent or not. We can now abbreviate the presentation because the underlying logic is clear from the earlier development.

14.7.1.   Type II Error Control: the Šidák Approach

The anticipated count G_1 of differentially expressed genes (i.e., true positives) when taken together with the power level $1 - \beta_1$ for any individual test for gene $g \in \mathcal{G}_1$ can be used to calculate either measure of familywise type II error control. Under the assumption of independent estimation errors among genes, the desired familywise type II error probability β_F and the type II error probability for an individual test β_1 for any gene are connected as follows for the gene index set $\mathcal{G}_1$.

$$
\begin{aligned}
P(A_1 = 0) &= P(\text{No false negative}) \\
&= (1 - \beta_1)^{G_1} = 1 - \beta_F \qquad (14.22)
\end{aligned}
$$

Also, under independence, the random variable R_1 follows a binomial distribution with parameters G_1 and $1 - \beta_1$. Thus, the expected number of true positives is given by

$$
E(R_1) = G_1(1 - \beta_1). \qquad (14.23)
$$

To illustrate these power calculations numerically, consider a microarray study for which G_1 is anticipated to be 50 genes and for which $1 - \beta_1 = 0.99$. In this case, $1 - \beta_F = (0.99)^{50} = .605$. Observe how high the power level must be for a single gene in index set $\mathcal{G}_1$, namely 0.99, in order to have even a moderate probability of discovering all 50 differentially expressed genes (0.605). In this same situation, the expected number of true positives among the 50 differentially expressed genes would be $G_1(1 - \beta_1) = 50(0.99) = 49.5$. In other words, 99 percent of the differentially expressed genes are expected to be declared as such.

14.7.2.   Type II Error Control: the Bonferroni Approach

Under an assumption of dependence for the estimated differential expression vectors, recourse to the Bonferroni inequality gives the following specification for the familywise power level $1 - \beta_F$ as a function of the individual type II error probability β_1 for testing any single gene.

$$
\begin{aligned}
P(V_g \in \mathcal{A} \text{ for one or more genes } g \in \mathcal{G}_1) &\leq \sum_{g \in \mathcal{G}_1} P(V_g \in \mathcal{A} \mid g \in \mathcal{G}_1) \\
&= G_1 P(V_g \in \mathcal{A} \mid g \in \mathcal{G}_1) \\
&= G_1 \beta_1 \qquad (14.24)
\end{aligned}
$$

Hence,

$$1 - \beta_F \geq \max(0, 1 - G_1 \beta_1) \tag{14.25}$$

As $1 - \beta_F$ must be non-negative, a minimum of zero is imposed in (14.25). Thus, the Bonferroni inequality gives us a lower bound on the familywise power level.

The expected number of true positives under the Bonferroni approach is given by

$$E(R_1) = G_1(1 - \beta_1). \tag{14.26}$$

For the previous numerical example, where $G_1 = 50$ and $1 - \beta_1 = 0.99$, the lower bound on the familywise power level is $1 - 50(0.01) = 0.50$. The expected number of true positives is $E(R_1) = 50(0.99) = 49.5$. Thus, again, 99 percent of the differentially expressed genes are expected to be declared as such. As is the case with false positives, there is no direct link between the familywise type II error probability β_F and the probability distribution for the number of true positives R_1 under dependence. The Bonferroni procedure controls the chance of incurring one or more false negatives but provides no probability statement about how many false negatives may be present if some do occur.

14.8. Contrast of Planning and Implementation in Multiple Testing

In this section, we emphasize the contrast between specifications for type I and type II error controls at the planning stage of a microarray study (before microarray experiments are conducted) and implementation algorithms used at the actual testing stage after gene expression data have been collected, such as step-down P-value methods.

For planning purposes, our methodology posits the index sets $\mathcal{G}_0$ and $\mathcal{G}_1$ for undifferentially and differentially expressed genes, respectively. Although the planning does not identify the members of each set, it does specify the cardinality of each. In this statistical setting, test implementation algorithms seek to maximize the power of detecting which genes are truly in the index set $\mathcal{G}_1$ while still controlling either the familywise type I error probability or a related quantity, such as the false discovery rate (discussed later in section 14.10.1).

As discussed in Chapter 11, many approaches have been proposed for actual test implementation once the microarray data are in hand. For example, step-down P-value algorithms and methods for controlling the false discovery rate have been widely adopted for error control in microarray studies. See, for instance, Efron *et al.* (2001) and Dudoit *et al.*[5] (2002). Observed P-values for the G hypothesis tests in a microarray study will be derived from the null p.d.f. $f_0(v)$ or its estimate, evaluated at the respective realizations v_g, $g = 1, \ldots, G$. The observed P-values, say $p_1, \ldots, p_G$, will vary from gene to gene because of inherent sampling variability and also because the null hypothesis may hold for some genes but not for others. The information content of the observed P-values is used in these testing procedures to assign genes to either the index set $\mathcal{G}_0$ or the index set $\mathcal{G}_1$ without knowing the size of either set. These approaches exploit information in the data themselves (specifically, the observed P-values) and, hence, are data-dependent. In contrast, in planning for power and sample size, we must anticipate the sizes of these two index sets, and control the two types of errors accordingly, without the benefit of having the observed P-values themselves. The P-values derived from the actual observed microarray data not only allow classification of individual genes as differentially expressed or not but also provide a report card on the study plan and whether its specifications were reasonable or not.

14.9. Power Calculations for Different Summary Measures

We now present power calculations for the two classes of functions $V_g = h(\hat{\boldsymbol{\mathcal{I}}}_g)$ mentioned earlier, both of which are important in microarray studies.

14.9.1. Designs with Linear Summary Measure

Consider a situation where the summary measure V_g is a linear combination of differential expression estimates $\hat{\mathcal{I}}_{gc}$ of the following form.

$$V_g = h(\hat{\boldsymbol{\mathcal{I}}}_g) = \boldsymbol{\lambda}' \hat{\boldsymbol{\mathcal{I}}}_g = \sum_{c \in C} \lambda_c \hat{\mathcal{I}}_{gc} \qquad (14.27)$$

where $\boldsymbol{\lambda}' = (\lambda_1, \ldots, \lambda_C)$ is a vector of design-related coefficients specified by the investigator. Examples of such linear combinations include any single differential expression estimate, say $\hat{\mathcal{I}}_{g1}$, or any difference of such estimates, say $\hat{\mathcal{I}}_{g1} - \hat{\mathcal{I}}_{g2}$. Frequently the linear combination of interest will be a contrast of interaction estimates that reflects, for example, the difference between treatment and control conditions.

As discussed earlier, we may assume that the vector $\hat{\mathcal{I}}_g$ has an approximate multivariate normal distribution with mean zero under the null hypothesis and covariance matrix $\boldsymbol{\Sigma}$. This assumption is reasonable, first, because of the application of the central limit theorem in deriving individual estimates $\hat{\mathcal{I}}_{gc}$ from repeated readings and, second, from a further application of the central limit theorem where the linear combination in (14.27) involves further averaging of the individual estimates. It then follows from this normality assumption that the null p.d.f. $f_0(v)$ is an approximate normal distribution with mean zero and null variance

$$\sigma_0^2 = \text{Var}(V_g|H_0) = \boldsymbol{\lambda}'\boldsymbol{\Sigma}\boldsymbol{\lambda} \qquad (14.28)$$

Under the alternative hypothesis H_1, we assume that the $\hat{\mathcal{I}}_{gc}$ have the same multivariate normal distribution but with mean $\boldsymbol{\Delta}$. In other words, that the null distribution is simply translated to a new mean position. In this case, the summary measure V_g has an approximate normal p.d.f. $f_1(v)$ with the same variance σ_0^2 and mean parameter

$$\mu_1 = E(V_g|H_1) = \boldsymbol{\lambda}'\boldsymbol{\Delta} = \sum_{c \in C} \lambda_c \Delta_c \qquad (14.29)$$

We consider only linear combinations for which μ_1 is non-zero.

Here are the steps for computing power.

1 Compute the null variance σ_0^2 in (14.28) from specifications for the vector $\boldsymbol{\lambda}$ and covariance matrix $\boldsymbol{\Sigma}$.

2 Compute μ_1 in (14.29) from specifications for the vectors $\boldsymbol{\lambda}$ and $\boldsymbol{\Delta}$.

3 Specify the desired familywise type I error probability α_F or the equivalent mean number of false positives $E(R_0)$.

The first step is the most difficult because it requires some knowledge of the inherent variability of the data in the planned microarray study. As we discuss later, this inherent variability is intimately connected with the experimental error in the scientific process, the experimental design and the number of replicates of the design used in the study.

We now present a brief numerical example of a power calculation based on the methodology for a linear function of differential expression.

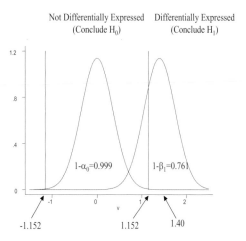

Figure 14.1.   Illustration of a power calculation in the non-central normal case. Source: Statistics in Medicine, Volume 21, page 3556, Copyright 2002, John Wiley & Sons Ltd. Reproduced by permission of John Wiley & Sons Limited.

14.9.2.   Numerical Example for Linear Summary

Consider a microarray study in which interest lies in the difference between two experimental conditions representing, say, two tissue types. Thus, differences in differential gene expression of the form $\hat{\mathcal{I}}_{g1} - \hat{\mathcal{I}}_{g2}$ are being considered. The null standard deviation for such differences is expected to be similar to that found in a previous study, namely, $\sigma_0 = 0.35$ on a log-scale with base 2. We suppose that a difference of $\mu_1 = 1.40$ (on a log-2 scale) is the difference specified under the alternative hypothesis by the investigator. Observe that this difference represents a $2^{1.40} = 2.64$ fold difference in gene expression and is four times the null standard deviation (i.e., $\mu_1/\sigma_0 = 1.40/0.35 = 4.0$).

The study involves a gene set of $G = 2100$ genes. It is anticipated that $G_0 = 2000$ genes will show no differential expression, while the the remaining $G_1 = 100$ will be differentially expressed at the level μ_1 specified under H_1. We assume statistical independence among the estimated differential expression vectors $\hat{\mathcal{I}}_g$ of the gene set. The order statistic rule (14.14) with $E(R_0) = k = 2$ will be used for setting the acceptance interval $\mathcal{A}$. It then follows that $\mathcal{A}$ is defined by $\pm z \sigma_0$ where z denotes the standard normal percentile $z(2000/2001) = z(0.9995) = 3.2905$. The resulting interval is (-1.152, 1.152) on a log-2 scale. In making this determination of $\mathcal{A}$, we have used the interval centered on zero as it is the shortest interval. The situation is illustrated in Figure 14.1. Observe that the area spanned by the acceptance interval under the null p.d.f. in this illustration corresponds to $1999/2001 = 0.999$ so $\alpha_F = 1 - (0.999)^{2000} = 0.8648$ and $E(R_0) = -\ln(1 - 0.8648) = 2$, as required.

Finally, the power for detecting a single differentially expressed gene is given by the area under the alternative p.d.f. in Figure 14.1, labelled $1 - \beta_1$. Reference to the standard normal distribution gives $1 - \beta_1 = 0.761$. Thus, any single gene with a 2.64-fold difference in expression between the two tissue types has probability 0.761 of being declared as differentially expressed in this study (i.e., of leading to conclusion H_1). This probability is the same whether the difference refers to an up- or down-regulated gene. This same power value implies that about 76 percent of the anticipated $G_1 = 100$ differentially expressed genes in the array will be correctly declared as differentially expressed. The probability of detecting all 100 of these genes is the familywise power level $1 - \beta_F$ given by (14.22). Here that familywise power level is 0.761^{100}, a vanishingly small value.

14.9.3.   Designs with Quadratic Summary Measure

We now consider a situation where the summary measure V_g is a quadratic form. To represent this quadratic form symbolically, we restrict vectors $\hat{\boldsymbol{\mathcal{I}}}_g$ and $\boldsymbol{\Delta}$ to their first $C - 1$ components and restrict matrix $\boldsymbol{\Sigma}$ to the principal submatrix defined by the first $C - 1$ interaction parameters. We denote these restricted forms by $\hat{\boldsymbol{\mathcal{I}}}_{g|R}$, $\boldsymbol{\Delta}_R$ and $\boldsymbol{\Sigma}_R$, respectively. These restricted forms are required by the interaction sum constraint which makes one component of each vector redundant. With this restricted notation, the quadratic form of interest is expressed as follows.

$$V_g = \hat{\boldsymbol{\mathcal{I}}}_{g|R}' \boldsymbol{\Sigma}_R^{-1} \hat{\boldsymbol{\mathcal{I}}}_{g|R} \tag{14.30}$$

We see that this measure implicitly takes account of all differential expression estimates and, hence, is responding to differential expression in any of the C experimental conditions in the study. Statistic V_g in (14.30) is larger whenever one of the interaction estimates in $\hat{\mathcal{I}}_g$ is larger. It is therefore a comprehensive measure of differential gene expression. Measure V_g in (14.30) can be interpreted as the squared statistical distance between the restricted interaction estimate vector $\hat{\mathcal{I}}_{g|R}$ and the zero vector $\mathbf{0}$ specified in the null hypothesis. It is also intimately connected to the sum of squares for the set of interaction effects $\hat{\mathcal{I}}_{gc}, c = 1, \ldots, C$, in the ANOVA model.

The quadratic measure in (14.30) is suitable for microarray studies which examine an assortment of experimental conditions with the simple aim of discovering genes that are differentially expressed *in any pattern* among the conditions. For example, a microarray study may examine tissues from C different tumors with the aim of seeing if there are genetic differences among the tumors. As another example, the experimental conditions may represent a biological system at C different time points and interest may lie in the time course of genetic change in the system, if any. Thus, measure (14.30) is suited to uncovering differential gene expression in a general set of experimental conditions where theory may provide no guidance about where the differential expression is likely to arise among the conditions.

To apply measure (14.30), we assume, as before, that the estimate vector $\hat{\mathcal{I}}_g$ is approximately multivariate normal with covariance matrix $\mathbf{\Sigma}$ and mean zero under the null hypothesis. The theory of quadratic forms then states that V_g follows an approximate chi-square distribution with $C - 1$ degrees of freedom. One degree of freedom is lost because of the interaction sum constraint. Under the alternative hypothesis, V_g has a non-central chi-square distribution with non-centrality parameter ψ_1 where

$$\psi_1 = \mathbf{\Delta}'_R \mathbf{\Sigma}_R^{-1} \mathbf{\Delta}_R. \tag{14.31}$$

We caution that the assumption of chi-square and non-central chi-square distributions for quadratic measure V_g in (14.30) is a little more sensitive to the assumed normality of the vectors $\hat{\mathcal{I}}_g$ than is the case with the linear summary measure (14.27). The reason is that the quadratic measure does not have the benefit of a secondary application of the central limit theorem from taking a linear combination of estimates.

The non-central chi-square p.d.f. can be used in (14.21) to calculate the power of the microarray study. The steps for computing power are as follows.

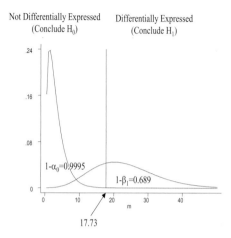

Figure 14.2. Illustration of a power calculation in the non-central chi-square case. Source: Statistics in Medicine, Volume 21, page 3558, Copyright 2002, John Wiley & Sons Ltd. Reproduced by permission of John Wiley & Sons Limited.

1 Compute the non-centrality parameter ψ_1 in (14.31) from specifications for vector $\mathbf{\Delta}$ and covariance matrix $\mathbf{\Sigma}$.

2 Specify the familywise type I error probability α_F or the equivalent mean number of false positives $E(R_0)$.

As before, the first step is the most difficult because it requires some knowledge of the inherent variability of the data in the planned microarray study, which depends on the experimental error in the scientific process, the experimental design and the number of replicates of the design used in the study.

14.9.4. Numerical Example for Quadratic Summary

As a brief numerical example of a power calculation for the quadratic summary measure, assume that the gene set contains $G = 2100$ genes and that the study includes $C = 4$ experimental conditions. Furthermore, assume statistical independence among the estimated differential expression vectors $\hat{\mathcal{I}}_g$ of the gene set. We suppose that the non-centrality parameter ψ_1 under the alternative hypothesis is calculated from (14.31) and equals 20.0.

It is anticipated that $G_0 = 2000$ of the 2100 genes will not be differentially expressed and $G_1 = 100$ genes will be differentially expressed at the level ψ_1 under the alternative hypothesis H_1. We will use the order statistic rule (14.14) with $E(R_0) = k = 1$ to set the acceptance interval $\mathcal{A}$. It then follows that $\mathcal{A}$ is defined by $\chi_3^2(2000/2001) = \chi_3^2(0.9995) = 17.73$. The resulting acceptance interval $\mathcal{A}$ under the null p.d.f. $f_0(v)$ is (0, 17.73) as shown in Figure 14.2. Finally, the power of this microarray study to detect a single differentially expressed gene is given by the area under the alternative p.d.f. $f_1(v)$, labelled $1 - \beta_1$ in Figure 14.2. Reference to the relevant non-central chi-square distribution gives this power value as $1 - \beta_1 = 0.689$. Thus, 69 percent of the 100 differentially expressed genes in the index set $\mathcal{G}_1$ are expected to be detected by this study.

14.10.   A Bayesian Perspective on Power and Sample Size

Lee *et al.*[6] (2000) and Efron *et al.* (2001) describe a mixture model for differential gene expression that provides a Bayesian posterior probability for the event that a given gene is differentially expressed. This mixture model has a useful interpretation in terms of our study of power and sample size. (A more extensive discussion of Bayesian modeling of microarray data appears in Chapter 13.).

In the multiple testing framework presented earlier, we defined p_1 and its complement $p_0 = 1 - p_1$ as the respective probabilities that a randomly selected gene would be differentially expressed (H_1) or not (H_0). We now take these probabilities as prior probabilities in a Bayesian model for the summary measure of gene expression V_g for gene g. The marginal p.d.f. for the summary statistic V_g under this model is

$$f(v) = p_0 f_0(v) + p_1 f_1(v). \tag{14.32}$$

This model simplifies reality in two respects. First, it assumes that the prior probabilities are the same for all genes, although this assumption can be relaxed easily. Second, it assumes that if a gene is differentially expressed then it is expressed at the level specified in the alternative hypothesis H_1.

From Bayes theorem, the posterior probabilities for any gene g having summary statistic $V_g = v_g$ can be calculated from the components of the mixture model (14.32) as follows.

$$P(H_0|v_g) = \frac{p_0 \, f_0(v_g)}{f(v_g)}, \quad P(H_1|v_g) = \frac{p_1 \, f_1(v_g)}{f(v_g)} \tag{14.33}$$

The posterior probabilities $P(H_1|v_g)$ and $P(H_0|v_g)$ are the respective probabilities that gene g is truly differentially expressed or not, given its summary measure has outcome v_g.

14.10.1.   Connection to Local Discovery Rates

For some classification cutoff value v_*, defined by an appropriate balancing of misclassification costs, each gene g can be *declared* as differentially expressed or not, depending on whether $v_g > v_*$ or not. Probability $P(H_1|v_g)$ is then the probability of a correct declaration of differential expression when an expression reading of $v_g > v_*$ is presented.

Efron *et al.*[7] (2001) interpret the posterior probability $P(H_0|v_g)$ in (14.33) as the false discovery rate for all genes sharing summary measure v_g, for given $v_g > v_*$. As $P(H_0|v_g)$ describes the FDR in the locality of v_g, Efron *et al.* (2001) refer to it as a *local false discovery rate* or local FDR for short. By analogy, the complementary posterior probability $P(H_1|v_g)$ might be interpreted as a *local true discovery rate* or local TDR. The local TDR is the proportion of genes that are truly differentially expressed among those genes sharing summary measure $v_g > v_*$.

14.10.2.   Representative Local True Discovery Rate

A representative value of the local TDR can be chosen to summarize the ability of a microarray study to correctly classify genes declared to be differentially expressed. As a representative value of the posterior probability $P(H_1|v_g)$, we suggest replacing v_g by the corresponding parameter $h(\mathbf{\Delta})$ under the alternative hypothesis H_1. The resulting probability is

$$P[H_1|h(\mathbf{\Delta})] = \frac{p_1 f_1[h(\mathbf{\Delta})]}{f[h(\mathbf{\Delta})]} \tag{14.34}$$

When h is the linear function in (14.29), parameter $h(\mathbf{\Delta})$ is μ_1. When h is the quadratic function in (14.31), parameter $h(\mathbf{\Delta})$ corresponds to ψ_1.

The representative local TDR that we have just defined is not directly comparable to the power level of a test although it does convey closely related information about the ability of a microarray study to correctly identify genes that are truly differentially expressed. As defined at the outset of the paper, classical power refers to the conditional probability of declaring a gene as differentially expressed when, in fact, that is true. In this Bayesian context, local TDR refers to the conditional probability that a gene is truly differentially expressed when, in fact, it has been declared as expressed by the test procedure. The conditioning events of these two probabilities are reversed in the classical and Bayesian contexts.

14.10.3.   Numerical Example for TDR and FDR

To give a numerical example of local TDR and FDR, we consider the demonstration depicted in section 14.9.4 and Figure 14.2. The gene set contains $G = 2100$ genes. Four experimental conditions are under study, so $C = 4$. It is anticipated that $G_0 = 2000$ of the 2100 genes will not be differentially expressed and $G_1 = 100$ genes will be differentially expressed. These counts correspond to prior probabilities of $p_0 = 2000/2100 = 0.952$ and $p_1 = 1 - 0.952 = 0.048$.

The non-centrality parameter ψ_1 has been specified as 20 under H_1. Setting v_g equal to $\psi_1 = 20$, the probability densities $f_0(20)$ and $f_1(20)$ for central and non-central chi-square distributions with $C - 1 = 3$ degrees of freedom are calculated as 0.00008096 and 0.04457, respectively. The marginal probability density $f(20)$ is then calculated from (14.32) as 0.002216. Finally, the desired posterior probabilities in (14.33) are calculated as 0.035 and 0.965, respectively. In other words, among genes having observed differential expression at level $v_g = 20$, about 96 percent will be truly differentially expressed and 4 percent will not. Assuming $v_g = 20$ is above the classification cutoff v_*, these respective probabilities are the local FDR and TDR. Observe how the probability that H_1 is true rises from a prior level of 0.048 to a posterior level of 0.965 if the gene has an observed differential expression level v_g equal to $\psi_1 = 20$.

14.11.   Applications to Standard Designs

We now show applications of the power methodology to some standard microarray designs and present representative sample size and power tables for these designs. The tables presented in this chapter are abbreviated and intended only to illustrate the methodology. More extensive power and sample size tables and a brief explanation of their use are provided in Appendices A and B.

14.11.1. Treatment-control Designs

Consider a microarray study in which treatment and control conditions are to be compared for differential expression. The study may involve a *completely randomized design* or a *matched-pairs design*. For example, in a toxicity study, $C = 2n$ mice may be randomly assigned in equal numbers to treatment and control groups, the n mice in the treatment group being exposed to a toxin and the n mice in the control group being unexposed. In contrast, in a study of liposarcoma, each treatment-control pair may consist of liposarcoma tissue from a patient and matching normal fat tissue taken from the same patient. The study has n such matched pairs and, therefore, $C = 2n$ tissue samples in total. In the case of a completely randomized design, we consider only the standard case of equal sample sizes for the treatment and control groups.

In what follows, we assume that there is a total of $C = 2n$ experimental conditions under consideration, where indices $c = 1, \ldots, n$ denote the treatment conditions and $c = n + 1, \ldots, 2n$ denote the control conditions. The assumption is made that a given gene g either has no difference in log-expression between the treatment and control conditions (null hypothesis H_0) or has a difference in log-expression equal to some non-zero value μ_1 (alternative hypothesis H_1). This assumption implies that the interaction parameters Δ_c have the following values under the alternative hypothesis.

$$\Delta_c = \left\{ \begin{array}{ll} \mu_1/2 & \text{for } c = 1, \ldots, n \quad\;\; \text{treatment conditions} \\ -\mu_1/2 & \text{for } c = n + 1, \ldots, 2n \quad \text{control conditions} \end{array} \right. \tag{14.35}$$

Observe that these parameter values follow the sum-to-zero constraint as required. The difference μ_1 in log-expression may be interpreted equivalently as a specified expression fold change. For instance, if logarithms to base 2 are used, then the corresponding specified alternative fold change is 2^{μ_1}.

We illustrate the methodology for a linear function of differential gene expression. The linear combination of interest in this context involves a contrast of gene expression under treatment and control conditions. We choose the convenient definition

$$\boldsymbol{\lambda}' = (1/n, \ldots, 1/n, -1/n, \ldots, -1/n)$$

where there are n coefficients of each sign. Thus, from (14.29) and (14.28), we have

$$E(V_g|H_1) = \sum_{c=1}^{n} \left(\frac{1}{n}\right)\left(\frac{\mu_1}{2}\right) + \sum_{c=n+1}^{2n} \left(\frac{-1}{n}\right)\left(\frac{-\mu_1}{2}\right) = \mu_1 \qquad (14.36)$$

$$\mathrm{Var}(V_g|H_0) = \sigma_d^2/n = \sigma_0^2 \qquad (14.37)$$

Here σ_d signifies the standard deviation of the difference in log-expression between treatment and control conditions.

Table 14.2 shows fragments of ANOVA tables for the completely randomized and matched pairs designs we have just described. The tables give degrees of freedom *df* for the sources of variance and the expected mean squares $E(MS)$ for the Treatments and Error sources. For a completely randomized design, the variance of the difference in log-expression between treatment and control is given by $\sigma_d^2 = 2\sigma^2$, where σ^2 is the experimental error variance of gene log-expression in Table 14.2 (a). For a matched pairs design, σ_d^2 is the variance of the difference in log-expression between treatment and control conditions for a matched pair. In terms of notation, again $\sigma_d^2 = 2\sigma^2$ but here σ^2 is the experimental error variance of gene log-expression in Table 14.2(b). The benefit of matching pairs of treatment and control conditions is indicated by the extend to which σ^2 in Table 14.2 (b) is smaller than σ^2 in Table 14.2(a). The ratio $|\mu_1|/\sigma_d$ can be interpreted as the statistical distance (i.e., the number of standard deviations) between the treatment and control log-expression levels under the alternative hypothesis.

14.11.2.   Sample Size for a Treatment-control Design

Table 14.3 gives the sample size n of the treatment and control groups required to achieve a specified individual power level $1 - \beta_1$ for the experimental design we have just described. A more extensive version of the table is provided in Appendix A. The table is entered based on the specified mean number of false positives $E(R_0)$, ratio $|\mu_1|/\sigma_d$, anticipated number of undifferentially expressed genes G_0 and desired individual power level $1 - \beta_1$. If G_0 is expected to be similar to the total gene count G, the table could be entered using G without introducing great error. To conserve space, only two individual power levels are offered in this illustrative table, namely, 0.90 and 0.99. The sample size shown

(a) Completely randomized design		
Source	df	E(MS)
Treatments	1	$\sigma^2 + n\mu_1^2/2$
Error	$2(n-1)$	σ^2
Total	$2n-1$	

(b) Matched pairs design		
Source	df	E(MS)
Treatments	1	$\sigma^2 + n\mu_1{}^2/2$
Blocks(pairs)	$n-1$	
Error	$n-1$	σ^2
Total	$2n-1$	

Table 14.2. Fragments of ANOVA tables for completely randomized and matched-pairs designs for a comparison of treatment and control conditions.

	Mean Number of False Positives											
	$E(R_0) = 0.5$				$E(R_0) = 1.0$				$E(R_0) = 2.0$			
	Distance $|\mu_1|/\sigma_d$				Distance $|\mu_1|/\sigma_d$				Distance $|\mu_1|/\sigma_d$			
	1.0	1.5	2.0	2.5	1.0	1.5	2.0	2.5	1.0	1.5	2.0	2.5
Genes G_0	Power $1 - \beta_1 = $ Proportion correctly declared as diff. expressed $= 0.90$											
1000	23	11	6	4	21	10	6	4	20	9	5	4
2000	25	11	7	4	23	11	6	4	21	10	6	4
Genes G_0	Power $1 - \beta_1 = $ Proportion correctly declared as diff. expressed $= 0.99$											
1000	34	15	9	6	32	15	8	6	30	14	8	5
2000	36	16	9	6	34	15	9	6	32	15	8	6

Table 14.3. Abbreviated sample-size table for treatment-control designs. The number listed in a cell is the sample size n required in each of the treatment and control groups to yield the specified individual power level $1 - \beta_1$, which is the expected proportion of differentially expressed genes that will be correctly declared as such by the tests. The requisite total sample size is $C = 2n$.

in the table is the smallest whole number that will yield the specified power. The total number of experimental conditions C is double the entry in the table, i.e., $C = 2n$. An examination of Table 14.3 shows that the required sample size is most sensitive to the ratio $|\mu_1|/\sigma_d$ and the required power level and least sensitive to the mean number of false positives $E(R_0)$. The required sample size is also moderately sensitive to the number of undifferentially expressed genes G_0 because of the effect of controlling for simultaneous inferences. The practical lesson to be drawn from this last observation is that the gene set $\mathcal{G}_0$ should be kept as small as possible, consistent with the scientific objective of the microarray study. Inclusion of superfluous genes in the analysis, possibly for reasons of data exploration or data mining, will have a cost in terms of power loss. Of course, housekeeping genes and genes included on the arrays as positive controls may be used for diagnostic and quality-control checks but do not enter the main analysis. Such monitoring genes should not be counted in the number G_0 used in power calculations.

A formula for the sample size n in this setting has the following simple form:

$$n = \left(\frac{z_a + z_b}{|\mu_1|/\sigma_d} \right)^2 \tag{14.38}$$

Here z_a and z_b are standard normal percentiles with $a = 1 - \alpha_0/2$ and $b = 1 - \beta_1$. The connection of α_0 to the expected number of false positives is $E(R_0) = G_0\alpha_0$, as noted earlier.

As one example of a reference to Table 14.3, consider a matched-pairs design for which $G_0 = 2000$ genes are not differentially expressed. The investigator wishes to control the mean number of false positives at $E(R_0) = 1.0$ and to detect a two-fold difference between treatment and control conditions with an individual power level of $1 - \beta_1 = 0.90$. Previous studies by the investigator may suggest that the experimental error standard deviation σ in Table 14.2(b) is anticipated to be 0.35 on a log-2 scale. The standard deviation of gene expression differences in matched pairs will therefore be $\sigma_d = \sqrt{2}\sigma = \sqrt{2}(0.35) = 0.495$. The two-fold difference represents a value of $|\mu_1| = \log_2(2) = 1.00$ on a log-2 scale. Thus, the ratio $|\mu_1|/\sigma_d$ equals $1.00/0.495 = 2.02$. Reference to Table 14.3 for these specifications shows that $n = 6$. Thus, 6 pairs of treatment and control conditions are required in the study. The specified individual power level of 0.90 indicates that 90 percent of the differentially expressed genes are expected to be discovered.

To illustrate the use of Table 14.3 for a completely randomized design with equal numbers of treatment and control conditions, suppose the experimental error standard deviation σ in Table 14.2(a) is anticipated to be 0.40. Furthermore, suppose that $|\mu_1| = 1.00$, $E(R_0) = 1.0$, $G_0 = 2000$ and the desired individual power level is $1 - \beta_1 = 0.90$ as specified before. As the standard deviation of the difference in log-expression between treatment and control is given by $\sigma_d = \sqrt{2}\sigma = \sqrt{2}(0.40) = 0.566$, reference is made to the ratio $|\mu_1|/\sigma_d = 1.00/0.566 = 1.768$ in the table. From Table 14.3, the required sample size can be seen to be somewhere between 6 and 11. The table in Appendix A gives a more precise value of $n = 8$.

We can calculate the exact value of n for this last illustration by using formula (14.38). As $E(R_0) = 1$ and $G_0 = 2000$, we have $\alpha_0 = 1/2000 = 0.0005$. Hence, $a = 1 - \alpha_0/2 = 0.99975$, $b = 1 - \beta_1 = 0.90$, $z_a = 3.481$ and $z_b = 1.282$. Substitution of these and the other values in the example into the formula gives:

$$n = \left(\frac{3.481 + 1.282}{1.768}\right)^2 = 7.3 \qquad (14.39)$$

Thus, a sample size of $n = 8$ is required.

Under the Šidák approach, in which estimated differential expression vectors $\hat{\boldsymbol{I}}_g$ are assumed to be mutually independent across genes, the familywise power level $1 - \beta_F$ and expected number of true positives $E(R_1)$ can be calculated from $1 - \beta_1$ using (14.22) and (14.23). Under the Bonferroni approach, in which estimated differential expression vectors $\hat{\boldsymbol{I}}_g$ may be dependent across genes, a lower bound on the familywise power level $1 - \beta_F$ and the expected number of true positives $E(R_1)$ can be calculated from $1 - \beta_1$ using (14.25) and (14.26). Recall from (14.19) that we have $E(R_0) = G_0\alpha_0 = \alpha_F$ in the Bonferroni approach. . Thus, in the Bonferroni case, the expected number of false positives is necessarily smaller than 1 and, hence, cannot be controlled at an arbitrary level. As a consequence, Table 14.3 is entered with reference to a fractional value of $E(R_0)$ (the only fractional value shown in this illustrative table is 0.5).

14.11.3.   Multiple-treatment Designs

In this section, we consider sample size and power calculations for a family of standard designs that involve multiple treatments. The number of treatments is denoted by T. The treatment-control designs considered in the previous section correspond to the case where $T = 2$. The multiple-treatment design involves either a *completely randomized design* or a *randomized block design*. In the completely randomized design, a set of $C = nT$ experimental units are randomly divided into n groups of T units each. The T treatments are then randomly assigned to the units in each group.

For example, consider a time-course experiment studying the effect of an opiate on mouse tissue. The experiment may involve an examination of gene expression at $T = 3$ time points: (1) baseline (0 hours), just prior to administration of an opiate, and (2) 12 hours and (3) 24 hours after administration of the opiate. Twenty-four mice of an identical age, gender, and strain ($C = 24$) are chosen and randomly divided into eight lots of three mice each (so $n = 8$). The three mice in each lot are then selected at random for sacrifice at the time points (0, 12, and 24 hours). A randomized block design is similar but in this design the $C = nT$ experimental units form n homogeneous groups or *blocks*, each of size T. To take advantage of the group homogeneity, the T treatments are assigned at random to the experimental units in each group. To continue the previous illustration, if the 24 mice consist of eight lots of three littermates then the litter is a natural blocking variable. Thus, the time-course treatments (at 0, 12 and 24 hours) would then be assigned randomly to the three littermates of each block.

As we now are considering a general case where there may be more than two treatments, we cannot speak of treatment-control differences but rather must consider more general patterns of differential gene expression among the T treatments. This fact brings us to use the quadratic summary measure considered earlier. Let $\mathbf{\Delta} = (\Delta_t, t = 1, \ldots, T)'$ denote the column vector of investigator-specified interaction parameters describing differential expression levels under the alternative hypothesis H_1. Recall the convention that requires these interaction parameters to sum to zero.

Table 14.4 shows fragments of the ANOVA tables for completely randomized design and randomized block designs for multiple treatments. The fragments show the sources of variance, degrees of freedom, and relevant expected mean squares that are required for later discussion.

For these multiple treatment designs, the noncentrality parameter ψ_1 defined in (14.31) takes the simple form

$$\psi_1 = \frac{n}{\sigma^2} \mathbf{\Delta}'\mathbf{\Delta} = \frac{n}{\sigma^2} \sum_{t=1}^{T} \Delta_t^2. \qquad (14.40)$$

Here σ^2 is the experimental error variance or expected mean square error for the completely randomized design or randomized block design, taken from Table 14.4(a) or Table 14.4(b) as the case may be.

The specification of the alternative differential expression vector $\mathbf{\Delta}$ requires some thought by the investigator. To illustrate the process, we return to the time-course illustration and a randomized block design with groups of littermates forming the blocks. Suppose that a gene of scientific

(a) Completely randomized design		
Source	df	E(MS)
Treatments	$T-1$	$\sigma^2 + (n\sum_{t=1}^{T}\Delta_t^2)/(T-1)$
Error	$(n-1)T$	σ^2
Total	$nT-1$	

(b) Randomized block design		
Source	df	E(MS)
Treatments	$T-1$	$\sigma^2 + (n\sum_{t=1}^{T}\Delta_t^2)/(T-1)$
Blocks	$n-1$	
Error	$(n-1)(T-1)$	σ^2
Total	$nT-1$	

Table 14.4. Fragments of ANOVA tables for completely randomized and randomized block designs for multiple treatments.

interest in this study is anticipated to be down-regulated 1.5-fold at baseline and up-regulated 1.5-fold at 24 hours, relative to the expression level at 12 hours. Using logarithms to base 2, the components of vector Δ will be $\{\log_2(1/1.5), \log_2(1), \log_2(1.5)\} = \{-0.585, 0, 0.585\}$. Observe that the up-regulation and down-regulation have been specified in a convenient form so the interaction parameters sum to zero. If σ for gene expression for litermates receiving the same treatment is anticipated to be 0.40 on a log-2 scale and $n = 4$ blocks are to be used then the noncentrality parameter in (14.40) equals

$$\psi_1 = \frac{4}{0.40^2}\left[(-0.585)^2 + (0)^2 + (0.585)^2\right] = 17.1.$$

Once the components of vector Δ, σ^2 and the type I error probability α_0 are specified then either a sample size n can be calculated for the desired power level $1 - \beta_1$ or the power level for a given sample size can be calculated. Each of these calculations is illustrated shortly.

A useful formulation of the alternative differential expression vector Δ is based on the idea of an *isolated treatment effect*. Many microarray studies anticipate the presence of differential gene expression somewhere among the T treatments, but the investigator does not know in advance where the differential expression will appear. The science underpinning these studies is often at a formative stage so they are essentially exploratory in nature. Therefore, consider a microarray study in which one treatment, which we refer to as the *distinguished treatment*, exhibits differential expression for a gene g relative to all other $T-1$ treatments under study. The latter $T - 1$ conditions are assumed to be uniform in their gene expression. Without loss of generality, we take this distinguished condition as $t = 1$ and assume that the hypothesized difference in expression between condition $t = 1$ and all other conditions is μ_1 on the log-intensity scale. This assumption implies that the interaction parameters have the following values under the alternative hypothesis H_1.

$$\Delta_t = \begin{cases} \mu_1(T-1)/T & \text{for } t = 1 \qquad \text{distinguished treatment} \\ -\mu_1/T & \text{for } t = 2,\ldots,T \quad \text{all other treatments} \end{cases} \tag{14.41}$$

Observe that these parameter values sum to zero as required by the interaction sum constraint. The differential expression in question may be either an up- or down-regulation, depending on the sign of the difference μ_1. The non-centrality parameter (14.40) for this pattern of gene expression has the following form.

$$\psi_1 = \frac{n(T-1)}{T} \left(\frac{|\mu_1|}{\sigma} \right)^2 \tag{14.42}$$

We note in (14.42) that the non-centrality parameter depends strongly on the number of replicates or blocks n and the statistical distance between the log-expression levels for the distinguished treatment condition and all other conditions, as measured by the ratio $|\mu_1|/\sigma$. The effect of the number of treatments T is less pronounced as the ratio $(T-1)/T$ approaches 1 as T increases.

14.11.4.   Power Table for a Multiple-treatment Design

In principle, tables of power levels and sample sizes can be computed for multiple-treatment designs. Recall that Table 14.3 gives sample sizes for the treatment-control design, which corresponds to the multiple-treatment design with $T = 2$ (treatment and control counting as two treatment conditions). Presentation of a full set of sample size tables, however, is not feasible as they must span a range of values for T. We shall therefore focus on power tables alone. A sample size can always be derived by reverse-engineering a power table, as we demonstrate shortly. We also mention that power tables have the added advantage of being applicable to unbalanced designs and a range of other designs. The only

	Mean Number of False Positives								
	$E(R_0) = 0.5$			$E(R_0) = 1.0$			$E(R_0) = 2.0$		
	Non-centrality ψ_1			Non-centrality ψ_1			Non-centrality ψ_1		
	20	30	40	20	30	40	20	30	40
Genes G_0	Number of treatments $T = 5$								
1000	.63	.90	.98	.70	.93	.99	.76	.95	.99
2000	.57	.87	.98	.63	.90	.98	.70	.93	.99
Genes G_0	Number of treatments $T = 10$								
1000	.44	.78	.94	.51	.83	.96	.58	.87	.97
2000	.37	.72	.92	.44	.78	.94	.51	.83	.96

Table 14.5. Abbreviated power table for multiple-treatment designs. The number listed in each cell is the individual power level $1 - \beta_1$. This value is the expected proportion of differentially expressed genes that will be correctly declared as differentially expressed by the tests. Quantity $E(R_0)$ denotes the mean number of false positives, ψ_1 the non-centrality parameter, T the number of treatments and G_0 the anticipated number of undifferentially expressed genes involved in the experiment.

requirement is that the non-centrality parameter ψ_1 in (14.31) must be specified.

If an investigator selects two treatments from among the T treatments in a multiple-treatment design and considers them to be of paramount interest, then the sample size table (Table 14.3) and its counterpart in Appendix A can be used for sample size determination. That sample size can then be extended to all treatments by arguing that interest may lie in studying any arbitrary pair of treatments. This approach is not equivalent to the method described in this section but is a practical alternative that is simple to apply.

Table 14.5 gives an abbreviated set of power levels for a multiple-treatment design. More extensive power tables are provided in Appendix B. The table is entered based on the number of treatments T, number of undifferentially expressed genes G_0, noncentrality parameter ψ_1 and the specified mean number of false positives $E(R_0)$. The noncentrality parameter can be calculated from the general formula (14.31) or one of the special cases in (14.40) or (14.42). If G_0 is expected to be close to the total number of genes G then G can be used in place of G_0 without introducing great error. The cell entry is the individual power level $1 - \beta_1$. From Table 14.5 it can be seen that power is most sensitive to the noncentrality parameter ψ_1 and the number of treatments T but less sensitive to the the mean number of false positives $E(R_0)$ and the number of undifferentially expressed genes G_0.

As one example of a reference to Table 14.5, consider a randomized block design involving $T = 5$ treatments and $G_0 = 2000$ undifferentially expressed genes. Assume that the investigator wishes to control the mean number of false positives at $E(R_0) = 1.0$ and to detect an isolated effect that amounts to a two-fold difference between the distinguished condition and all others. The experimental error standard deviation is anticipated to be about $\sigma = 0.40$ on a log-2 scale for the randomized block design. The two-fold difference represents a value of $|\mu_1| = \log_2(2) = 1.00$ on a log-2 scale. Thus, the ratio $|\mu_1|/\sigma$ equals $1.00/0.40 = 2.5$. Six blocks are to be used ($n = 6$). For these specifications, the non-centrality parameter (14.42) equals

$$\psi_1 = \frac{n(T-1)}{T}\left(\frac{|\mu_1|}{\sigma}\right)^2 = \frac{6(5-1)}{5}(2.5)^2 = 30.0$$

Reference to the cell corresponding to $E(R_0) = 1$, $\psi_1 = 30$, $T = 5$ and $G_0 = 2000$, shows an individual power level of $1 - \beta_1 = .90$. Thus, 90 percent of differentially expressed genes are expected to be discovered with this study design. The table can be used iteratively to explore the effect on power of specific design changes. For example, if $n = 7$ blocks were to be used in lieu of $n = 6$, then recalculation of the non-centrality parameter gives $\psi_1 = 35$ and the individual power level is seen to rise to between .90 and .98. Reference to the more extensive tables in Appendix B gives an interpolated power level for this case of .96.

To illustrate how a power table such as Table 14.5 can be used to obtain a sample size, the preceding illustration provides an immediate example. Suppose that the investigator wants to know the number of blocks for this design that will ensure a power level of 0.98. Reference to the cell containing power level .98 for $E(R_0) = 1$, $T = 5$ and $G_0 = 2000$, shows that the noncentrality parameter must be $\psi_1 = 40$. From the isolated-effect formula (14.42), we have

$$\psi_1 = \frac{n(T-1)}{T}\left(\frac{|\mu_1|}{\sigma}\right)^2 = \frac{n(5-1)}{5}(2.5)^2 = 40$$

Solving for n gives $n = 8$, i.e., eight blocks are required in the randomized block design where an isolated effect is anticipated.

Under the Šidák approach, in which estimated differential expression vectors $\hat{\mathcal{I}}_g$ are assumed to be mutually independent across genes, the familywise power level $1 - \beta_F$ and expected number of true positives $E(R_1)$ can be calculated from $1 - \beta_1$ using (14.22) and (14.23). Under the Bonferroni approach, in which estimated differential expression vectors $\hat{\mathcal{I}}_g$ may be dependent across genes, a lower bound on the familywise power level $1 - \beta_F$ and the expected number of true positives $E(R_1)$ can be calculated from $1 - \beta_1$ using (14.25) and (14.26). Recall from (14.19) that we have $E(R_0) = G_0 \alpha_0 = \alpha_F$ in the Bonferroni approach. Thus, in the Bonferroni case, the expected number of false positives is necessarily smaller than 1 and, hence, cannot be controlled at an arbitrary level. As a consequence, Table 14.5 is entered with reference to a fractional value of $E(R_0)$ (the only fractional value shown in this illustrative table is 0.5).

14.11.5. Time-course and Similar Multiple-treatment Designs

A special case of a multiple-treatment design is the time-course experiment that was discussed in Chapter 9. A typical design was illustrated in Table 9.11 and a more elaborate one in Table 9.14. The design in Table 9.11 can be seen to involve six ratios of time-point expressions, using time zero as the common reference (the common reference being in the opposite dye). Viewed this way, the design involves $T = 6$ treatments. The methods described earlier in this section can be used to choose the number of replications n of this design that are required to attain a given power. The design in Table 9.14 involves a two-way structure consisting of two treatments (A, B) and eight time points t0,…,t24, including time zero, for a total of $T = 16$ treatment-time combinations. Again, the methods described earlier in this section can be used to determine the power associated with $n = 2$ replicates of this two-way structure as called for in the design of Table 9.14. The case example on opioid dependence that appears later in section 14.13.2 illustrates the numerical calculation of power for such a two-way structure. The time-course in this case example concerns morphine time dependence versus placebo in mice.

The time-course structure also arises in other microarray experimental contexts. An important example is a *dilution-series experiment*, in which interest lies in gene expression changes over successive dilutions of a biological sample. For example, the design in Table 9.11 might be copied for a dilution series with t_k denoting the kth dilution of a biological sample and t_0 being the undiluted sample. The calculational methods for sample size and power with multiple treatments would apply here exactly as for the time-course experiment.

Time-course experiments are prototypical contexts for using the non-centrality formulations in both (14.40) and (14.42). For example, where a trend in gene expression is expected under the research hypothesis H_1, the Δ_t in (14.40) can be given a linear form $\Delta_t = b_0 + b_1\tau_t$ where τ_t denotes the time value of treatment level t and b_0 and b_1 are the intercept and slope of the time trend. The slope and intercept would be chosen so the interaction sum constraint $\sum_t \Delta_t = 0$ is satisfied. Similarly, where the investigator is uncertain if and where differential expression may arise during the time course, the formula (14.42) for the isolated treatment effect may be appropriate. The isolated effect in this instance may take the form of a *change point* in the time course of gene expression.

14.12.   Relation Between Power, Replication and Design

With given specifications for the familywise type I error probability α_F and the vector $\boldsymbol{\Delta}$ in the alternative hypothesis, the power is determined by the properties of the distribution of $\hat{\boldsymbol{\mathcal{I}}}_g$ and, in particular, its covariance matrix $\boldsymbol{\Sigma}$. In section 14.9, it was shown how the covariance matrix affects the variance of the null p.d.f. $f_0(v)$ in (14.28) and the non-centrality parameter in (14.31). Both the sample size and experimental design influence this covariance matrix.

14.12.1.   Effects of Replication

If a given microarray design is repeated so that there are r independent repetitions of the design then $\boldsymbol{\Sigma}$ for a single replicate is reduced by a multiple of $1/r$. Thus, σ_0^2 in (14.28), for example, is reduced by the factor $1/r$ and the non-centrality parameter ψ_1 in (14.31) is multiplied by r. These reductions are illustrated in the previous standard designs we used to demonstrate the power and sample size methodology. For instance, referring to (14.37) for the matched-pairs design, we can see that σ_d^2 is the variance of a single matched pair and n plays the role of the number of replicates (the number of matched pairs in this instance). Note that σ_d^2 is reduced by the multiplier $1/n$ with n replications. Similarly, with the multiple-treatment design, the number of replications n appears as a multiplier in the formula for the non-centrality parameter in (14.40).

The replication discussed here refers to the simple repetition of a basic experiment and, hence, we are considering a pure statistical effect that is captured by the number of repetitions r. This parameter does not reveal if the replicated design is a sound one in terms of the scientific question of interest.

We do want to point out, however, that the nature of replication is an important issue in terms of the overall study plan. Contrast the following two situations. First, imagine an experiment in which a single small tissue fragment is cut from a tumor core and then used to prepare six arrays. Next, imagine an alternative experiment in which six small tissue fragments are cut from six different regions of the same tumor in a spatially randomized fashion and then used to prepare six arrays, one array being prepared from each of the tissue fragments. Both imaginary designs yield six arrays of data. The first design allows inferences to be made only about the single core tissue fragment based on a sample of size six. The ANOVA model for this design describes the population of all arrays that could be constructed from this single tissue fragment. The second design allows inferences about the whole tumor based on a sample of size six. The ANOVA model for this design describes the population of all arrays that could be prepared from the whole tumor. The respective sets of inferences clearly relate to different biological populations (the single tumor core fragment and the whole tumor, respectively). Both designs involve six replications ($r = 6$) but the replications have different elemental designs. The variance structures of the two designs will differ, of course, making them essentially incommensurate. The method of calculating power, however, based on the two designs follows the logic we explained earlier. The choice of design here depends on which population is of scientific interest to the investigator. Is it the single core tissue fragment or the whole tumor?

14.12.2. Controlling Sources of Variability

The choice of experimental design, as opposed to simple replication, has a more complicated influence on power. A good design will be one that takes account of important sources of variability in the microarray study and reduces the experimental error variance of the expression data. As discussed in Chapter 9, various experimental designs for microarray studies have been proposed that aim to be more efficient. Experience with different designs gives some indication of the correlation structure and the magnitudes of variance parameters that can be expected in co-variance matrix Σ. These expectations, in turn, can be used to compute the anticipated power.

To give a concrete illustration, suppose that an ANOVA model is modified by adding a main effect for the subarray in which each spot is located, the aim being to account for regional variability on the surface of the slide for the microarray. Furthermore, suppose that incorporation of this main effect would reduce the error variance by 15%, other factors remaining unchanged. Then, the covariance levels in Σ are reduced by a multiple of 0.85. The direct effect of this refinement on the numerical example in section 14.9.4, for instance, is to increase the non-centrality parameter ψ_1 by a factor of $1/0.85 = 1.1765$ from 20 to 23.53. This change increases individual power level $1 - \beta_1$ from 0.689 to 0.806, a worthwhile improvement.

14.13.   Assessing Power from Microarray Pilot Studies

The purpose of a power calculation is to assess the ability of a proposed study design to uncover a differential expression pattern having the specification Δ under the alternative hypothesis H_1. Thus, our methodology should find its main application at the *planning stage* of microarray studies. As part of this planning process, investigators sometimes wish to calculate the power of a pilot study in order to decide how the pilot study should be expanded to a full study or to decide on the appropriate scale for a new and related study. Power calculation for a pilot study involves an application of the same methodology but with the benefit of having estimates of relevant parameters needed for the calculation from the pilot study data. For instance, power calculations need estimates of inherent variability. The pilot study data can provide those estimates. As illustrations of power calculations from pilot studies and as demonstrations of real applications of our methodology, we now consider two microarray case studies involving mice.

14.13.1.   Example 1: Juvenile Cystic Kidney Disease

The Mouse Juvenile Cystic Kidney Data Set was introduced in section 6.2.2. Table 14.6 again presents the experimental design for this data set for ready reference. The table shows how the tissue types (mutant or wild type) were assigned to the four arrays and two color channels of each array. A total of $G = 1728$ genes were under investigation.

The scientists in this study were interested in differential gene expression for the two tissue types, mutant (type 1) and wild type (type 2). Thus, the difference $\hat{\mathcal{I}}_{g1} - \hat{\mathcal{I}}_{g2}$ was the summary measure of interest for gene g. The alternative hypothesis for which power was to be calculated was $H_1 : \mu_1 = 1.00$. This specification corresponds to a 2.72-fold difference between mutant and wild type tissues on the natural log scale. The study data gave an estimate of 0.2315 for σ_0, the standard deviation of the summary measure on the same log-scale. The expected number of false positives was to be controlled at $E(R_0) = 2$. We let the total gene count G stand in for G_0. Using the methodology presented in section 14.9.1, the individual power level for the study was calculated to be

	Channel 1 Cy3 (Green)	Channel 2 Cy5 (Red)
Array 1	mutant	mutant
Array 2	mutant	wild type
Array 3	wild type	mutant
Array 4	wild type	wild type

Table 14.6.   Experimental design for the Mouse Juvenile Cystic Kidney Data Set.

$1 - \beta_1 = 0.858$, which suggests that 86 percent of differentially expressed genes are expected to be declared as such.

14.13.2.   Example 2: Opioid Dependence

Lee *et al.*[8] (2003) describe an experiment designed to investigate how morphine dependence in mice alters gene expression. The study involved two treatments (morphine, placebo) and four time points corresponding to consecutive states of opioid dependence, classified as *tolerance, withdrawal, early abstinence,* and *late abstinence.* In the experiment, mice received either morphine (treatment) or placebo (control). Treatment mice were sacrificed at four time points corresponding to the tolerance, withdrawal, early abstinence, and late abstinence states. Control mice were sacrificed at the same time points, with the exception of the withdrawal state, which was omitted on the assumption that the tolerance and withdrawal states are identical with placebo. The microarray data resulted from hybridization of mouse spinal cord samples to a custom-designed array of 1728 cDNA sequences. At each time point (i.e., in each dependence state), for both the treatment and control groups, three mice were sacrificed, for a total of 21 mice. The paucity of spinal column mRNA in any single mouse required that the mRNA of the three mice sacrificed together be combined and blended into a single sample. The treatment and control samples were labeled with red dye. Other control samples, derived from mouse brain tissue, were labeled with green dye. The green readings were not used in the analysis reported here.

The experimental design is shown in Table 14.7. The natural logarithm of the raw red intensity reading, without background correction, was used as the response variable. As noted above, no array was created for the placebo-withdrawal combination (marked '*' in Table 14.7). The original intention was to place the spinal column sample on two spots of

Treatment	Dependence State (Time Point)			
	1. Tolerance	2. Withdrawal	3. Early Abstin.	4. Late Abstin.
1. Placebo	array 1	*	array 2	array 3
2. Morphine	array 4	array 5	array 6	array 7

Table 14.7. Experimental design for a study of opioid dependence in mice. An asterisk * denotes an omitted array.

the same slide, yielding a replicated expression reading. This attempt was not entirely successful. The replicate was missed in the morphine-tolerance combination because of administrative error. Also, the array for the morphine-late abstinence combination had a large number of defective spots. Finally, several dozen spots in other arrays were faulty. The final microarray data set contains readings for only $G = 1722$ genes out of the original set of 1728. Six genes were dropped because they had defective readings for the morphine-tolerance combination, the unreplicated treatment combination in the design in Table 14.7. The fact that the replicated spots were nested within the same array was not taken into account in the analysis.

The aim of the study was to identify genes that characterize the tolerance, withdrawal, and two abstinence states and to describe how gene expression is altered as a mouse moves from one state to the next. As this aim is somewhat broad, it was decided to evaluate power on the assumption that a differential expression would appear in only one treatment combination, with all other combinations having a uniform expression level. This assumption is exactly what characterizes the isolated effect formulation of the multiple-treatment design. Hence, we shall use the quadratic summary measure and noncentrality parameter (14.42) for the power calculation, recognizing that this power value will slightly overstate the power achieved in this actual study because of the failure to replicate one of the seven treatment combinations and the nesting of duplicate spots within the same arrays.

The alternative hypothesis H_1, for which power was to be calculated, had the specified differential expression pattern given in (14.41) with $|\mu_1| = 0.693$, which corresponds to a two-fold differential expression on the natural log-scale. Thus, the specification under H_1 calls for a single treatment combination to exhibit a two-fold up- or down-regulation relative to all other treatment combinations. We assume there are $n = 2$ replicates for each of the $T = 7$ treatment combinations. The study data gave an estimate of 0.1513 for σ, the standard deviation of the ANOVA error variance. The expected number of false positives was controlled at $E(R_0) = 1$. The non-centrality parameter, calculated from (14.42), was $\psi_1 = 36.00$. We let the total gene count G stand in for G_0. Now, using the methodology presented in section 14.9.3, the individual power level was calculated to be $1 - \beta_1 = 0.944$, which implies that 94 percent of differentially expressed genes are expected to be declared as such. The same power value can be found by rough interpolation in Table 14.5 and with further refinement in the power table in Appendix B.

Acknowledgement: Several sections in this chapter appeared in *Statistics in Medicine*, volume 21, pages 3543-3570, Copyright 2002, John Wiley & Sons Limited. The contents are reproduced by permission of John Wiley & Sons Limited.

Notes

[1] Simon, R., Radmacher, M.D., Dobbin, K. (2002). *Genetic Epidemiology*, **23**, 21-36.

[2] Lee, M.-L.T., Whitmore, G.A. (2002c). *Statistics in Medicine*, **21**, 3543-3570.

[3] Benjamini, Y., Hochberg, Y. (1995). *Journal of Royal Statistical Society*, **B 57**, 289-300.

[4] Hochberg, Y. and Tamhane A.C. (1987). *Multiple Comparison Procedures*. John Wiley and Sons, New York.

[5] Dudoit, S., Yang, Y.H., Callow, M.J., Speed, T.P. (2002). *Statistica Sinica*, **12**, 111-139.

[6] Lee, M.-L.T., Kuo, F.C., Whitmore, G.A., Sklar, J. (2000). *Proceedings of the National Academy of Sciences*, **97**, 9834-9839.

[7] Efron, B., Tibshirani, R., Storey, J.D., Tusher, V. (2001). *Journal of American Statistical Association*, **96**, 1151-1160.

[8] Lee, M.-L.T., Whitmore, G.A., Yukhananov, R.Y. (2003). *Journal of Data Science*, **1**, 103-121.

III

UNSUPERVISED EXPLORATORY ANALYSIS

Chapter 15

CLUSTER ANALYSIS

Statistical learning methods can be divided into two general classes, namely, supervised and unsupervised learning. In supervised learning, objects in the training set are classified with respect to known reference labels. In unsupervised learning, no predefined reference labels are used and classifications are made independent of prior knowledge. In this chapter we discuss unsupervised clustering methods.

Unsupervised cluster analysis is concerned with grouping previously unclassified objects. That is, the composition of the classes is unknown at the beginning of the investigation. Cluster analysis has been used in different disciplines, including anthropology, archaeology, behavior sciences, biology, market research, social services, medicine and psychiatry. For a review of cluster analysis methods, see Jain and Dubes[1] (1988), Kaufman and Rousseeuw[2] (1990), Afifi and Clark[3] (1990) Jobson[4] (1992), Everitt[5] (1993), Johnson and Wichern[6](1998), and Hastie, Tibshirani, and Friedman[7] (2001), among others.

Cluster analysis methods are familiar to biologists through their application in phylogenetic and sequence analysis. Relationships among objects (genes) are represented by a tree whose branch lengths reflect the degree of similarity between the objects. In sequence comparison, cluster analysis methods are used to infer the evolutionary history of sequences being compared. By computing the degrees of similarity among groups of closely related genes, cluster analysis can be used to identify features in gene expression patterns that appear to reflect molecular signatures of the tissue from which the cells originated. Unsupervised methods are considered as useful approaches where there is little prior knowledge of the expected gene expression patterns for any condition.

Depending upon the objective of the research study, interest may focus on either finding clusters of genes having similar expression patterns across specimen samples or finding clusters of specimen samples sharing similar expression patterns across the gene set.

In order to cluster genes, each gene can be represented by a row vector across N specimen samples (cell lines or experiment conditions). Specifically, let the row vector

$$\boldsymbol{x}^{(g)} = (x_{g1}, \ldots, x_{gN})$$

represent expression levels for gene g measured across specimens such that x_{gn} denotes the intensity level measured from specimen sample n, for $n = 1, \ldots, N$. The vector $\boldsymbol{x}^{(g)}$ might be described as a point in an N-dimensional *biological specimen space*. Clusters of genes in this space might be viewed as *co-expressed*.

On the other hand, to cluster specimen samples, such as in cancer classification, each specimen can be represented by a G-dimensional column vector across the genes. For each specimen sample n, the column vector of expression levels is of the form

$$\boldsymbol{x}_n = (x_{1n}, \ldots, x_{gn}, \ldots, x_{Gn})',$$

where x_{gn} denotes the intensity of gene g measured from specimen n and the prime denotes a vector or matrix transpose. The vector $\boldsymbol{x}_n$ might be described as a point in a G-dimensional *gene space*. The clusters in this space may be identifiable subclasses of tumors, for instance.

15.1. Distance and Similarity Measures

Cluster analysis is based on measures of *distance* between the objects being clustered, whether these be genes or specimens. The idea is that objects that share many features will lie close to each other and, hence, may belong to the same group or cluster. In contrast, objects that are widely separated in their features are unlikely to belong to the same group. Selecting the correct measurement of distance for a feature space is one of the challenges of cluster analysis. Many possible distance measures can be defined, as will be demonstrated shortly.

Cluster analysis may also be based on measures of *similarity*. A similarity measure indicates the strength of the relationship between two objects and, in some sense, is the inverse of distance. Highly similar objects are close to each other while dissimilar objects lie a distance from each other; thus, larger distances correspond to smaller similarity.

As this last sentence demonstrates, reference is also made to measures of *dissimilarity* and these equate in some rough sense to measures of distance. An extensive list of dissimilarity measures is given in Gower[8] (1985). Clustering algorithms usually offer a selection of similarity measures to conduct the iterative selection procedures.

Distances between objects (genes, specimens) may be summarized by a *distance matrix* $\mathbf{D} = (d_{ij})$, where d_{ij} is the distance between objects i and j. Correspondingly, a *similarity matrix* $\mathbf{S} = (s_{ij})$ may be defined to summarize similarities between all pairs of objects i and j. Matrices $\mathbf{D}$ and $\mathbf{S}$ are also called *proximity matrices*. It is always possible to construct similarities from distances. For example, if $d_{g_1 g_2}$ denotes the distance between two genes g_1 and g_2 then the similarity of g_1 and g_2 might be defined by the measure

$$s_{g_1 g_2} = \frac{1}{1 + d_{g_1 g_2}}. \tag{15.1}$$

With appropriate care for technical considerations, one can work readily with either kind of proximity matrix. As an example of a technical consideration, Gower[9] (1967) has shown that one can only construct a proper distance measure from a similarity measure if the similarity matrix $\mathbf{S}$ is nonnegative definite.

15.2.   Distance Measures

A number of distance measures are available for comparing proximities of expression vectors in gene or specimen space. The discussion of these measures will be in terms of distances between genes, but they apply with a corresponding change in notation to distances between specimens. The formal properties of distance measures are considered first before describing particular measures.

15.2.1.   Properties of Distance Measures

A distance measure is usually required to satisfy a set of logical properties or conditions so that it will produce sensible conclusions. These properties are the following. Here, P, Q, and R are any three points in the multidimensional feature space of interest.
(1) $d(P,Q)=d(Q,P)$,
(2) $d(P,Q) > 0$ if $P \neq Q$ and $d(P,Q) = 0$ if $P = Q$,
(3) $d(P,Q) \leq d(P,R) + d(R,Q)$, for any other point R

The last property is known as the *triangle inequality.*

Although it is usually advisable to use distance measures that satisfy the preceding set of properties, many clustering algorithms will accept subjectively assigned measures that may not satisfy all of the distance properties (for example, the triangle inequality may be violated).

15.2.2.   Minkowski Distance Measures

Assume that genes g_1 and g_2 are represented by their expression vectors $\boldsymbol{x}^{(g_1)} = (x_{g_1 1}, \ldots, x_{g_1 N})$ and $\boldsymbol{x}^{(g_2)} = (x_{g_2 1}, \ldots, x_{g_2 N})$ across N specimens. A common family of distance measures has the following general form and is known as the *Minkowski family.*

$$d(\boldsymbol{x}^{(g_1)}, \boldsymbol{x}^{(g_2)}) = \left(\sum_{n=1}^{N} |x_{g_1 n} - x_{g_2 n}|^k \right)^{1/k} \qquad \text{for some integer } k$$

The following are three special cases of Minkowski measures that find application in clustering.

Euclidean distance

$$d(\boldsymbol{x}^{(g_1)}, \boldsymbol{x}^{(g_2)}) = \left(\sum_{n=1}^{N} (x_{g_1 n} - x_{g_2 n})^2 \right)^{1/2}$$

City-block distance

$$d(\boldsymbol{x}^{(g_1)}, \boldsymbol{x}^{(g_2)}) = \sum_{n=1}^{N} |x_{g_1 n} - x_{g_2 n}|$$

Maximum distance

$$d(\boldsymbol{x}^{(g_1)}, \boldsymbol{x}^{(g_2)}) = \max_{n} |x_{g_1 n} - x_{g_2 n}|$$

The maximum distance measure is the limiting form of the Minkowski family that is produced as parameter k tends to ∞.

Euclidean distance is scale dependent and, hence, distance rankings will not be preserved after a scale change. Therefore, variables x_{gn} are often standardized before calculating Euclidean distance. Standardizing each variable separately ignores possible correlations between the variables.

Dunn and Everitt[10] (1982) used the city block distance measure as the natural distance measure in comparing amino acids in homologous proteins.

15.2.3. Mahalanobis Distance

The Minkowski distance measures implicitly assume that the variables are uncorrelated with one another. In many applications this assumption will not be justified. Moreover, when the coordinates of points represent measurements that are subject to random fluctuations of differing magnitudes, or when the components of points may be dependent, it is often desirable to weight coordinates subject to a great deal of variability less heavily than those that are not highly variable.

The *Mahalanobis distance measure* for genes g_1 and g_2 is a distance measure that takes into account the correlation between vectors. It is defined as a weighted distance in which the weighting matrix[11] is the inverse of the covariance matrix of vectors $x^{(g_1)}$ and $x^{(g_2)}$, denoted here by Σ.

$$d(x^{(g_1)}, x^{(g_2)}) = \sqrt{(x^{(g_1)} - x^{(g_2)})\Sigma^{-1}(x^{(g_1)} - x^{(g_2)})'} \qquad (15.2)$$

Notice that Mahalanobis distance reduces to Euclidean distance if the covariance matrix is replaced by an identity matrix.

In practice, the matrix Σ is often taken to be the pooled within-groups covariance matrix. The observed overall covariance matrix is substituted when Σ is unknown.

15.3. Similarity Measures

A number of similarity measures are available for comparing similarities in expression vectors between genes or between specimens.

15.3.1. Inner Product

The *inner product*, also called the *dot product*, is a measure of similarity. For example, let $x^{(g_1)} = (x_{g_1 1}, \ldots, x_{g_1 N})$ and $x^{(g_2)} = (x_{g_2 1}, \ldots, x_{g_2 N})$ denote two gene vectors across specimens, then the inner product distance between $x^{(g_1)}$ and $x^{(g_2)}$ is defined by

$$s_{g_1 g_2} = \langle x^{(g_1)} \cdot x^{(g_2)} \rangle = x^{(g_1)} \, x^{(g_2)'} = \sum_{n=1}^{N} x_{g_1 n} \, x_{g_2 n}$$

15.3.2.   Pearson Correlation Coefficient

If the variables x_{g_1n} and x_{g_2n} are standardized, i.e., centered on their means and scaled by their standard deviations, then the inner product corresponds to the Pearson correlation coefficient, which is a similarity measure.

$$s_{g_1g_2} = \rho_{g_1g_2} = \sum_{n=1}^{N} z_{g_1n} \, z_{g_2n} \qquad (15.3)$$

Here z_{g_1n} and z_{g_2n} are the standardized forms of variables x_{g_1n} and x_{g_2n}, respectively.

Eisen *et al.*[12] (1998) pointed out that a good choice of a similarity measure for comparing microarray expressions from two genes should conform well to the intuitive biological notion of what it means for two genes to be "coexpressed". They consider the Pearson correlation coefficient as a similarity measure. Alon *et al.*[13] (1999) also note that the intensity of each gene across the tissues can be thought of as a pattern that can be correlated with expression patterns of other genes.

The correlation coefficient captures similarity in the "shapes" of expression profiles between two genes but places no emphasis on the magnitude of the expression levels. The correlation coefficient can be high between two genes that are affected by the same process, even if each has a different magnitude of response to the process. Positive correlation between two highly expressed genes is much more significant than the same value between two poorly expressed genes. By using correlations, one ignores this dependence of the reliability on the absolute expression level.

The Pearson correlation coefficient, like some other proximity measures, is sensitive to outliers. Hence, care must be taken to investigate outliers before using a Pearson correlation coefficient as a similarity measure. Moreover, the correlation coefficient has a range $-1 \leq \rho_{g_1g_2} \leq 1$, and negative values must be handled with care. If the coefficients are calculated from samples and, hence, subject to sampling variability, then small negative coefficients might be set to zero. Alternatively, negative correlation coefficients may be replaced by their absolute values, as is done optionally in some clustering algorithms. The implicit logic of reversing the sign is that if two variables x_{g_1n} and x_{g_2n} are negatively correlated then x_{g_1n} and $-x_{g_2n}$ will be positively correlated with the same absolute coefficient. The two variables (with one having reversed sign) are therefore more similar the larger their absolute correlation coefficient.

The investigator should be aware when an algorithm reverses the sign of a negative correlation coefficient because the fact that two genes have opposite expression profiles across specimens may be an important feature that should not be lost in a default calculation. For example, two genes may be highly (negatively) correlated across specimens because one is strongly upregulated in tumor tissue while the other gene is strongly upregulated in normal tissue.

Given a non-negative correlation coefficient $\rho_{g_1 g_2}$, a corresponding distance (dissimilarity) measure can be defined by

$$d_{g_1 g_2} = 1 - \rho_{g_1 g_2}.$$

This distance measure is symmetric and nonnegative and has the property that, if the expression vectors $x^{(g_1)}$ and $x^{(g_2)}$ of genes g_1 and g_2 are linearly independent (uncorrelated), then $\rho_{g_1 g_2} = 0$ and $d_{g_1 g_2} = 1$. If the expression vectors are linearly dependent (perfectly correlated), then $\rho_{g_1 g_2} = 1$ and $d_{g_1 g_2} = 0$.

15.3.3.   Spearman Rank Correlation Coefficient

Where expression levels are first converted to ranks, a rank correlation coefficient such as the Spearman rank correlation coefficient

$$s_{g_1 g_2} = \rho^*_{g_1 g_2}$$

may be appropriate as a similarity measure. The formula for $\rho^*_{g_1 g_2}$ is exactly the same as (15.3) except the standardized variables $z_{g_1 n}$ and $z_{g_2 n}$ are derived from the ranks $r_{g_1 n}$ and $r_{g_2 n}$ of variables $x_{g_1 n}$ and $x_{g_2 n}$. The ranks are computed across specimens for each gene as follows.

$$r_{g_1 n} = \text{Rank}_{1 \leq n \leq N}(x_{g_1 n}), \quad r_{g_2 n} = \text{Rank}_{1 \leq n \leq N}(x_{g_2 n})$$

15.4.   Inter-cluster Distance

As cluster analysis creates either clusters of genes or clusters of specimens, the choice of a clustering solution requires the notion of the proximity of clusters. Clusters that are 'close' in some appropriate sense may be candidates for being merged into a single larger cluster. Clusters that are well-separated may be judged as truly distinct groupings of objects. As clusters are generally collections of objects (genes or specimens) as opposed to singletons, the definition of distance or similarity must be appropriate for sets having more than one object.

The following sections describe a number of inter-cluster distance measures of importance in clustering genes or samples in microarray studies.

15.4.1.   Mahalanobis Inter-cluster Distance

This measure considers the separation of the mean coordinates or *centroids* of a pair of clusters. Specifically, consider two clusters of genes denoted by A and B. Assume that individual vectors in gene cluster A have mean $\bar{\boldsymbol{x}}_A = (\bar{x}_{A1}, \dots, \bar{x}_{AN})$ and that individual vectors in cluster B have mean $\bar{\boldsymbol{x}}_B = (\bar{x}_{B1}, \dots, \bar{x}_{BN})$, where $\bar{x}_{An}$ denotes the average of expressions x_{gn} for all $g \in A$ for the nth specimen and similarly for $\bar{\boldsymbol{x}}_{Bn}$. The Mahalanobis distance between two clusters of objects A and B is defined as

$$d(A, B) = \sqrt{(\bar{\boldsymbol{x}}_A - \bar{\boldsymbol{x}}_B)\boldsymbol{\Sigma}^{-1}(\bar{\boldsymbol{x}}_A - \bar{\boldsymbol{x}}_B)'}, \qquad (15.4)$$

When the covariance matrix in (15.4) is replaced by an identity matrix, the inter-cluster distance reduces to the Euclidean distance between the centroids of the clusters.

15.4.2.   Neighbor-based Inter-cluster Distance

Several inter-cluster distance measures for pairs of clusters are defined in terms of the pair-wise proximity of their members. The *nearest-neighbor distance* between two clusters is calculated as the distance between their closest members, one from each cluster. The *furthest-neighbor distance* between two clusters is calculated as the distance between the most remote pair of members, one from each cluster. The *cluster average distance* is the average of all the distances between pairs of members from different clusters. The average referred to here is usually the arithmetic mean but a more robust alternative is the use of median distances between pairs of points in different clusters as the inter-cluster distance measure.

15.5.   Hierarchical Clustering

In *hierarchical clustering*, data points are forced into a strict hierarachy of nested subsets. Hierarchical clustering techniques can be either agglomerative (bottom-up) or divisive (top-down). An *agglomerative clustering method* begins by considering each individual point as a cluster by itself. The procedure begins with G clusters (in the case of clustering genes) and successively combines the two closest clusters and thereby reduces the number of clusters by one in each step. In contrast, a *divisive clustering method* begins with one cluster containing all G genes and successively splits the least homogeneous cluster into two successor clusters that are each more uniform than the parent cluster. The splitting can continue until G singleton clusters (individual genes) are formed. In order to have a solution with an 'optimal' number of clusters, the investigator will need to decide on a particular stage at which to stop the iterative procedure.

Listed below are some commonly used linkage methods which allow us to specify the type of joining algorithm used to amalgamate clusters. Results of hierarchical methods can be shown in a tree diagram, known as a *dendrogram*. It must be borne in mind that using different linkage methods or encountering small changes in the data set can lead to very different dendrograms.

15.5.1.   Single Linkage Method

The *single linkage clustering* method uses the nearest-neighbor distance measure for inter-cluster distances. The clustering proceeds hierarchically, each level being obtained by the merger of two clusters from the previous level. When the two clusters have a small number of points lying in close proximity, referred to as noise points by Wishart[14] (1969), the single linkage method has a tendency to join these two clusters together. This produces a *chaining effect*. Johnson[15] (1967) notes that single linkage clustering is invariant under monotonic transformations of elements of the proximity matrix. In other words, the clusters formed by the single linkage method will remain unchanged by any assignment of distances (similarities) that gives the same ordering as the initial distances (similarities).

15.5.2.   Complete Linkage Method

The *complete linkage clustering* method uses furthest-neighbor distance for inter-cluster distances. Complete linkage clustering, like single linkage clustering, is invariant under monotonic transformations of proximities.

15.5.3.   Average Linkage Clustering

The *average linkage clustering* method uses the (arithmetic) average distance measure for inter-cluster distances. For average linkage clustering, changes in the assignment of distances (similarities) that preserve the ordering can affect the final configuration of clusters. Eisen *et al.* (1998) apply this hierarchical clustering method to the analysis of gene expression data.

15.5.4. Centroid Linkage Method

The *centroid linkage method* uses the average value of all points in a cluster (i.e., the cluster centroid) as the reference point for distances to other points or clusters. The distance between two clusters is defined as the Euclidean distance between the centroids of the cluster pair. The process proceeds by combining clusters according to the distance between their centroids, the clusters with the shortest distance being combined first. A disadvantage of the centroid method is that if the sizes of the two clusters to be considered are very different, then the centroid of the new cluster will be very close to that of the larger cluster.

15.5.5. Median Linkage Clustering

The *median linkage method* uses the median distances between pairs of points in different clusters as the inter-cluster distance measure. See Gower (1967) for a discussion of median clustering methods.

15.5.6. Ward's Clustering Method

Ward[16] (1963) proposed a clustering procedure that minimizes the *information loss* associated with clustering. Ward used an error sum-of-squares criterion to define information loss. At each step, union of every possible pair of clusters is considered and the two clusters whose fusion results in the smallest increase in 'information loss' are combined.

15.5.7. Applications

For microarray studies involving G genes from N specimens, one is interested in finding meaningful clusters among the G genes or N specimens. Examples of applications of hierarchical clustering in gene expression studies may be found in Wen *et al.*[17] (1998), Eisen *et al.* (1998), and Spellman *et al.*[18] (1998), among others. Also, Chu *et al.*[19] (1998) employed hierarchical clustering algorithms to organize genes into a phylogenetic tree, reflecting similarity in expression patterns.

15.5.8.   Comparisons of Clustering Algorithms

Milligan[20] (1980) examined the effect of six types of error perturbation on fifteen clustering algorithms and concluded that no single method could be claimed superior for all types of data. How to decide on the appropriate number of clusters for the data is an important question. Milligan and Cooper[21] (1985) reported an examination of procedures for determining the number of clusters in a data set. Kaufman and Rousseeuw[22] (1990) commented that hierarchical clustering suffers from the defect that it can never repair what was done in previous steps. Morgan and Ray[23] (1995) showed that hierarchical clustering suffers from lack of robustness, nonuniqueness, and inversion problems that complicate interpretation of the hierarchy.

15.6.   *K*-means Clustering

MacQueen[24] (1967) introduced a non-hierarchical clustering technique called the *K*-means method. This method assigns each object to the cluster having the nearest centroid. In applying the *K-means clustering method*, the total number of clusters, K, is specified in advance of applying the clustering procedure. Because a proximity matrix does not have to be built and the basic data do not have to be stored during the computer run, the *K*-means method can be applied to much larger data sets than hierarchical techniques. The basic steps in this clustering method are:

Step 1: Select a set of K points as *cluster seeds*. These seeds represent a first guess at the centroids of the K clusters.

Step 2: Assign each individual observation to the cluster whose centroid is nearest. The Euclidean distance is usually used as the distance measure with either standardized or unstandardized observations. The centroids are recalculated for the cluster receiving the new object and for the cluster losing the object.

Step 3: Repeat step 2 until no further changes occur in the cluster compositions.

K-means clustering does not give an ordering of objects within a cluster. The final assignment of clusters will be somewhat dependent on the initial selection of seed points. As the number of clusters K is changed, the cluster memberships can also change in arbitrary ways. For example, the solution for $K = 4$ clusters, may not be nested within the $K = 3$ cluster solution.

K-means clustering is an unstructured approach, which proceeds in a local fashion and produces an unorganized collection of clusters that is not always conducive to interpretation[25]. The K-means clustering method assigns each gene to only one cluster. This may not necessarily be biologically appropriate. Moreover, genes assigned to the same cluster may not necessarily have similar expression patterns. It is important to check if the cluster solution makes biological sense.

15.7.   Bayesian Cluster Analysis

Binder[26] (1978) considered a parametric model for partitioning individuals into mutually exclusive groups. The problem of cluster analysis was reformulated in a Bayesian context allowing for the possibility of an unknown number of groups. Bayesian methods for cluster analysis were also discussed in Bernado and Giron[27] (1988).

15.8.   Two-way Clustering Methods

In several earlier reports on cluster analysis of microarray data, such as Eisen *et al.* (1998) and Alizadeh *et al.*[28] (2000), specimens and genes were clustered completely independently. Also, if the methods of self-organizing maps or K-means clustering are used, the number of clusters is externally prescribed. As an improvement, Getz *et al.*[29] (2000) introduce a coupled two-way clustering (CTWC) method where the number of clusters is determined by the algorithm itself.

Getz *et al.* (2000) believe that only a small subset of the genes participate in any cellular process of interest, which takes place only in a subset of the specimens. By focusing on small subsets, they lower the noise induced by the other specimens and genes. They note that, in clustering specimens, correlations do not always capture their similarity. Consider two specimens, taken at different stages of some process, with the absolute expression levels of a family of genes much below average in one specimen and much higher in the other. Even if the expression levels of the two specimens over these genes are correlated, one would like to assign them to different clusters. They use a normalization scheme such that data values in each column are divided by their column (specimen) mean, and each resulting row (gene) is then normalized so that each normalized row has a norm value equal to one. After the data are standardized, Euclidean distance is used as the dissimilarity measure. They look for pairs of a relatively small subset F_i of features (genes or specimens) and of objects O_k (specimens or genes), such that when the objects in O_k are represented using only the features from F_i, the clustering yields stable and significant partitions. The CTWC method produces pairs (O_k, F_i) in an iterative clustering process. They apply CTWC to two experiments and demonstrate a possible diagnostic use of the approach. They show that one can identify different responses to treatment and the groups of genes to be used as the appropriate probe.

15.9. Reliability of Clustering Patterns for Microarray Data

Because clustering methods are essentially data driven, clustering algorithms always produce clusters, even on random data. Hence, it is important to check the reliability of clustering results. Kerr and Churchill[30] (2001a) propose a bootstrap method to check whether the resulting clusters are statistically reliable relative to the noise in the data. Their approach begins with estimation of relative expression by using the ANOVA model, instead of ratios. Then, the authors used the residuals from the fitted ANOVA model to provide an empirical estimate of the error distribution for the following bootstrapping step. They created 499 bootstrap data sets and, for each simulated data set, they construct a bootstrap pattern for each gene. They repeat the filtering and clustering steps with these bootstrap estimates. The advantage of their model-based approach is that it separates systematic variations from noise and uses an empirical estimate of error that is free of distributional assumptions.

McShane *et al.*[31] (2002) propose a global test for clustering and a method to examine cluster-specific reproducibilities. They note that one minus Pearson correlation distance is proportional to the square of Euclidean distance computed on the standardized expression profiles. The authors present statistical methods for testing for the existence of meaningful cluster patterns based on examination of the Euclidean distances between specimens in principal components space. By adjusting for the effects of correlations among genes, they transform the log ratios to a space based on the first three principal components to simplify calculations. If the gene expression data come from a single multivariate Gaussian distribution, there should be no meaningful clusters. If the gene expression data follow an approximate Gaussian distribution, then the principal components will also have similar distributions. Hence, they generate simulated data under a Gaussian null distribution and examine the distribution of the nearest neighbor distance to quantify the clustering patterns in the real and simulated data. Then, the approach of data perturbation was used to assess clustering stability.

Notes

[1] Jain, A.K. and Dubes, R.C. (1988). *Algorithms for Clustering Data*, Prentice Hall, Englewood Cliff, New Jersey.

[2] Kaufman, L. and Rousseeuw, P.J. (1990). *Finding Groups in Data*, John Wiley and Sons, New York.

[3] Afifi, A.A. and Clark, V. (1990). *Computer-aided Multivariate Analysis*. 2nd edition, Chapman and Hall, New York.

[4] Jobson, J.D. (1992). *Applied Multivariate Data Analysis*, Springer-Verlag, New York.

[5] Everitt, B.S. (1993). *Cluster Analysis*. Edward Arnold, New York.

[6] Johnson, R.A., Wichern, D.W. (1998). *Applied Multivariate Statistical Analysis*, 4th edition, Prentice-Hall, Inc., New Jersey.

[7] Hastie, T., Tibshirani, R., and Friedman, J. (2001). *The Elements of Statistical Learning*, Springer, New York.

[8] Gower, J.C. (1985). In *Encyclopedia of Statistical Sciences*, **V.5**, (S. Kotz, N.L. Johnson and C.B. Read, eds.), John Wiley & Sons, New York.

[9] Gower, J.C. (1967). *Biometrics*, **23**, 623-628.

[10] Dunn, G. and Everitt, B.S. (1982). *An Introduction to Mathematical Taxonomy*, Cambridge University Press, Cambridge.

[11] Kohonen, T. (1997). *Self-Organizing Maps*, Springer, Berlin.

[12] Eisen, M., Spellman, P.T., Brown, P.O. and Botstein, D. (1998). *Proceedings of the National Academy of Sciences, USA*, **95**, 14863-14868.

[13] Alon, U., Barkai, N., Notterman, D.A., Gish, K., Ybarra, S.,Mack, D., Levine, A.J., (1999). *Proceedings of the National Academy of Sciences, USA*, **96**, 6745-6750.

[14] Wishart, D. (1969). *Biometrics*, **25**, 165-170.

[15] Johnson, S.C. (1967). *Psychometrika*, **32**, 241-254.

[16] Ward, J. H. (1963). *Journal of the American Statistical Association*, **58**, 236-244.

[17] Wen, X., Fuhrman, S., Michaels, G.S., Carr, D.B., Smith, S., Baker, J.L., and Somogyi, R. (1998). *Proceedings of the National Academy of Sciences, USA*, **95**, 334-339.

[18] Spellman, P.T., Sherlock, G., Zhang, M.Q., Iyer, V.R., Anders, K. , Eisen, M.B., Brown, P.O., Botstein, D. and Futcher, B. (1998). *Molecular Biology of the Cell*, **9**, 3273-329.

[19] Chu, S., DeRisi, J., Eisen, M., Mulholland, J., Botstein, D., Brown, P.O., Herskowitz, I. (1998). *Science*, **282**, 699-705.

[20] Milligan, G.W. (1980). *Psychometrika*, **45**, 325-342.

[21] Milligan, G.W. and Cooper, M.C. (1985). *Psychometrika*, **50**, 159-179.

[22] Kaufman, L. and Rousseeuw, P.J. (1990). *Finding Groups in Data*, John Wiley and Sons, New York.

[23] Morgan, B.J.T. and Ray, A.P.G. (1995). *Applied Statistics*, **44**, 117-134.

[24] MacQueen, J.B. (1967). In: *Proceedings of 5th Berkeley Symposium on Mathematical Statistics and Probability*, **1**, 281-297, University of California Press, Berkeley, California.

[25] Tamayo *et al.* (1999), *Proceedings of the National Academy of Sciences, USA*, 2907-2912.

[26] Binder, D. A. (1978). *Biometrika*, **65**, 31-38.

[27] Bernardo, J.M., Giron, J. (1988). A Bayesian approach to cluster analysis, *Questiio*, **12**, 97-112.

[28] Alizadeh, A., Eisen, M.B., Davis, R.E., Ma, C., *et al.* (2000). *Nature*, **403**, 503-511.

[29] Getz, G., Levine, E., Domany, E. (2000). *Proceedings of the National Academy of Sciences, USA*, **97**, 12079-12084.

[30] Kerr, M.K., and Churchill, G.A. (2001a). *Proceedings of the National Academy of Sciences, USA*, **98**, 8961-8965.

[31] McShane, L.M., Radmacher, M. D., Freidlin, B., Yu, R., Li, M.-C., and Simon, R. (2002). *Bioinformatics*, **18**, 1462-1469.

Chapter 16

PRINCIPAL COMPONENTS AND SINGULAR VALUE DECOMPOSITION

16.1. Principal Component Analysis

Pearson[1] (1901) and Hotelling[2] (1933) are pioneers in considering the method of *principal component analysis* (PCA). Hotelling (1933) used principal component analysis as a dimension reduction technique in the field of educational testing. Because of the high-dimensional nature of microarray data, the method of principal components is well suited for exploring gene expression data.

Given any specimen with index n, a G-dimensional column vector $x_n = (x_{1n}, \ldots, x_{gn}, \ldots, x_{Gn})'$ is observed, consisting of possibly correlated expression variables from G genes. To simplify notation, we suppress the specimen index n in this section and focus on reducing the dimension of the space of gene variables $\{x_1, \ldots, x_g, \ldots, x_G\}$. The technique of principal component analysis is a method of transforming the original set of variables into a smaller set of $r < G$ uncorrelated composite variables called principal components. The r principal components must together account for most of the variation in the G original variables.

For any given specimen, let $x = (x_1, \ldots, x_G)'$ denote a random vector consisting of expression levels of G genes, with mean $E(x) = 0$ and variance $E(xx') = \Sigma$. Then there exists an orthogonal linear transformation

$$L = B'x \tag{16.1}$$

such that the covariance matrix of L is $E(LL') = \Lambda$ and

$$
\boldsymbol{\Lambda} = \begin{pmatrix} \lambda_1 & 0 & \cdot & 0 \\ 0 & \lambda_2 & \cdot & 0 \\ \cdot & & \cdot & \\ \cdot & & \cdot & \\ \cdot & & \cdot & \\ 0 & 0 & \cdot & \lambda_G \end{pmatrix}, \tag{16.2}
$$

where $\lambda_1 \geq \lambda_2 \ldots \geq \lambda_G \geq 0$ are *eigenvalues* of the covariance matrix $\boldsymbol{\Sigma}$.

The k-th column of $\boldsymbol{B}$, denoted by $\boldsymbol{\beta}^{(k)}$, is the *eigenvector* corresponding to the k-th eigenvalue λ_k. That is, $\boldsymbol{\beta}^{(k)}$ satisfies the equation

$$
\boldsymbol{\Sigma} \, \boldsymbol{\beta}^{(k)} = \lambda_k \, \boldsymbol{\beta}^{(k)}. \tag{16.3}
$$

Then $L_k = \boldsymbol{\beta}^{(k)'} \boldsymbol{x}$ is called the k-th principal component of $\boldsymbol{x}$, with

$$
\mathrm{Var}(L_k) = \boldsymbol{\beta}^{(k)'} \boldsymbol{\Sigma} \, \boldsymbol{\beta}^{(k)} = \lambda_k.
$$

The principal components are orthogonal because

$$
\mathrm{Cov}(L_k, L_l) = \boldsymbol{\beta}^{(k)'} \boldsymbol{\Sigma} \, \boldsymbol{\beta}^{(l)} = 0 \ \text{ if } \ k \neq l.
$$

Hence, L_1 is the normalized linear combination with maximum variance. Similarly, L_k, the k-th component of $\boldsymbol{L}$, has maximum variance among all normalized linear combinations uncorrelated with $L_1, \ldots, L_{k-1}$. The sum of the variances of the principal components is the sum of the variances of the original variates. That is,

$$
\sum_{k=1}^{G} \mathrm{Var}(L_k) = \lambda_1 + \ldots + \lambda_G = \mathrm{Trace} \, (\boldsymbol{\Sigma}). \tag{16.4}
$$

The proportional of total variance explained by the k-th principal component is given by

$$
\frac{\lambda_k}{\lambda_1 + \ldots + \lambda_G}. \tag{16.5}
$$

If $r = \mathrm{rank}(\boldsymbol{\Sigma})$ then only the first r eigenvalues of $\boldsymbol{\Sigma}$ are nonzero. That is, $\lambda_1 \geq \lambda_2 \geq \ldots \lambda_r > 0$, and $\lambda_{r+1} = \ldots = \lambda_G = 0$. Let $P_k = \lambda_k^{-1/2} \boldsymbol{\beta}^{(k)'} \boldsymbol{x}$. The set of components $P_1, \ldots, P_r$ forms an orthonormal basis of the space spanned by random variables $\boldsymbol{x}_1, \ldots, \boldsymbol{x}_G$.

The method of principal components analysis maps the original dataset to a new *feature space* in which the features are formed by linear functions of the original attributes. These new features are sorted by the amount of variance that the data exhibit in each direction. Dimensionality reduction is achieved by removing features corresponding to dimensions in which the data have low variance. There is no guarantee, however, that these removed features are not essential for understanding the data.

Principal component analysis is based on the covariance matrix Σ of the random vector x. If the components of random vector $x = (x_1, \ldots, x_G)$ are measured on scales with wide ranges, or in incommensurate units, then the variables should probably be standardized before undertaking the principal component analysis. Principal component analysis can also be based on the correlation matrix of x. When standardized variables are used, the total variance is simply G, and the proportion of total variance explained by the k-th principal component is λ_k/G, for $k = 1, \ldots, G$. Principal components derived from the covariance matrix, however, are different from those derived from the correlation matrix. The two sets of principal components are not related in a simple fashion. Hence, a decision to standarize the variables requires careful consideration.

The geometry of the principal component method does not require a multivariate normal assumption. Inferences about the principal component structure require a random sample of specimen samples $x_n, n = 1, \ldots, N$. The parameters of the structure can then be inferred from the sample data. With the additional assumption of multivariate normality of x, large sample inferences can be made for the resulting eigenvalue estimates $\hat{\lambda}_k$ and eigenvector estimates $\hat{\beta}_k$. Since the method of principal component analysis yields a small number of uncorrelated variables, it is useful as an exploratory analysis of gene expression data that are collected under a mixture of conditions. When the number of samples is small, however, there is a potential risk that some principal components might be identified as important solely because of excessive noise in the data. Detailed descriptions of the method of principal component analysis can be found in Anderson[3] (1958), Seal[4] (1964), Morrison[5] (1976), Joliffe[6] (1986), Flury[7] (1988), Krzanowski[8] (1988), Dunteman[9] (1989), Jackson[10] (1991), Jobson[11] (1992), and Johnson and Wichern[12] (1998), among others.

16.1.1. Applications of Dominant Principal Components

Khan *et al.*[13] (2001) used microarray gene expression data to model cancers in each of the four categories of the small, round blue cell tumors (SRBCTs) of childhood, including neuroblastoma (NB), rhabdomyosarcoma (RMS), non-Hodgkin lymphoma (NHL), and the Ewing family of tumors (EWS). These cancers belong to four distinct diagnostic categories but have similar appearance on routine histology. Hence, they often present diagnostic dilemmas in clinical practice.

The authors first reduced the number of genes from 6567 to to 2308 by filtering for a minimal level of expression. Then, they conducted a principal component analysis and found that 10 dominant principal components contained 63% of the variance in the data matrix. The remaining components contained variance unrelated to separating the four cancers. As linear combinations of genes, these 10 dominant principal components might be referred to as *eigengenes*. The analysis thus reduced the dimensionality of the gene space from 2308 to 10. Hence, instead of using the high-dimensional original data, the authors used these 10 dominant components for each sample as inputs for further analyses.

16.2.   Singular-value Decomposition

When the dimension G of the vector space of gene expression levels is large, the matrix computations for principal components discussed in the previous section can be time consuming. The method of *singular-value decomposition* (SVD) proves to be very useful in solving this problem. SVD is also known as the Karhunen-Loéve expansion in the field of pattern recognition.

For each specimen sample n, let $\boldsymbol{x}_n = (x_{1n}, x_{2n}, \ldots, x_{Gn})'$ denote the column vector of intensity levels of the G genes. Let $\boldsymbol{X} = (\boldsymbol{x}_1, \ldots, \boldsymbol{x}_N)$ denote the $G \times N$ matrix of expression data from N specimen samples.

The *Singular-value Decomposition Theorem*[14] states that for any $G \times N$ dimensional matrix $\boldsymbol{X}$ of rank r, there exists a $G \times G$ orthogonal transformation matrix $\boldsymbol{U}$, an $N \times N$ orthogonal transformation matrix $\boldsymbol{V}$, and a $r \times r$ diagonal matrix $\boldsymbol{D}$ such that $\boldsymbol{X}$ can be decomposed as follows

$$\boldsymbol{X} = \boldsymbol{U} \begin{pmatrix} \boldsymbol{D} & \boldsymbol{0} \\ \boldsymbol{0} & \boldsymbol{0} \end{pmatrix} \boldsymbol{V}', \tag{16.6}$$

where $r = \text{rank}(\boldsymbol{X}) \leq \min(G, N)$, $\boldsymbol{D} = (\delta_{jk})$ is a $r \times r$ diagonal matrix with

$$\delta_{11} \geq \delta_{22} \geq \ldots \geq \delta_{rr} > 0 \quad \text{and} \quad \delta_{jk} = 0, \text{ for } j \neq k. \tag{16.7}$$

The $\boldsymbol{0}$s are zero matrices of suitable dimensions. Note that both matrices $\boldsymbol{X}\boldsymbol{X}'$ and $\boldsymbol{X}'\boldsymbol{X}$ are nonnegative definite and symmetric and that they have the same nonzero eigenvalues, which are all positive. Diagonal values $\delta_{jj}, j = 1, \ldots, r$, are called the singular values of $\boldsymbol{X}$, they are the positive square roots of the nonzero eigenvalues of the matrix $\boldsymbol{X}'\boldsymbol{X}$ (or, equivalently, of $\boldsymbol{X}\boldsymbol{X}'$); that is,

$$\boldsymbol{U}'\boldsymbol{X}\boldsymbol{X}'\boldsymbol{U} = \begin{pmatrix} \boldsymbol{D}^2 & \boldsymbol{0} \\ \boldsymbol{0} & \boldsymbol{0} \end{pmatrix}_{G \times G} \text{ and } \boldsymbol{V}'\boldsymbol{X}'\boldsymbol{X}\boldsymbol{V} = \begin{pmatrix} \boldsymbol{D}^2 & \boldsymbol{0} \\ \boldsymbol{0} & \boldsymbol{0} \end{pmatrix}_{N \times N}. \quad (16.8)$$

The above theorem can be written in the following simpler form. If rank $\boldsymbol{X} = r$, let $U^{(k)}$ denote the k-th column of the matrix $\boldsymbol{U}$, and $V^{(k)}$ denote the k-th column of the matrix $\boldsymbol{V}$. Let $\boldsymbol{U}^*$ denote the $G \times r$ dimensional submatrix of $\boldsymbol{U}$ consisting of its first r columns $U^{(1)}, \ldots, U^{(r)}$. Let $\boldsymbol{V}^*$ denote the $n \times r$ dimensional submatrix of $\boldsymbol{V}$ consisting of its first r columns $V^{(1)}, \ldots, V^{(r)}$. Then equation (16.6) can be simplified as

$$\boldsymbol{X} = \boldsymbol{U}^*\boldsymbol{D}\boldsymbol{V}^{*'}, \quad (16.9)$$

and equation (16.8) can be simplified as

$$\boldsymbol{U}^{*'}\boldsymbol{X}\boldsymbol{X}'\boldsymbol{U}^* = \boldsymbol{V}^{*'}\boldsymbol{X}'\boldsymbol{X}\boldsymbol{V}^* = \boldsymbol{D}^2. \quad (16.10)$$

16.3.   Computational Procedures for SVD

Since thousands of genes are usually involved in a microarray study, the number of genes is often much larger than the number of sample arrays, i.e., $G >> N$. One advantage of the SVD method is that, instead of finding eigenvalues and eigenvectors for the genes using a large $G \times G$ dimensional matrix $\boldsymbol{X}\boldsymbol{X}'$, one will first find eigenvalues and eigenvectors for specimen samples using the much smaller $N \times N$ dimensional matrix $\boldsymbol{X}'\boldsymbol{X}$. Therefore, the method of SVD also has the advantage that it can find principal components for both the genes and specimen samples simultaneously.

To begin, one first computes the matrix $\boldsymbol{X}'\boldsymbol{X}$, and finds the matrix $\boldsymbol{V}^*$ and a diagonal matrix $\boldsymbol{D}^2$ such that

$$\boldsymbol{X}'\boldsymbol{X} = \boldsymbol{V}^*\boldsymbol{D}^2\boldsymbol{V}^{*'}. \quad (16.11)$$

Then, one projects the resulting matrices $\boldsymbol{V}$ and $\boldsymbol{D}$ onto $\boldsymbol{X}$ and obtains

$$\boldsymbol{U}^* = \boldsymbol{X}\boldsymbol{V}^*\boldsymbol{D}^{-1}, \quad (16.12)$$

where $\boldsymbol{D}^{-1}$ is the inverse of the diagonal matrix $\boldsymbol{D}$.

A useful consequence of the singular-value decomposition is that one can compute

$$X^{-1} = V \begin{pmatrix} D^{-1} & 0 \\ 0 & 0 \end{pmatrix} U' \qquad (16.13)$$

as the Moore-Penrose inverse of the matrix X.

16.4. Eigengenes and Eigenarrays

We have already referred to a linear combination of genes in principal component analysis as an *eigengene*. The corresponding linear combination of arrays (specimens) might be called an *eigenarray*.

Alter, Brown, and Botstein[15] (2000) describe the use of singular value decomposition in analyzing genome-wide expression data. They show that SVD is a linear transformation of the expression data from the $G \times N$ dimensional "gene $\times$ array" space to a reduced $r \times r$ dimensional "eigengene $\times$ eigenarray" space. In equation (16.6), the G-dimensional vector in the jth column of the matrix U consists of the genome-wide expression in the jth eigenarray corresponding to the positive eigenvalue δ_{jj} of $X'X$. The N-dimensional vector in the j-th row of the matrix V' consists of the expression of the j-th eigengene across the different arrays. The eigengenes and eigenarrays are orthonormal superpositions of the genes and arrays such that the transformations U^* and V^* are both orthogonal, i.e.,

$$U^{*\prime}U^* = V^{*\prime}V^* = I. \qquad (16.14)$$

where I is the identity matrix of dimension $r \times r$. Hence the expression of each eigengene (or eigenarray) is uncorrelated with that of other eigengenes (or eigenarrays).

16.5. Fraction of Eigenexpression

Alter, Brown, and Botsein (2000) use the "fraction of eigenexpression",

$$f_j = \frac{\delta_{jj}{}^2}{\sum_{k=1}^{r} \delta_{kk}{}^2}, \qquad (16.15)$$

to indicate the relative significance of the j-th eigengene and eigenarray in terms of the fraction of the overall expression that they capture. *Shannon entropy*, defined here as

$$\eta = -\frac{1}{\log(r)} \sum_{k=1}^{r} f_k \log(f_k), \qquad (16.16)$$

where $0 \leq \eta \leq 1$, measures the complexity of the microarray dataset in terms of the distribution of the eigen expression across the different eigengenes (and eigenarrays). Entropy $\eta = 0$ corresponds to an ordered and redundant data set in which all expression is captured by a singular eigengene (and eigenarray), and $\eta = 1$ corresponds to a disordered and random data set where all eigengenes (and eigenarrays) are equally expressed.

SVD provides a useful mathematical framework for processing and modeling genome-wide expression data, in which both the mathematical variables and operations may be assigned biological meaning.

16.6. Generalized Singular Value Decomposition

In comparing yeast and human cell-cycle gene expression, Alter *et al.*[16] (2003) use generalized singular value decomposition (GSVD) to provide a comparative mathematical framework for two genome-scale expression data sets. Using GSVD, the authors transform the data sets from two gene-by-array spaces to two reduced and diagonalized genelet-by-arraylet spaces where the genelets are shared by both data sets. *Genelet* and *arraylet* are their terms for eigengenes and eigenarrays. The authors found biological similarity in the two disparate organisms in terms of their mRNA expression during their cell-cycle programs as well as experimental dissimilarity in terms of yeast and human mRNA expression during their different synchronization-response program. They also found differential gene expression in the yeast and human cell-cycle programs versus their synchronization-response programs. They note that the method of GSVD can be used in many other applications, such as comparing two different types of genomic information (e.g., mRNA expression or protein abundance) collected from the same set of samples to elucidate the molecular composition of the overall biological signal in these samples.

16.7. Robust Singular Value Decomposition

The conventional SVD, being a least-squares method, is highly susceptible to outlier values that are common to microarray data. On the basis of the Gabriel-Zamir alternating least squares (ALS) algorithms[17] and the alternating robust fitting[18] methods, Liu *et al*[19] (2003) proposed a robust analysis using a variant of the SVD method. Their approach uses all of the observed data and does not require imputation of missing data. Hence, the strength of thier method is that it does not require complete information and is not affected by a minority of outliers.

Notes

[1] Pearson, K. (1901). *Phil. Mag.* (6), **2**, 559-572.

[2] Hotelling, H. (1933). *Journal of Educational Psychology*, **24**, 417-441.

[3] Anderson, T.W., (1958), *An Introduction to Multivariate Statistical Analysis*. Wiley, New York.

[4] Seal, H. (1964). *Multivariate Statistical Analysis for Biologists*, New York, Wiley.

[5] Morrison, D.F. (1976). *Multivariate Statistical Methods*, 2nd ed., New York, McGraw-Hill.

[6] Joliffe, I.T. (2002). *Principal Component Analysis*, 2nd ed., Springer-Verlag, New York.

[7] Flury, B. (1988), *Common Principal Components and Related Multivariate Models*, Wiley, New York.

[8] Krzanowski, W.J. (1988). *Principles of Multivariate Analysis: A User's Perspective*, Oxford University Press, Oxford.

[9] Dunteman, G.H. (1989), *Principal Component Analysis*, Sage University Papers, Sage, Newbury Park, California.

[10] Jackson, J.E. (1991). *A User's Guide to Principal Components*, John Wiley & Sons, New York.

[11] Jobson, J.D. (1992). *Applied Multivariate Data Analysis*, Springer-Verlag, New York.

[12] Johnson, R.A., Wichern, D.W. (1998). *Applied Multivariate Statistical Analysis*, 4th ed., Prentice-Hall, Inc., New Jersey.

[13] Khan, J., Wei, J.S., Ringner, M., Saal, L.H., Ladanyi, M., Westermann, F., Berthold, F., Schwab, M., Antonescu, C.R., Peterson, C., and Meltzer, P.S. (2001). *Nature Medicine*, **7**, 673-679.

[14] Searle, S.R. (1982), *Matrix Algebra Useful for Statistics*, John Wiley and Sons, pages 316-317.

[15] Alter, O., Brown, P.O. and Botstein, D. (2000), *Proceedings of the National Academy of Sciences, USA*, **97**, 10101-10106.

[16] Alter, O., Brown, P.O., and Botstein, D. (2003) *Proceedings of the National Academy of Sciences, USA*, **100**, 3351-3356.

[17] Gabriel, K.R. and Zamir, S. (1979) *Technometrics*, **21**, 489-498.

[18] Croux, C., Filzmoser, P., Pison, G. and Rousseeuw, P.J. (2002) *Statistical Computations*, **13**, 23-36.

[19] Liu, L., Hawkins, D.M., Ghosh, S., Young, S.S. (2003). *Proceedings of the National Academy of Sciences, USA*, **100**, 13167-13172.

Chapter 17

SELF-ORGANIZING MAPS

In unsupervised learning, not only do we have the disadvantage of having data of unknown classification, often we do not even know how many categories exist. Kohonen[1] demonstrated that a *self-organizing map* (SOM) can organize data without knowing the correct classification of input patterns beforehand. Using the language of cluster analysis, the user has the option of specifying the number of clusters to be identified in a SOM. The SOM algorithm then finds an optimal set of centroids around which the data points of the input space appear to aggregate. It partitions the data set, with each centroid defining a cluster consisting of the data points nearest to it. The SOM algorithm builds a low-dimensional display or map (the output space) that associates similar input data patterns with contiguous map locations. SOM can also be considered as a nonlinear projection of the probability density function of the high-dimensional input data onto the map. The basic method of SOM is an unsupervised neural network algorithm that has been used in data mining, image processing and speech voicing because it has many desirable mathematical properties, including scaling well to large data sets.

17.1.   The Basic Logic of a SOM

As its name suggests, a SOM algorithm organizes a set of items into a map in such a way that similar items (in terms of attributes) appear close to each other on the map. It is intended that the map yield useful information about associations among the items.

Consider a data set consisting of a finite set of items, each having a number of characteristics or *attributes*. Attributes may be variables with discrete values or continuous measurements. An attribute may also be a classification variable, such as a binary variable, having values of 1 or 0 say, that indicate the presence or absence of a characteristic, respectively. Distances between items or their similarities can be quantified in various ways, as described earlier in the chapter on cluster analysis (Chapter 15). The distance or similarity measures assess the proximity of items in terms of their component attributes. For example, the similarity of two items with several binary attributes might be defined in terms of the number of attributes common to both sets, i.e., as the dot product of their respective attribute vectors. Euclidean distances can also be used for measuring the proximity of items.

As we are concerned with the use of SOMs in analyzing microarray data, a SOM can be used to either cluster genes or cluster specimens. We present the methods of SOM in terms of clustering genes. For microarray studies involving a total of N specimens and G genes, let the row vector $\boldsymbol{x}^{(g)} = (x_{g1}, x_{g2}, \ldots, x_{gN})$ denote the expression levels for gene g across N specimens, where x_{gn} denotes the expression level of gene g for specimen n, $g = 1, \ldots, G$, $n = 1, \ldots, N$. The set of vectors $\mathcal{G} = \{\boldsymbol{x}^{(g)}, g = 1, \ldots, G\}$ is the data set. These G data vectors lie in an *input space* of dimension N that we will denote by $\mathcal{X}$. Some variants on this data structure may be of interest. For example, in some applications, x_{gn} might be defined as a classification outcome rather than an intensity measurement. The classification might be coded -1, 0, or 1 according to whether gene g is down-regulated, equally regulated, or up-regulated in specimen n.

A SOM is a mapping from the N-dimensional input space $\mathcal{X}$ onto a lower-dimensional output space or *map*, which we denote by $\mathcal{M}$. The map $\mathcal{M}$ is usually two-dimensional but may be one-dimensional (as a later example shows), three-dimensional, or of higher order. The analytical elements required for a SOM are as follows:

1. Grid. The map $\mathcal{M}$ has a lattice of *grid points* $\mathcal{J} = \{1, \ldots, j, \ldots, J\}$. The grid points may be arrayed along a line in one dimension, over a plane in two dimensions or, in similar fashion, over a higher dimensional space. The two-dimensional map is the most common. Grid points are sometimes referred to as *nodes*.

2. Reference vectors. Associated with each grid point in $\mathcal{M}$ is a reference attribute vector $\boldsymbol{w}_j = (w_{j1}, w_{j2}, \ldots, w_{jN})$ in the N-dimensional input space $\mathcal{X}$. Thus, $\boldsymbol{w}_j \in \mathcal{X}, j = 1, \ldots, J$. The reference vector at a grid

point describes the attribute profile of that point on the map. In clustering terminology, a reference vector is sometimes referred to as a *centroid*.

3. Attribute distance function. A distance function $d(\boldsymbol{u}, \boldsymbol{v})$ is defined for all pairs of attribute vectors $\boldsymbol{u}, \boldsymbol{v} \in \mathcal{X}$. This function evaluates the proximity of two attribute profiles; for instance, the proximity of gene expression profiles for two specimens.

4. Map neighborhood function. As noted earlier, the aim of a SOM is to organize the data so similar items (in terms of attributes) appear close to each other on the map. Being 'close on the map' requires a judgment about what defines the neighborhood of a grid point. This judgment uses another proximity function called a neighborhood function $h_j(k)$ for all $j, k \in \mathcal{J}$. Here $h_j(k)$ describes the extent or degree to which grid point k lies in the neighborhood of grid point j. Larger values of $h_j(k)$ indicate that point k is closer to point j. For example, the neighborhood of a grid point might simply include any immediately adjacent grid point but no other points. The neighborhood function in this case might have $h_j(k) = 1$ if k is adjacent to j and equal 0 otherwise. As another example, the neighborhood function may be a function whose value declines the farther apart are grid points k and j. The neighborhood function must usually be adjusted to take account of the borders of the map.

5. Updating algorithm. The SOM methodology requires an algorithm that will revise (organize) the reference vectors at the grid points so neighboring points (as defined by the map neighborhood function) have similar reference vectors (as defined by the attribute distance function). We return to the technical aspects of the algorithm later. For the moment, we wish to convey the basic idea that the SOM procedure starts with an arbitrary set of reference vectors and then proceeds to revise them using a training sample from the data set until the final map is produced.

6. Cluster formation. Once the SOM is prepared, each data vector $\boldsymbol{x}_g$ in the gene expression data set can be assigned to its closest reference vector and, thereby, becomes associated with a particular grid point or position on the map. The scientist can then examine the clusters of genes for valuable information about patterns, associations, gaps, etc.

Figure 17.1 shows a schematic that illustrates these elements for the case of a one-dimensional map (i.e., an arrangement of reference vectors

on a line). The schematic is simplified but is realistic nonetheless because it may be of interest to see how G genes might array themselves
along a line in terms of the similarity of their expression profiles for N
specimens. The figure shows an array of J grid points along a line. The
map neighborhood function represented in the scheme shows that the
neighborhood of a point (say, point 3 in the figure) includes its adjacent
points 2 and 4. The attribute distance function can be used in two ways.
First, it can measure the distances between reference vectors; for exam-
ple, between $\boldsymbol{w}_2$ and $\boldsymbol{w}_3$ as shown in the figure. Neighboring reference
vectors in a SOM should be separated by small attribute distances in the
final map. Second, the attribute distance function decides to which grid
point on the map a data vector $\boldsymbol{x}_g$ should be assigned, based on the distance between the data vector and the reference vector. The schematic
shows $\boldsymbol{x}_g$ being assigned to grid point 3 because $\boldsymbol{x}_g$ is closest to $\boldsymbol{w}_3$. The
attainment of the final map depends on applying the SOM algorithm
which we now describe.

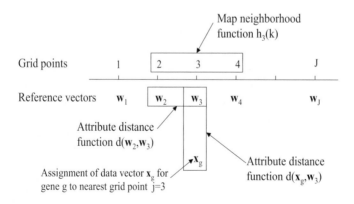

Figure 17.1.   Schematic showing the elements of a one-dimensional SOM

17.2.   The SOM Updating Algorithm

Many algorithms have been proposed for generating a SOM. All have the following basic features. Detailed mathematics and variations of the SOM algorithm can be found in Kohonen (1997), for example.

- A SOM algorithm is often applied first to a *training set*, a subset of the full data set. We denote this subset by $\mathcal{G}_T \subseteq \mathcal{G}$.

- The algorithm is iterative. We use t to denote the iteration, with $t = 0, 1, 2, \ldots$.

- Experience shows that algorithms have better convergence properties if the neighborhood function is adjusted with each iteration. In essence, the neighborhood of a reference vector is generally shrunk as t increases. Hence, we will use $h_j^{(t)}(k)$ to denote the form of the neighborhood function used at the tth iteration.

- The algorithm adjusts the reference vectors at each iteration. We shall denote the set of reference vectors at the tth iteration by $\{w_j^{(t)}, j = 1, \ldots, J\}$. Thus, the algorithm begins with an initial set of reference vectors $\{w_j^{(0)}, j = 1, \ldots, J\}$. These J vectors may be chosen arbitrarily or chosen judiciously from the training set $\mathcal{G}_T$.

- At the tth iteration, one of the data vectors in the training set, say $x_{g(t)} \in \mathcal{G}_T$, is presented to the algorithm. The specific grid point $j \in \mathcal{J}$ is found for which $d(x_{g(t)}, w_j^{(t)})$ is a minimum. This grid point closest to $x_{g(t)}$ is denoted by $j(t)$. Thus, $w_{j(t)}^{(t)}$ is the reference vector that is closest to $x_{g(t)}$.

- All reference vectors are then updated as follows:

$$w_k^{(t+1)} = w_k^{(t)} + \lambda^{(t)} h_{j(t)}^{(t)}(k)(x_{g(t)} - w_k^{(t)}), \quad \text{for all } k \in \mathcal{J}. \quad (17.1)$$

The scalar $\lambda^{(t)}$ is a learning rate coefficient that generally varies with the iteration t. The updating in (17.1) can be seen to involve moving each reference vector $w_k^{(t)}$ in the neighborhood of the grid point $j(t)$ closer to $x_{g(t)}$. The movement is scaled by the magnitude of the neighborhood function $h_{j(t)}^{(t)}(k)$ as well as the learning coefficient $\lambda^{(t)}$. Grid points k for which $h_{j(t)}^{(t)}(k)$ is zero are not adjusted because they are not in the neighborhood of the grid point $j(t)$.

- Data vectors from the training set $\mathcal{G}_T$ are successively submitted to the SOM algorithm, usually in random order and recycled as required until convergence is achieved.

Figure 17.2 carries on from the schematic shown in Figure 17.1. It illustrates the updating of reference vectors that occurs at one iteration. We leave the iteration t unspecified. The figure shows the reference vectors $\boldsymbol{w}_1, \boldsymbol{w}_2, \boldsymbol{w}_3, \boldsymbol{w}_4, \ldots, \boldsymbol{w}_J$ positioned in the input space $\mathcal{X}$. This space is shown as two-dimensional here for simplicity. The two dimensions correspond to gene expression outcomes for $N^* = 2$ samples (referred to as samples 1 and 2 in the figure). The data vector $\boldsymbol{x}_g$ for gene g is presented to the algorithm at this iteration and falls closest to reference vector $\boldsymbol{w}_3$ in this case (as measured by Euclidean distance). The vectors $\boldsymbol{w}_2$ and $\boldsymbol{w}_4$ lie in the designated neighborhood of $\boldsymbol{w}_3$ as shown in Figure 17.1. Thus, all three of these reference vectors (circled by a halo) are drawn closer to $\boldsymbol{x}_g$ in the updating process. The other reference vectors are left unchanged in this iteration. Observe that in early iterations, as the map is organizing itself, the neighboring grid points may not necessarily be the closest points in the input space. Notice in Figure 17.2, for instance, that $\boldsymbol{w}_1$ lies slightly closer to $\boldsymbol{w}_3$ than to $\boldsymbol{w}_4$ but is not updated by $\boldsymbol{x}_g$. The reason is that grid point 1 is not one of the immediate neighbors of grid point 3 in Figure 17.1.

To summarize, the algorithm operates as follows. Each iteration involves randomly selecting an expression vector $\boldsymbol{x}_g$ for one gene in the training set $\mathcal{G}_T$ and moving the reference vectors of the SOM in the direction of this vector. The closest reference vector to $\boldsymbol{x}_g$ and its neighbors is moved by amounts that depend on the values of their neighborhood function and the learning rate coefficient. Hence, neighboring grid points on the map tend to have reference vectors that lie close to each other in the N-dimensional input space $\mathcal{X}$, which contains the expression vectors of the genes. Well separated grid points tend to have reference vectors that are quite distinct. In the end, each gene is assigned to that grid point which has a reference vector closest to its expression profile. Hence, neighboring grid points tend to define related clusters of genes. The algorithm continues for many iterations (many thousands if needed) until the map stabilizes. By repeating the process, selecting a different number of grid points, one can explore the underlying structure and vary the geometry of the SOM. If only a few points are chosen, distinct patterns may not be seen and there is large within-cluster scatter. Adding more grid points might help meaningful cluster patterns to emerge. On the other hand, having too many grid points will produce sparse clusters.

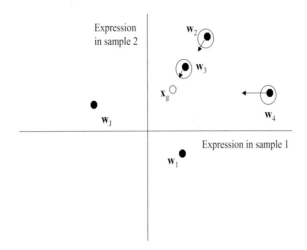

Figure 17.2. Schematic showing the updating of positions of reference vectors in the neighborhood defined by the data vector $\boldsymbol{x}_g$ for gene g. The input space $\mathcal{X}$ here is two-dimensional.

17.3.   Program GENECLUSTER

Tamayo *et al.*[2] (1999) developed a SOM method for microarray data in a publicly available computer package, GENECLUSTER, which is written in the computer language C, runs under UNIX, and requires a Web browser. This program is available from the Center of Genome Research of the Whitehead Institute at the Massachusetts Institute of Technology. The center's website address is http://www-genome.wi.mit.edu/.

The program GENECLUSTER accepts an input file of expression levels from either oligonucleotide arrays or spotted arrays. The user specifies a *range* for the number of clusters in order to find the cluster structure and number of iterations that optimizes the grouping of data objects. To improve the ability to detect meaningful patterns, the program begins with two preprocessing steps. To begin, the data are passed through a variation filter to eliminate those genes with little differential expression across the samples. This screening prevents reference vectors from gravitating toward large numbers of invariant genes. Next, the expression level of each gene is normalized across experiments so that the map will be based on patterns of expression rather than absolute levels of expression. A SOM is computed such that each cluster of genes is represented by its average expression pattern or centroids. The centroids are used to discern similarities and differences among the patterns. The variation around the pattern can be visualized by means of error bars (standard deviations).

17.4.   Supervised SOM

The SOM method was originally used in an unsupervised process like classical clustering methods. Classification accuracy, however, can be improved greatly if information about class identity is taken into account in the training phase. In order to make the SOM supervised, the input vectors x_g are extended to include a class component vector ξ_g, where ξ_g is a unit vector with a 1 assigned to the class of gene g and 0s assigned elsewhere. Since the vector ξ_g is the same for all genes that share the same class but different for genes from different classes, clustering of vectors x_g, incorporating their class unit vectors, leads to improved class-separation.

17.5.   Applications

17.5.1.   Using SOM to Cluster Genes

Tamayo *et al.* (1999) applied SOM to hematopoetic differentiation in four models (HL-60, U937, Jurkat and NB4 cells). Expression patterns of some 6,000 human genes were assayed. A total of 567 genes passed a variation filter and their expression levels were normalized.

For their updating algorithm, they chose a learning rate coefficient in (17.1) of the form $\lambda^{(t)} = 0.02T/(T + 100t)$ (using the earlier notation), where t denotes the iteration, $t = 1, \ldots, T$ and T is the maximum number of iterations. The authors' neighborhood distance function $h_{j(t)}^{(t)}(k)$ was a circular region (a "bubble neighborhood") that started with an initial radius of 3 and shrank with each iteration t in a linear fashion until it became zero. For their attribute distance measure, the authors presumably used Euclidean distance although this is not mentioned explicitly in the article.

Use of a hierarchical clustering method would have resulted in hundreds, or even thousands, of nested clusters that could be difficult to interpret. The advantage of the SOM method is that it extracts the J

most prominent patterns (where J is the optional number of grid points or nodes specified in the SOM algorithm) and arranges them so that similar patterns occur as neighbors in the SOM.

Figure 17.3 shows the resulting SOM. It is arranged in a 4×3 grid with $J = 4(3) = 12$ grid points, referred to in the figures as clusters (numbered 0 to 11). The components of the reference vectors are plotted as line graphs. The similar profiles of neighboring graphs is apparent. Error bars are also shown to illustrate the variation of expression profiles for genes assigned to each cluster. The variation is reasonably small, indicating relatively homogeneous clusters.

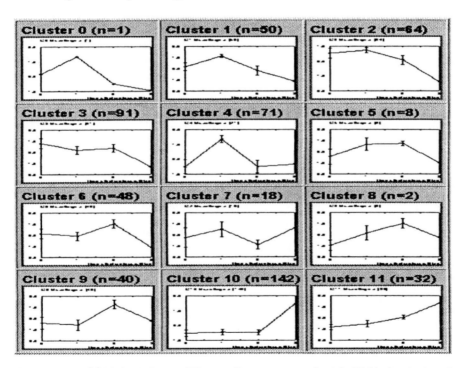

Figure 17.3. SOM for HL-60. HL-60 cells were treated with PMA for 0, 0.5, 4, or 24 hours, and expression levels of more than 6,000 genes were measured at each time point. The 567 genes passing the variation filter were grouped by a 4×3 SOM. Source: Proceedings of the National Academy of Sciences (PNAS), volume 96, Copyright (1999) National Academy of Sciences, U.S.A. Reprinted with permission from PNAS.

17.5.2. Using SOM to Cluster Tumors

Golub, Slonim, Tamayo *et al.*[3] (1999) used a SOM in a classification of tumors. They applied a two-cluster SOM to automatically group the

38 initial leukemia samples into two classes on the basis of the expression pattern of all 6817 genes. The SOM was constructed using the GENECLUSTER software, with a variation filter excluding genes with less than five-fold variation across the collection of samples. By comparing the clusters obtained by SOM to the known AML-ALL classes, the evaluation results show that the SOM paralleled the known classes closely: Class A_1 contained mostly ALL (24 of 25 samples), and class A_2 contained mostly AML (10 of 13 samples). The SOM was thus quite effective in discovering the two types of leukemia.

17.5.3.   Multiclass Cancer Diagnosis

Ramaswamy, Tamayo, Rifkin *et al.*[4] (2001) considered multiclass cancer diagnosis using tumor gene expression signatures. To determine whether the diagnosis of common adult malignancies could be achieved purely by molecular classification, they subjected 218 tumor samples, spanning 14 common tumor types, and 90 normal tissue samples to oligonucleotide microarray gene expression analysis. The expression levels of 16,063 genes and expressed sequence tags were used to evaluate the accuracy of a multiclass classifier.

To exclude genes showing minimal variation across the specimens, a variation filter was used to exclude any gene that exhibited less than 5-fold and 500 units of absolute variation across the dataset. A lower threshold of 20 units and ceiling of 16,000 units was applied. Of 16,063 expression values that they considered, a total of 11,322 passed this filter and were used for clustering. The dataset was then normalized by standardizing each gene to mean zero and unit variance. Using the CLUSTER and TREEVIEW[5] software, the authors conducted average-linkage hierarchical clustering. They constructed a SOM by using the GENECLUSTER analysis package.

Figure 17.4 shows the results of both hierarchical and SOM clustering of this dataset. In panel (c), the columns represent 190 primary human tumor specimens ordered by class. Rows represent 10 genes, including the known cancer markers prostate-specific antigen (PSA), carcinoembryonic antigen (CEA), and estrogen receptor (ER). Red indicates high relative level of expression and blue represents low level. Specimen classes considered include breast adenocarcinoma (BR), prostate adenocarcinoma (PR), lung adenocarcinoma (LU), colorectal adenocarcinoma (CR), lymphoma (LY), bladder transitional cell carcinoma (BL), melanoma (ML), uterine adenocarcinoma (UT), leukemia (LE), renal cell carcinoma (RE), pancreatic adenocarcinoma (PA), ovarian adenocarcinoma (OV), pleural mesothelioma (ME), and central nervous system (CNS). Although some tumor types, such as lymphoma (LY), leukemia (LE), and central nervous system (CNS), formed relatively discrete clusters with both methods, others, in particular the epithelial tumors, were largely intermixed. Their finding indicates that unsupervised learning might not adequately capture the distinctions in the tissue of origin among these molecularly complex tumors. Hence, the authors also analyzed the data by the method of support vector machines, which will be discussed in a later chapter.

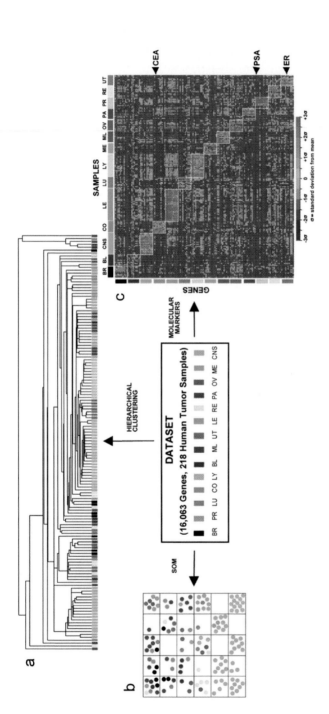

Figure 17.4. Clustering of tumor gene expression data and identification of tumor-specific molecular markers. Hierarchical clustering (a) and a 5x5 self organizing map (b) were used to cluster 144 tumors spanning 14 tumor classes according to their gene expression patterns (c) gene expression values for class-specific OVA markers. Source: Proceedings of the National Academy of Sciences (PNAS), volume 98, Copyright (2001) National Academy of Sciences, U.S.A. Reprinted with permission from PNAS.

Notes

[1] Kohonen, T. (1997). *Self-Organizing Maps*, Springer, Berlin.

[2] Tamayo, P., Slonim, D., Mesirov, J., Zhu, Q., Kitareewan, S., Dimitrovsky, E., Lander, E.S., Golub, T.R. (1999). *Proceedings of the National Academy of Sciences, USA*, **96**, 2907-2912.

[3] Golub, T.R., Slonim, D.K., Tamayo, P., *et al.* (1999). *Science*, **286**, 531-537.

[4] Ramaswamy, S., Tamayo, P., Rifkin, R. *et al.* (2001). *Proceedings of the National Academy of Sciences, USA*, **98**, 15149-15154.

[5] Eisen, M., Spellman, P.T., Brown, P.O. and Bostein, D. (1998). *Proceedings of the National Academy of Sciences, USA*, **95**, 14863-14868.

IV

SUPERVISED
LEARNING METHODS

Chapter 18

DISCRIMINATION AND CLASSIFICATION

Unlike cluster analysis, which is considered as unsupervised learning because it involves grouping previously unclassified objects, discriminant analysis, classification trees, neural networks, and support vector machines are considered as supervised learning because one begins the analysis with a set of training data (*the training set*) with known classes or groups. In analyzing microarray data, use of a supervised learning framework allows a researcher to start with a training set of specimens, for example, and to ask questions such as: How well does the gene set discriminate among classes of specimens? Which specimens may be misclassified? What is the typical expression profile of a specimen of a specified type, say A_1?

This chapter considers discrimination and classification methods. The method of discrimination analysis involves classifying items using multivariate data when class memberships are known in advance. We begin with a training set of M specimens composed of L mutually exclusive classes (or groups) with known labels represented by integers ranging from 1 to L. The m-th observation in the training set can be represented as a pair $(\boldsymbol{x}_m, y_m)$, where $\boldsymbol{x}_m = (x_{1m}, \ldots, x_{gm}, \ldots, x_{Gm})'$ consists of expression levels of G genes from specimen m and y_m denotes the known group label for $\boldsymbol{x}_m$. Let M_l denote the number of observations belonging to group l so that $\sum_l M_l = M$. The main objective of discriminant analysis is to derive an effective classification rule from data in the training set. To validate the derived classification rule, it is applied to a fresh data set referred to as the *test set* to see how well it performs. Discrimination and classification methods differ in their assumptions about the structure and distribution pattern of the data, the availability of prior information and the form of the classification rule. We begin with a classical method of discriminant analysis proposed by R. A. Fisher.

18.1.   Fisher's Linear Discriminant Analysis

First, we shall consider the case where the training samples can be divided into two known categories, denoted by A_1 and A_2. Two groups A_1 and A_2 are said to be *linearly separable* if there exists a constant d and a linear function $w'x_i + b$ such that

$$\begin{aligned} w'x_i + b &= \textstyle\sum_{g=1}^{G} w_g x_{gi} + b > d, & \text{if } x_i \in \text{group } A_1 \\ w'x_i + b &= \textstyle\sum_{g=1}^{G} w_g x_{gi} + b < d, & \text{if } x_i \in \text{group } A_2 \quad (18.1) \end{aligned}$$

The set of points $\{x: w'x + b = d\}$ is called a separating hyperplane.

For simplicity of notation, we omit the specimen index i in the remainder of this section. Assume that groups A_1 and A_2 have means $\mu_1 = E(x|A_1)$ and $\mu_2 = E(x|A_2)$, respectively. Also, assume that the covariance matrix is the same for both groups, i.e.,

$$\Sigma_l = E[(x - \mu_l)(x - \mu_l)'] = \Sigma, \quad \text{for } l = 1, 2.$$

Fisher's linear discriminant analysis is based on finding the linear transformation $w'x$ which maximizes the ratio of the between-groups sum of squares and the within-groups sum of squares. This ratio can be expressed in the following form

$$\frac{(w'\mu_1 - w'\mu_2)^2}{w'\Sigma w} \quad (18.2)$$

which equals the squared distance between linear combinations of means divided by the variance of the linear combinations.

The maximum of the above ratio is attained at $w_*' = (\mu_1 - \mu_2)'\Sigma^{-1}$, with

$$\max_{w} \frac{(w'\mu_1 - w'\mu_2)^2}{w'\Sigma w} = (\mu_1 - \mu_2)'\Sigma^{-1}(\mu_1 - \mu_2). \quad (18.3)$$

The linear transformation

$$w_*'x = (\mu_1 - \mu_2)'\Sigma^{-1}x \quad (18.4)$$

is called Fisher's linear discriminant function.

Let

$$\mu_* = \frac{1}{2}\left[E(w_*'x|A_1) + E(w_*'x|A_2)\right] = \frac{(\mu_1 - \mu_2)'\Sigma^{-1}(\mu_1 + \mu_2)}{2}(18.5)$$

denote the midpoint between the means $w_*'\mu_1$ and $w_*'\mu_2$ of the two groups A_1 and A_2. It can be easily shown that $E(w_*'x|A_1) - \mu_* > 0$ and $E(w_*'x|A_2) - \mu_* < 0$ when $\mu_1 > \mu_2$. If an expression vector x is from group A_1, then the linear transformation $w_*'x$ is, on the average, expected to be larger than the midpoint μ_*. Similarly, if x is from group A_2, then the linear transformation $w_*'x$ is expected to be smaller than the midpoint μ_*. Hence Fisher's discrimination function can be used as a classification device for the test set as follows. An expression vector x in the test set is allocated to group A_1 if $w_*'x \geq \mu_*$, and to group A_2 otherwise. Or, equivalently, expression vector x in the test set is allocated to group A_1 if

$$(\mu_1 - \mu_2)'\Sigma^{-1}\left[x - \frac{(\mu_1 + \mu_2)}{2}\right] \geq 0, \tag{18.6}$$

and to group A_2 otherwise.

In practice, mean μ_l can be replaced by the sample mean $\bar{x}_l$ for group $A_l, l = 1, 2$. The covariance matrix Σ can also be replaced by the sample covariance matrix S pooled from both groups A_1 and A_2. An observation x in the test set is then allocated to group A_1 if

$$(\bar{x}_1 - \bar{x}_2)'S^{-1}\left[x - \frac{(\bar{x}_1 + \bar{x}_2)}{2}\right] \geq 0, \tag{18.7}$$

and to group A_2 otherwise.

Fisher's linear discriminant analysis can be extended to discriminating among more than two classes. Detailed descriptions on discriminant analysis can be found in Mardia, Kent, and Bibby[1] (1979), Johnson and Wichern[2] (1998), McLachlan[3] (1992), and Hastie, Tibshirani, and Friedman[4] (2001).

18.2. Maximum Likelihood Discriminant Rules

Assume that data in the training set contains L groups, denoted by $A_1, \ldots, A_L$. Suppose that random vector $x \in A_l$, for $l = 1, \ldots, L$, has a multivariate normal distribution with probability density function

$$f_l(x) = \frac{1}{(2\pi)^{G/2}|\Sigma_l|^{1/2}}e^{-\frac{1}{2}(x-\mu_l)'\Sigma_l^{-1}(x-\mu_l)} \tag{18.8}$$

An observation x in the test set is assigned to group A_{l*} if the function

$$(x - \mu_l)'\Sigma_l^{-1}(x - \mu_l) + \log|\Sigma_l| \tag{18.9}$$

achieves its minimum when $l = l^*$. When the class parameters $\boldsymbol{\mu}_l$ and $\boldsymbol{\Sigma}_l$ are unknown, they can be estimated from a training set. As shown by equation (18.9), the maximum likelihood discriminant rule is a quadratic discriminant rule in the normal case.

Notice that, for the special case when $L = 2$ and $\boldsymbol{\Sigma}_1 = \boldsymbol{\Sigma}_2 = \boldsymbol{\Sigma}$, i.e., when discriminating between two groups with the same covariance matrix, the maximum likelihood discriminant rule (18.9) assigns an observation $\boldsymbol{x}$ to group A_1 if

$$(\boldsymbol{x} - \boldsymbol{\mu}_1)'\boldsymbol{\Sigma}^{-1}(\boldsymbol{x} - \boldsymbol{\mu}_1) \leq (\boldsymbol{x} - \boldsymbol{\mu}_2)'\boldsymbol{\Sigma}^{-1}(\boldsymbol{x} - \boldsymbol{\mu}_2),$$

and assigns $\boldsymbol{x}$ to group A_2 otherwise. As the quadratic terms $\boldsymbol{x}'\boldsymbol{\Sigma}^{-1}\boldsymbol{x}$ cancel on both sides of the inequality, this rule corresponds to Fisher's linear discriminant rule (18.6).

If one assumes that the two groups have a common diagonal covariance matrix, which ignores correlations between genes, i.e.,

$$\boldsymbol{\Sigma} = \begin{pmatrix} \sigma_1{}^2 & 0 & \cdot & 0 \\ 0 & \sigma_2{}^2 & \cdot & 0 \\ \cdot & \cdot & & \cdot \\ \cdot & \cdot & & \cdot \\ \cdot & \cdot & & \cdot \\ 0 & 0 & \cdot & \sigma_G{}^2 \end{pmatrix} \tag{18.10}$$

then, from equation (18.7), the maximum likelihood discriminant rule assigns a gene expression vector $\boldsymbol{x}_i = (x_{1i}, \ldots, x_{gi}, \ldots, x_{Gi})$ for specimen i to group A_1 if

$$\sum_{g=1}^{G} \frac{(\bar{x}_{A_1}{}^{(g)} - \bar{x}_{A_2}{}^{(g)})}{\hat{\sigma}_g{}^2} \left[x_i - \frac{(\bar{x}_{A_1}{}^{(g)} + \bar{x}_{A_2}{}^{(g)})}{2} \right] \geq 0, \tag{18.11}$$

and assigns $\boldsymbol{x}_i$ to group A_2 otherwise. Here, $\bar{x}_{A_l}{}^{(g)}$ denotes the sample mean for gene g for specimens in group A_l, $l = 1, 2$, in the training set and $\hat{\sigma}_g{}^2$ is the pooled sample variance for gene g for the training set.

18.3.   Bayesian Classification

Assume that the training set contains L groups $A_1, \ldots, A_L$. Let π_l be the prior probability that a specimen belongs to group l, with $\sum_{l=1}^{L} \pi_l = 1$. Let $\boldsymbol{x}$ denote an expression vector for a specimen in the test set. Assume that $\boldsymbol{x} \in A_l$ follows a probability density function $f_l(\boldsymbol{x})$, for $l = 1, \ldots, L$. Then, $\boldsymbol{x}$ in the test set is assigned to group A_{l*} if the posterior probability function

$$\frac{f_l(\boldsymbol{x})\pi_l}{\sum_{l=i}^{L} f_l(\boldsymbol{x})\pi_l} \tag{18.12}$$

attains its maximum when $l = l*$.

If we further assume that each group has a multivariate normal density $f_l(\boldsymbol{x})$ as defined in (18.8), and that the covariance matrix $\boldsymbol{\Sigma}_l$ is the same for $l = 1, \ldots, L$, then the log-ratio of the posterior probabilities for any pair of groups l_1 and l_2 reduces to the following linear function

$$\log \pi_{l_1} - \log \pi_{l_2} + (\boldsymbol{\mu}_{l_1} - \boldsymbol{\mu}_{l_2})'\boldsymbol{\Sigma}^{-1}\boldsymbol{x} - \frac{(\boldsymbol{\mu}_{l_1} - \boldsymbol{\mu}_{l_2})'\boldsymbol{\Sigma}^{-1}(\boldsymbol{\mu}_{l_1} + \boldsymbol{\mu}_{l_2})}{2}$$

The above function can be used to compare assignments between groups l_1 and l_2. For a good review of Bayesian classification, see Denison *et al.*[5] (2002).

18.4. *k*-Nearest Neighbor Classifier

Early applications of the *k*-nearest neighbor classifier can be found in Fix and Hodges[6] (1951) and Cover and Hart[7] (1967). On the basis of a given training set and a chosen distance measure for pairs of observations, such as the Euclidean distance or correlation coefficient, the *k*-nearest neighbor method classifies an observation in the test set by the following steps:

(1) Select an integer k
(2) Find the k closest observations in the training set
(3) Classify this test observation using majority vote; that is, choose the class that is most common among these k neighbors in the training set

For a given value of k, the algorithm begins by sequentially omitting one observation from the training set and using this observation as a test observation. It next computes the distance of this test observation to all of the other observations left in the training set and then classifies it by the nearest neighbor rule. This procedure is repeated until every observation is selected and omitted once as the test case. The cross-validation error rate can then be computed for the original training set. The above steps are repeated for different k values, and the k value with the smallest cross-validation error rate can be applied to the test data set.

18.5.   Neighborhood Analysis

Let $\mu_l{}^{(g)}$, and $\sigma_l{}^{(g)}$ denote the sample mean and standard deviation of the logarithms of expression levels of gene g for specimens in class A_l of the training set, for $l = 1, 2$. Golub *et al.*[8] (1999) considered a measure w_g defined by

$$w_g = \frac{\mu_1{}^{(g)} - \mu_2{}^{(g)}}{\sigma_1{}^{(g)} + \sigma_2{}^{(g)}} \qquad (18.13)$$

to emphasize the *signal-to-noise ratio* in using gene g as a class predictor. They argued that large absolute values of w_g indicate a strong correlation between expression level and class assignment (which the authors called *class distinction*). The sign of w_g being positive or negative corresponds to gene g being more highly expressed in class A_1 or class A_2. Unlike a standard Pearson correlation coefficient, this measure w_g is not confined to the range $[-1, +1]$.

Let M denote the number of specimens in the training set. Each gene g can be represented by an expression vector $\boldsymbol{x}^{(g)}$ of dimension M, consisting of its expression levels on all specimens. A class distinction is represented by an idealized expression pattern denoted by the index vector $\boldsymbol{c} = (c_1, c_2, ..., c_M)$, where $c_i = 1$ if the i-th specimen belongs to class A_1 and $c_i = 0$ if the i-th specimen belongs to class A_2. This idealized expression pattern corresponds to a gene that is uniformly high in one class and uniformly low in the other. Define neighborhoods $N_1(\boldsymbol{c}, r)$ and $N_2(\boldsymbol{c}, r)$ of radius r around class A_1 and class A_2 to be the sets of genes such that $w_g = r$ and $w_g = -r$, respectively. Neighborhood analysis then involves counting the number of genes having various levels of correlation with the idealized expression pattern $\boldsymbol{c}$.

A high density of genes in the neighborhoods of $\boldsymbol{c}$ indicates that there are many more genes that are correlated with the pattern than can be expected by chance. Golub *et al.* (1999) used a permutation test to infer whether the density of genes in a neighborhood was significantly higher than expected. They compared the number of genes in the neighborhood to the number of genes in similar neighborhoods around idealized expression patterns corresponding to random class distinctions, obtained by randomly permuting the coordinates of $\boldsymbol{c}$. The neighborhood analysis selects a set of *informative genes*, consisting of genes closest to class A_1 with signal-to-noise ratio w_g as large as possible and genes closest to class A_2 with $-w_g$ as large as possible.

18.6. A Gene-casting Weighted Voting Scheme

To predict tumor type, Golub *et al.* (1999) used the methods of supervised learning and derived discriminant decision rules based on the magnitude and threshold of prediction strength. They considered a *weighted voting scheme* that uses a weighted linear combination of relevant "marker" or "informative" genes obtained in the training set to provide a classification scheme for new specimens. They divided cancer classification into two challenges: *class discovery* and *class prediction*. Class discovery refers to defining previously unrecognized tumor subtypes. Class prediction refers to assigning specimens to already-defined classes.

The selection of informative genes is accomplished by computing the signal-to-noise ratio statistic w_g as defined in (18.13). The class predictor is uniquely defined by the initial set of specimens and informative genes. The prediction of a new specimen is based on "weighted votes" of the set of informative genes. Each informative gene g casts a weighted vote for either class A_1 or A_2, with the weight of each vote dependent on its log-expression level in the new specimen and the degree of that gene's correlation with the class distinction. The number of informative genes is a free parameter in defining the class predictor.

The class predictor is formed from specimens in the training set and the set of informative genes. Let w_g denote the weighting factor defined in (18.13) which reflects the correlation between the expression levels of gene g and the class distinction. Let $b_g = (\mu_1{}^{(g)} + \mu_2{}^{(g)})/2$ denote the average of the mean log-expression values in the two classes. Parameters (w_g, b_g) can be defined for each informative gene.

Consider a new specimen to be predicted having gene expression vector $\boldsymbol{x}$. Let $x^{(g)}$ denote the normalized log-expression level of gene g in this new specimen. To predict the class of this test specimen, each gene in the informative gene set casts a vote. The vote of gene g is

$$v_g = w_g(x^{(g)} - b_g) = \frac{(\mu_1{}^{(g)} - \mu_2{}^{(g)})}{\sigma_1{}^{(g)} + \sigma_2{}^{(g)}} \left[x_g - \frac{(\mu_1{}^{(g)} + \mu_2{}^{(g)})}{2} \right] \qquad (18.14)$$

Note that the form of v_g is very similar to the special case of the linear maximum likelihood discriminant rule described in (18.11), except that the sum of standard deviations $\sigma_1{}^{(g)} + \sigma_2{}^{(g)}$ is used in the denominator of the weight instead of the variance $\hat{\sigma}_g^2$. A positive value of v_g indicates a vote for class 1 and a negative value indicates a vote for class 2. The total vote $V_1 = \sum_g \max(v_g, 0)$ for class 1 is obtained by summing the absolute values of the positive votes over the informative genes. Analogously, the total vote $V_2 = \sum_g \max(-v_g, 0)$ for class 2 is obtained by summing the absolute values of the negative votes. The sample is assigned to the class with the higher vote total, provided that the prediction strength exceeds a predetermined threshold.

The gene-casting weighted models can be evaluated by leave-one-out cross validation with the training set being used to predict the class of a randomly withheld specimen. This procedure can be repeated for all specimens and the cumulative error rate recorded. Thereafter, the total number of prediction errors in cross-validation can be calculated and a final model chosen which minimizes cross-validation error.

18.7.   Example: Classification of Leukemia Samples

Golub *et al.* (1999) noted that cancer classification has been difficult in part because it has historically relied on specific biological insights, rather than systematic and unbiased approaches for recognizing tumor subtypes. Therefore, they considered a generic approach to cancer classification based on gene expression monitoring by DNA microarrays.

To demonstrate the methods, they applied their methods to human acute leukemias as a test case. An initial collection of samples belonging to known classes was used to create a "class predictor" to classify new, unknown samples. Their initial leukemia data set consisted of 38 bone marrow samples (27 acute lymphoblastic leukemia (ALL), 11 acute myeloid leukemia (AML)) obtained from acute leukemia patients at the time of diagnosis. RNA prepared from bone marrow mononuclear cells was hybridized to high-density oligonucleotide microarrays, produced by Affymetrix and containing probes for 6817 human genes. The 6817 genes were sorted by their degree of correlation. Neighborhood analysis and permutation tests showed that about 1100 genes were more highly correlated with the AML-ALL class distinction than would be expected by chance.

To test the validity of the class predictors, a two-step procedure was conducted. The accuracy of the predictors was first tested by leave-one-out cross-validation on the initial data set. Then a final predictor was built based on the initial data set and its accuracy was accessed using a test set of samples.

The two-step validation approach was applied to the 38 leukemia samples. The set of informative genes to be used in the predictor was chosen to be the 50 genes most closely correlated with AML-ALL distinction in the training set. The parameters of the predictor were determined by the expression levels of these 50 genes in the training samples. The predictor was then used to classify new samples, by applying it to the expression levels of these 50 genes in the new samples. A sample was assigned to the class with the greater prediction strength, provided that strength exceeded a predetermined threshold. Otherwise, the sample was given an 'uncertain' classification. On the basis of previous analysis, Golub *et al.* (1999) used a threshold value of 0.3. The 50-gene predictors derived in cross-validation tests assigned 36 of the 38 samples as either AML or ALL and the remaining two as uncertain. They then applied this 50-gene predictor to an independent collection of 34 leukemia samples that consisted of 24 bone marrow and 10 peripheral blood samples. The predictor made strong predictions for 29 of these 34 test samples.

To test the hypothesis that class discovery could be tested by class prediction, Golub *et al.* evaluated the clusters A_1 and A_2 found by the SOM methods as discussed earlier in section 17.5.2. They constructed predictors to assign new samples as "type A_1" or "type A_2". Predictors that used a wide range of different numbers of informative genes performed well in cross-validation. The results suggest an iterative procedure for refining clusters, in which a SOM is used initially to cluster the data, a predictor is constructed, and samples not correctly predicted in cross-validation are removed. The edited data set could then be used to generate an improved predictor to be tested on an independent data set.

The procedure for selecting a threshold of prediction strength is an essential ingredient for classification. Heuristic selection rules can be used, but with a certain unavoidable level of subjectivity.

Notes

[1] Mardia, K.V., Kent, J.T., Bibby, J.M. (1979). *Multivariate Analysis*, Academic Press, Inc., San Diego.

[2] Johnson, R.A., Wichern, D.W. (1998). *Applied Multivariate Statistical Analysis*, 4th Edition, Prentice-Hall, Inc., New Jersey.

[3] McLachlan, G.J. (1992). *Discriminant Analysis and Statistical Pattern Recognition.* Wiley, New York.

[4] Hastie, T., Tibshirani, R., and Friedman, J. (2001). *The Elements of Statistical Learning*, Springer, New York.

[5] Denison, D.G.T., Holmes, C.C., Mallick, B.K., and Smith, A.F.M. (2002). *Bayesian Methods for Nonlinear Classification and Regression*, John Wiley & Sons, West Sussex, U.K.

[6] Fix, E., and Hodges, J. (1951). Technical report, Randloph Field, Texas, USAF School of Aviation Medicine.

[7] Cover, T. and Hart, P. (1967). *Proc. IEEE Trans. Inform. Theory*, **11**, 21-27.

[8] Golub, T.R., Slonim, D.K., Tamayo, P., *et al.* (1999). *Science*, **286**, 531-537.

Chapter 19

ARTIFICIAL NEURAL NETWORKS

Artificial neural networks (ANN) are computer-based learning algorithms that are modeled on the neural activity of the brain and can be trained to recognize and categorize complex patterns[1]. The three basic features of a neural network are its nodes, its architecture, and its training algorithm[2]. The nodes are generally arranged in layers with the network connections running in a directed fashion between layers, as shown in Figures 19.1 and 19.2. The details of these figures will be explained later. Continuing with the brain analogy, each node represents a neuron and the connections between them represent synapses. A neuron fires when the total signal passed to it exceeds a certain threshold. The firing is its output. The brain learns by adapting the strength of the synaptic connections. Likewise, the synaptic weights in neural networks are adjusted to solve the problem presented to the network. *Learning* or *training* are terms used to describe the process of finding the appropriate values of these weights.

Mathematically, the nodes of a neural network are simple computational units. Each node combines input values received from its incoming connections and outputs a value to its outgoing connections. The output value is generating from the input values by a smooth mathematical function that gives appropriate weights to each input. Neural networks permit development of a well-fitted model through the learning algorithm if sufficient data are available. Mathematical descriptions of neural networks are plentiful in the literature, including Hertz, Krogh and Palmer[3] (1991), Ripley[4] (1996), Vapnik[5] (1998), and Hastie, Tibshirani, and Friedman[6] (2001), among others.

Neural networks have been applied to clinical problems such as classifying arrhythmias from electrocardiograms[7] and diagnosing myocardial infarctions[8]. Hudson and Cohen[9] (2000) described the development of neural networks in medical applications. In analysis of microarray data, the method of neural networks can be used for classifying either genes or sample specimens. Khan *et al.*[10] (2001) applied the method of artificial neural networks to decipher gene-expression signatures of small, round blue-cell tumors and used them for diagnostic classification. In this chapter, we present the methods in terms of classifying sample specimens. The methods, however, can be easily adapted to applications in classifying genes.

19.1.   Single-layer Neural Network

To introduce neural networks, it is instructive to begin with one that consists of a layer of input nodes and an output layer consisting of one output node, as illustrated in Figure 19.1. This kind of neural network is called a *single-layer network* because the input layer is not counted. In this section, it will be explained how this simple network can classify biological specimens based on their gene expression data.

Let $x = (x_0, x_1, \ldots, x_G)'$ denote a vector of input variables. The leading variable x_0 is, in fact, fixed to the constant value 1, i.e., $x_0 = 1$. The remaining G input variables in vector x are gene expression readings for a given sample specimen in the microarray data set. The augmentation of the vector of gene expression variables by constant x_0 helps to simplify later notation. In Figure 19.1, there are $G = 3$ input variables plus constant x_0.

19.1.1.   Separating Hyperplanes

Assume that the learning set for the neural network consists of n biological specimens $x_1, x_2, \ldots, x_n$ from two classes A_1 and A_2. If the two classes are linearly separable, then there exists a *hyperplane*, also called a *decision boundary*, denoted by $\mathcal{D}$, such that $w'x = 0$ for any x on the hyperplane $\mathcal{D}$, and

$$w'x > 0, \quad \text{if} \quad x \in A_1$$
$$w'x < 0, \quad \text{if} \quad x \in A_2$$

where $w = (w_0, w_1, \ldots, w_G)'$ is called the *weight vector*. Now the role of constant $x_0 = 1$ can be seen as adding a constant term w_0 to the inner product $w'x$. Constant w_0 is often referred to as the *bias*. Alternatively, $-w_0$ may be called the *threshold* (Observe that w_0 plays the role of constant b in a support vector machine in Chapter 20). The weights of

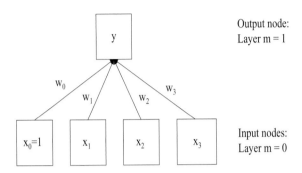

Figure 19.1. An illustration of the architecture of a single-layer neural network or perceptron. Each box denotes a neuron or node. The bottom row is a layer of input variables, including a constant term. One node defines the output layer. Each directed arrow is a synaptic connection representing an influence of an input node on the output node.

vector w are displayed in Figure 19.1 on the arrows connecting the input nodes to the output node.

19.1.2.   Class Labels

Next, the true class labels are added to the separable learning set of n samples to form the set of pairs $S = \{(x_1, y_1), (x_2, y_2), \ldots, (x_n, y_n)\}$. Here y_i denotes the true class label corresponding to x_i. As the learning set contains only two classes, A_1 and A_2, it is convenient to adopt the following number codes for the labels.

$$y_i = 1 \quad \text{if the sample point } x_i \text{ belongs to class } A_1,$$
$$y_i = -1 \quad \text{if the sample point } x_i \text{ belongs to class } A_2.$$

In Figure 19.1, the output node is labelled by the true class label y. The objective of the neural network is to have the output node predict the correct class label y_i for each sample i from knowledge of the input vector x_i. The success of the objective depends on a careful selection of weight vector w.

19.1.3.   Decision Rules

Given any sample point x, a decision rule $\hat{y} = \varphi(x)$ regarding the predicted label for x can be defined in terms of a linear function of the input x as follows

$$\hat{y} = \varphi(x) = \text{sign}(w'x), \tag{19.1}$$

where $\text{sign}(s)$ denotes the sign of quantity s. Thus, predicted class label $\hat{y}$ takes value 1 or -1 according to whether $w'x$ is positive or negative.

19.1.4.   Risk Functions

To choose weight vector w, a criterion of success (an objective function) is required. That role is played by the *risk function*. Let $R(w)$ denote the risk function for decision rule $\varphi(x)$ in (19.1). Risk function $R(w)$ measures the success of a decision rule by comparing the true labels y_i with the predicted labels $\hat{y}_i = \varphi(x_i)$. In a neural network that is used for classification, as here, the weight vector w is chosen to minimize the risk function. One commonly used risk function is the sum of squared errors:

$$R(w) = \sum_{i=1}^{n}(y_i - \hat{y}_i)^2. \tag{19.2}$$

Other risk functions are also used as will be demonstrated shortly.

19.1.5.   Gradient Descent Procedures

A *gradient descent procedure*, also called a *steepest descent procedure*, is a recursive numerical procedure to find a local minimum of a function. If $R(w)$ denotes the risk function of a decision rule $\varphi(x)$ of the type shown in (19.1), the goal is to find the vector w that minimizes $R(w)$. The gradient descent procedure can be used to find optimum weights w to achieve the minimum risk. Let $\frac{\partial}{\partial w}R(w)$ denote the gradient of $R(w)$ for the decision rule φ. Then $\frac{\partial}{\partial w}R(w)$ gives the direction of steepest descent for function $R(w)$ at point w. Starting with an initial approximation w_0 to the minimum, the procedure steps toward the minimum. At each step, the weight vector is adjusted according to the following rule:

$$w_{k+1} = w_k - \tau \frac{\partial}{\partial w} R(w_k), \quad \text{for } k = 0, 1, \dots \tag{19.3}$$

where τ is a specified scalar or step size.

19.1.6. Rosenblatt's Perceptron Method

To complete the development of this single-layer neural network, we link it to an historical idea attributed to F. Rosenblatt and often called *Rosenblatt's perceptron*. Extending the symbolic logic models of Mc-Culloch and Pitts[11] (1943), Rosenblatt[12] (1958) introduced early neural models[13] based on probability models that he called *perceptrons*. The photoperceptron contains an input sensory array corresponding to the retinal structure. Each point in the array responds to light in an on/off manner. Input is then transmitted to an association area. The connections have three possible weights: 1 (excitatory), -1 (inhibitory), or 0. When an optical pattern is presented to the sensory array, a unit in the association area becomes active and it generates an output response depending on whether the weighted input signals exceed a predetermined threshold. Similar to the gradient descent procedure, the basic perceptron model is an example of a statistical algorithm for learning linear classification.

To build this kind of model here, let us continue with the separable learning set that contains only two classes, A_1 and A_2, with class labels $y_i = 1$ and -1. If the decision function $\hat{y}_i = \varphi(x_i) = \text{sign}(w'x_i)$, as defined in (19.1), predicts a correct label for x_i, then the product $y_i \varphi(x_i) > 0$. If the predicted label is incorrect then $y_i \varphi(x_i) \leq 0$. Hence, for each pair (x_i, y_i) in the learning set, the following quantity can be defined

$$\gamma_i = y_i (w'x_i) \tag{19.4}$$

Value γ_i is called the *functional margin* of x_i with respect to a hyperplane defined by the set of weights w. The functional margin γ_i is positive if and only if the the assigned label $\varphi(x_i)$ and the true label y_i have the same sign. In this case the algorithm made a correct classification of x_i, and the weight vector w remains unchanged. If the functional margin γ_i is less than or equal to zero, then the classification $\varphi(x_i)$ is incorrect and the weight vector w needs to be adjusted.

The perceptron model adopts a risk function $R(w)$ that equals the total of the functional margins γ_i of incorrect decisions. Let J denote the set of points $\boldsymbol{x}_i$ which are misclassified by the decision rule φ defined in (19.1). Then,

$$R(\boldsymbol{w}) = \sum_{i \in J}(-\gamma_i) = -\sum_{i \in J} y_i \, (\boldsymbol{w}'\boldsymbol{x}_i). \tag{19.5}$$

For the perceptron model, the gradient of $R(\boldsymbol{w})$ can be easily computed.

$$\frac{\partial}{\partial \boldsymbol{w}} R(\boldsymbol{w}) = -\sum_{i \in J} y_i \, \boldsymbol{x}_i. \tag{19.6}$$

Hence, the gradient descent procedure can be applied as described in the next paragraph to find a numerical solution that minimizes the risk function (19.5).

In the context of learning algorithms, the steepest descent step size $\tau > 0$ is called the *learning rate*. Now, choose τ. Next, set the initial weight vector $\boldsymbol{w}_0 = \boldsymbol{0}$ and iteration counter $k = 0$. The perceptron algorithm works by adding misclassified positive training samples (i.e., class A_1) or subtracting misclassified negative ones (class A_2), to the initial weight vector. That is, if the functional margin $y_i \, (\boldsymbol{w}'\boldsymbol{x}_i) \leq 0$ for some i then, with the gradient defined as in (19.6), the perceptron algorithm updates the weight vector as follows.

$$\boldsymbol{w}_{k+1} = \boldsymbol{w}_k - \tau \, \frac{\partial}{\partial \boldsymbol{w}} R(\boldsymbol{w}_k) = \boldsymbol{w}_k + \tau \sum_{i \in J} y_i \boldsymbol{x}_i \tag{19.7}$$

The iterations continue until no classification errors are made within the loop for each $\boldsymbol{x}_i$. The algorithm computes a linear combination of the variables and returns the signs of the class predictions. This procedure is guaranteed to converge if there exists a hyperplane that correctly classifies (separates) the training data[14].

19.2. General Structure of Multilayer Neural Networks

A single-layer neural network (i.e., a linear perceptron learning machine) has limited computational power. From the preceding development, it is clear that a perceptron having only the input layer of variables can only separate categories that are linearly separable. The output classification is determined solely by a linear combination of the input variables. However, by adding one or more layers between the input and output layers, a multilayer neural network is formed and it has a much richer response structure. This idea is now explained by reference to the illustrative neural network in Figure 19.2.

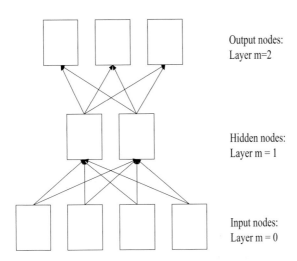

Output nodes:
Layer m=2

Hidden nodes:
Layer m = 1

Input nodes:
Layer m = 0

Figure 19.2. An illustration of the architecture of a multilayer neural network, with one hidden layer of neurons. Each box denotes a neuron or node. Each directed arrow is a synaptic connection representing an influence of a predecessor node on its successor in the next layer.

- Input Layer: The input layer (the bottom layer in Figure 19.2) consists of input nodes, where each node represents an input variable.

- Output Layer: The last layer (the top layer in Figure 19.2) contains the output nodes corresponding to response variables. For a two-category classification problem, the output layer needs only one node (as is the case in Figure 19.1). Two or more output nodes are used in neural network classification applications where there are more than two classes. The three output nodes in Figure 19.2, for example, would be appropriate for a classification problem having up to eight classes.

- Hidden Layer: A *hidden,* or *interactive layer,* consists of derived nodes, also called *features.* Figure 19.2 shows one such hidden layer, although other hidden layers may be added. The derived nodes in a hidden layer are unobservable and are formed by *activation functions* based on linear combinations of outputs from nodes in the preceding layer. Thus, for example, if $z_j^{(m-1)}$ denotes the output from node j in the $m-1$th layer then the output from the kth node in the mth layer is given by

$$z_k^{(m)} = f\left(\sum_{j=1}^{g_{m-1}} w_{kj}^{(m)} z_j^{(m-1)}\right), \quad \text{for} \ \ k = 1, \dots, g_m. \qquad (19.8)$$

Here g_m denotes the number of nodes in the mth layer and $w_{kj}^{(m)}$ is the weight given to the output from the jth node at the $m-1$th level in the activation function for the kth node at the mth level. The function $f(\cdot)$ represents some smooth monotone function. The input layer corresponds to level $m = 0$ and the output layer (the top layer) to level $m = M$. In Figure 19.2, $M = 2$.

The learning algorithm determines the optimal weights of the activation functions. The output of one layer forms the input for the next higher layer. The final output of the neural network are the values $z_k^{(M)}, k = 1, \dots, g_M$, of the output layer. Hence, mathematically, a neural network is a nested set of linear combinations of smooth monotone functions, starting with a linear combination of input variables. Linear functions, step functions or sigmoid functions are used as activation functions, as will be discussed shortly.

19.3.   Training a Multilayer Neural Network

This section looks at the mathematical methods required to train a multilayer neural network. It begins with an introduction to sigmoid functions. Next, the mathematical formulation for the multilayer neural network is presented. Finally, a solution algorithm is described.

19.3.1.   Sigmoid Functions

The activation function $f(\cdot)$ of a neural network that is being used for classification should be, ideally, an indicator function so that the output will be an on/off response. Yet, indicator functions are problematic when deriving the gradient for an application of the gradient descent procedure. Indicator functions (and, more generally, sign functions and step functions), however, can be well approximated by smooth sigmoid functions.

A *sigmoid function* ς is a smooth monotonic function such that $\varsigma(-\infty) = a$ and $\varsigma(+\infty) = b$, with $b > a$, where a and b are the asymptotes of the function. For example, the hyperbolic tangent (tanh) function

$$\varsigma(u) = \tanh(u) = \frac{\exp(u) - \exp(-u)}{\exp(u) + \exp(-u)} \tag{19.9}$$

is a sigmoid function for which $a = -1$ and $b = 1$. The logistic function

$$\varsigma(u) = \frac{1}{1 + \exp(-u)} \tag{19.10}$$

is also a sigmoid function with $a = 0$ and $b = 1$. Simple rescaling of any sigmoid function can adapt it to arbitrary values of the asymptotes a and b.

The sigmoid functions are easily differentiable and, hence, are widely used in constructing multilayer neural networks.

19.3.2.   Mathematical Formulation

The multilayer neural network consists of the following elements, which are described in a microarray context.

- An input layer with nodes representing the variables of a gene expression vector x_i for biological specimens $i = 1, \ldots, n$. Each vector also includes a leading constant $x_0 = 1$. For consistency with the notation that follows, x_i may be represented notationally by $z_i^{(0)}$.

- One or more hidden layers, indexed by $m = 1, \ldots, M - 1$, with the mth hidden level having l_m derived nodes. The feature variables for the nodes of the mth layer are represented by vector $z_i^{(m)}$ for sample $i, i = 1, \ldots, n$, where

$$z_i^{(m)} = \left(z_{1i}^{(m)}, \ldots, z_{l_m i}^{(m)} \right)'$$

.

For $m = 1, \ldots, M$, the $(m-1)$th layer is connected to the mth layer through a $l_m \times l_{m-1}$ dimensional weight matrix $\boldsymbol{W}^{(m)} = (w_{kj}^{(m)})$ and a sigmoid transformation ς. Specifically, the layer-to-layer connection can be written in the simple matrix form

$$z_i^{(m)} = \varsigma \left(\boldsymbol{W}^{(m)} z_i^{(m-1)} \right), \quad \text{for } i = 1, \ldots, n; \; m = 1, \ldots, M \tag{19.11}$$

where for any column vector $\boldsymbol{u} = (u_1, \ldots, u_l)'$, $\varsigma(\boldsymbol{u}) = (\varsigma(u_1), \ldots, \varsigma(u_l))'$ denotes the sigmoid transformation of $\boldsymbol{u}$.

- The Mth layer is the output layer. For the case of classifying samples into two classes with labels $y_i = 1$ or $y_i = -1$, the output layer has a single node with predicted output $\hat{y}_i = z_i^{(M)}$. The vector $z_i^{(M)}$ here reduces to a single scalar component, i.e., $l_M = 1$.

19.3.3.   Training Algorithm

One standard training algorithm for the multilayer neural network is called the back-propagation method. Using gradients, one can iteratively modify the coefficients (weights) of a neural network on the basis of standard gradient-based procedures.

For multilayer neural networks sigmoid approximations are used for the activation functions at the stage of estimating weight coefficients and then indicator or sign functions, evaluated with the resulting weight coefficients, are used at the stage of recognition so that one can evaluate the gradient for an entire set of neurons.

Consider the multilayer neural network defined in (19.11). Let the risk function of this network be defined by the sum of squared errors

$$
\begin{aligned}
R(\boldsymbol{W}^{(1)}, \ldots, \boldsymbol{W}^{(M)}) \; &= \sum_{i=1}^{n} \; (y_i - \hat{y}_i)^2 \\
&= \sum_{i=1}^{n} \; (y_i - z_i^{(M)})^2 .
\end{aligned} \tag{19.12}
$$

To find a local minimum for the risk function (19.12) under the set of equality type constraints $z_i^{(m)} = \varsigma\,(\boldsymbol{W}^{(m)} z_i^{(m-1)})$, for $m = 1 \ldots, M$, the method of Lagrange multipliers can be applied as follows.

Define the Lagrange function $\mathcal{L}$ such that

$$
\mathcal{L}(\boldsymbol{W}^{(1)}, .., \boldsymbol{W}^{(M)}, z_1^{(1)}, .., z_i^{(m)}, .., z_n^{(M)}, \boldsymbol{\alpha}_1^{(1)}, .., \boldsymbol{\alpha}_i^{(m)}, .., \boldsymbol{\alpha}_n^{(M)})
$$
$$
= \sum_{i=1}^{n} (y_i - z_i^{(M)})^2 + \sum_{i=1}^{n} \sum_{m=1}^{M} \boldsymbol{\alpha}_i^{(m)\prime} [z_i^{(m)} - \varsigma(\boldsymbol{W}^{(m)} z_i^{(m-1)})],
$$

where $\boldsymbol{\alpha}_i^{(m)} = (\alpha_{1i}^{(m)}, \ldots, \alpha_{g_m i}^{(m)})$ is a vector of Lagrange coefficients. To minimize the Lagrange function as listed above, it is necessary to find the gradient of $\mathcal{L}$ with respect to all parameters $\boldsymbol{W}^{(1)}, \ldots, \boldsymbol{W}^{(M)}$, $z_1^{(1)}, \ldots, z_i^{(m)}, \ldots, z_n^{(M)}, \boldsymbol{\alpha}_1^{(1)}, \ldots, \boldsymbol{\alpha}_n^{(M)}$ and set these gradients to zero. This leads to three sets of equations.

(1) *The Forward Dynamic:*

Set the partial derivatives

$$
\frac{\partial}{\partial \boldsymbol{\alpha}_i^{(m)}} \mathcal{L}(\boldsymbol{W}^{(1)}, \ldots, \boldsymbol{W}^{(M)}, z_1^{(1)}, \ldots, z_n^{(M)}, \boldsymbol{\alpha}_1^{(1)}, \ldots, \boldsymbol{\alpha}_i^{(m)}, \ldots, \boldsymbol{\alpha}_n^{(M)})
$$

equal to zero for all i, m. It then gives the following set of equations

$$
z_i^{(m)} = \varsigma\,(\boldsymbol{W}^{(m)} z_i^{(m-1)}), \quad i = 1, \ldots, n, \quad m = 1, \ldots, M, \tag{19.13}
$$

with initial conditions $z_i{}^{(0)} = x_i, i = 1, \ldots, n$.

Equation (19.13) iteratively defines values of the vector $z_i{}^{(m)}$ for all levels $m = 1, \ldots, M$ of the neural net and, hence, is called the *forward dynamic*.

(2) *The Backward Dynamic:*

Set the partial derivatives

$$\frac{\partial}{\partial z_i{}^{(m)}} \mathcal{L}(\boldsymbol{W}^{(1)}, \ldots, \boldsymbol{W}^{(M)}, z_1{}^{(1)}, \ldots, z_i{}^{(m)}, \ldots, z_n{}^{(M)}, \boldsymbol{\alpha}_1{}^{(1)}, \ldots, \boldsymbol{\alpha}_n{}^{(M)})$$

equal to zero for all i, m. It then gives the following equations.

- For the Mth layer, i.e., the output layer, we have

$$\boldsymbol{\alpha}_i{}^{(M)} = 2(y_i - \hat{y}_i) = 2(y_i - z_i{}^{(M)}), \quad i = 1, \ldots, n. \qquad (19.14)$$

- For the hidden layers, for all $i = 1, \ldots, n$, and $m = 1, \ldots, M - 1$, we have

$$\boldsymbol{\alpha}_i{}^{(m)} = \boldsymbol{W}^{(m+1)\prime} \Delta_\varsigma(\boldsymbol{W}^{(m+1)} z_i{}^{(m)}) \boldsymbol{\alpha}_i{}^{(m+1)}, \qquad (19.15)$$

where the function $\Delta_\varsigma(\boldsymbol{W}^{(m+1)} z_i{}^{(m)})$ is a $l_{m+1} \times l_{m+1}$ diagonal matrix with diagonal elements $\frac{d}{du}\varsigma(u)$.

Equation (19.15) iteratively defines Lagrange multipliers $\boldsymbol{\alpha}_i{}^{(m)}$ in terms of $\boldsymbol{\alpha}_i{}^{(m+1)}$, and, hence, is called the *backward dynamic*.

(3) *Estimating the Weight Matrix by the Gradient Descent Method:*

The condition

$$\frac{\partial}{\partial \boldsymbol{W}^{(m)}} \mathcal{L}(\boldsymbol{W}^{(1)}, \ldots, \boldsymbol{W}^{(M)}, z_1{}^{(1)}, \ldots, z_n{}^{(M)}, \boldsymbol{\alpha}_1{}^{(1)}, \ldots, \boldsymbol{\alpha}_n{}^{(M)}) = 0,$$

for all m, gives the following iterative equations.

$$\boldsymbol{W}^{(m)} \leftarrow \boldsymbol{W}^{(m)} - \tau_t \frac{\partial \mathcal{L}}{\partial \boldsymbol{W}^{(m)}}, \quad \text{for all } m = 1, \ldots, M, \qquad (19.16)$$

where τ_t is the step size for iteration t.

19.3.4.   Discussion

A multilayer neural network trained using back-propagation can solve a problem that is not linearly separable. Because a neural network can approximate any continuous function to any degree of accuracy, a neural network is useful when one does not have any idea of the functional relationship between the input and the output variables, i.e., for a 'black-box' problem. A neural network, however, cannot reveal the true functional relation that is buried in the summing of the sigmoidal functions. The solution usually provides a very good fit to a training sample because of the abundance of weight coefficients being fitted in the model. The fitted model may perform less well with predicting outcomes for a holdout sample. As a neural network does not embody any random error terms, it does not have a built-in inference mechanism. Continuing research is attempting to remedy this limitation.

The empirical risk function may have more than one minimum. The gradient descent procedure will converge to one of them. Neural networks typically have a slow convergence rate. The quality of the resulting solution depends on many characteristics of the network formulation and algorithmic implementation, including the initial starting weights, the number of hidden layers, the risk function, and the learning rate or step size τ.

Hence, neural networks are not well-controlled learning machines. In many practical applications, however, neural networks demonstrate good results[15].

19.4.   Cancer Classification Using Neural Networks

Khan *et al.* (2001) developed a method of classifying cancers to specific diagnostic categories. They studied the small, round blue cell tumors (SRBCTs) of childhood, which include neuroblastoma (NB), rhabdomyosarcoma (RMS), non-Hodgkin lymphoma (NHL), and the Ewing family of tumors (EWS). They used 63 training samples that included both tumor biopsy material (13 EWS and 10 RMS) and cell lines (10 EWS, 10 RMS, 12 NB, and 8 Burkitt lymphomas (BL); a subset of NHL). For two samples, ST486 (BL-C2 and C4) and GICAN (NB-C2 and C7), they performed two independent microarray experiments to test the reproducibility of the experiments and these were subsequently treated as separate samples.

As was discussed in section 16.1.1, Khan *et al.* first performed a principal component analysis and then reduced the dimension from 2308 genes to 10 eigengenes, formed by the 10 dominant principal components, for each sample. With these 10 eigengenes as the inputs and the four types of tumors, EWS, RMS, NB, or BL as outputs, they constructed neural network models. The authors used a three-fold cross-validation procedure to produce a total of 3750 neural networks. Using these neural networks, all of the 63 training samples were correctly assigned/classified to their respective categories, having received the highest committee vote (average output) for that category. The contribution of each gene to the classification by the neural network models can be determined by measuring the sensitivity of the classification to a change in the expression level of each gene, using the 3750 calibrated models. The authors ranked the genes according to their significance for the classification and then determined the classification error rate using increasing numbers of these ranked genes. The classification error rate reached a minimum of 0% at 96 genes. The 10 dominant principal components for these 96 genes contained 79% of the variance in the data matrix. Using only these 96 genes, they recalibrated the neural network models and again correctly classified all 63 samples.

Notes

[1] Bishop, C.M. (1995). *Neural Networks for Pattern Recognition.* Clarendon Press, Oxford.

[2] Stern, H.S. (1996). Neural networks in applied statistics (with discussion), *Technometrics,* **38**, 205-220

[3] Hertz, J., Krogh, A., Palmer, R.G. (1991). *Introduction to the Theory of Neural Computation,* Addison-Wesley, Redwood City, CA.

[4] Ripley, R.B. (1996). *Pattern Recognition and Neural Networks,* Cambridge University Press, Cambridge, U.K.

[5] Vapnik, V. (1998). *Statistical Learning Theory,* John Wiley & Sons, New York.

[6] Hastie, T., Tibshirani, R., and Friedman, J. (2001). *The Elements of Statistical Learning,* Springer, New York.

[7] Silipo, R., Gori, M., Taddei, A., Varanini, M., and Marchesi, C. (1995). *Computational Biomedical Research,* **28**, 305-318.

[8] Heden, B., Ohlin, H., Rittner, R., and Edenbrandt, L. (1997), 1798-1802.

[9] Hudson, D.L. and Cohen, M.E. (2000). *Neural Networks and Artificial Intelligence for Biomedical Engineering,* The Institute of Electric and Electronics Engineers, Inc., New York.

[10] Khan, J., Wei, J.S., Ringner, M., Saal, L.H., Ladanyi, M., Westermann, F., Berthold, F., Schwab, M., Antonescu, C.R., Peterson, C., and Meltzer, P.S. (2001). *Nature Medicine,* **7**, 673-679.

[11] McCulloch, W.S. and Pitts, W. (1943). *Bull. Math. Biophys,* **5**, 115-137.

[12] Rosenblatt, F. (1958). *Psychological Review,* **65**, 386-408. Reprinted in Shavlik & Dietterick (1990).

[13] Rosenblatt, F. (1962). *Principles of Neurodynamics: Perceptrons and the Theory of Brain Mechanisms.* Spartan Books, Washington D.C.

[14] Christiani, N. and Shawe-Taylor, J. (2000), *Support Vector Machines,* pages 11-18.

[15] Vapnik (1998), *Statistical Learning Theory,* Wiley, New York.

Chapter 20

SUPPORT VECTOR MACHINES

A *support vector machine* (SVM) is a supervised learning technique that has proven useful in classification problems encountered in working with microarray data. The presentation of the basic concepts will assume that a training set with two groups, say group A_1 and group A_2, is specified in advance with clear information about which item belongs to which group. After an SVM has learned from the expression data based on this training set to discriminate between members of group A_1 and group A_2, the SVM could then be used to classify new items (that are not in the training set) as members of group A_1 or group A_2. In the context of cancer classification, the items may be biological specimens. In the context of finding functional classes of genes, the items may be genes. Detailed background and mathematical derivations for SVMs can be found in Vapnik[1] (1998), and Cristianini and Shawe-Taylor[2] (2000).

The basic idea behind SVMs is to construct an optimal separating hyperplane for the two groups by mapping the gene expression data to a higher-dimensional space. This involves finding a hyperplane defined by a weight vector $\boldsymbol{w}$ and a bias b such that the separation of the two groups is maximized in a specific sense. Using kernel representations, linear separation in this higher-dimensional space corresponds to a nonlinear decision boundary in the original space.

20.1.   Geometric Margins for Linearly Separable Groups

Assume that $S = \{(\boldsymbol{x}_1, y_1), (\boldsymbol{x}_2, y_2), \ldots, (\boldsymbol{x}_n, y_n)\}$ is a training set of n specimen samples, where $\boldsymbol{x}_i$ denotes the input vector from the ith sample and the sign function y_i denotes the *true class label* corresponding to x_i. If the training set contains only two classes, say A_1 and A_2, we use the conventional labels that $y_i = 1$ if the sample point x_i actually belongs to class A_1 and $y_i = -1$ if the sample point x_i actually belongs to class A_2.

If the two groups A_1 and A_2 are linearly separable, there exists a hyperplane $\mathcal{D}$ defined by a weight vector w, having unit norm

$$\| w \| = \langle w \cdot w \rangle^{1/2} = 1$$

and some bias b, such that $\langle w \cdot x \rangle + b = 0$, for any $x \in \mathcal{D}$, and that

$$\begin{aligned}
\langle w \cdot x \rangle + b \ &> 0, \quad \text{if } x \in A_1, \\
\langle w \cdot x \rangle + b \ &< 0, \quad \text{if } x \in A_2.
\end{aligned} \tag{20.1}$$

Here, we use the notation $\langle w \cdot x \rangle = w'x = \sum_k w_k x_k$ to denote the inner or dot product of vectors w and x. The geometric *margin* of a sample point x with reference to the hyperplane $\mathcal{D}$ is its distance $\langle w \cdot x \rangle + b$ from the hyperplane. The unit weight vector w that maximizes the following *margin*

$$\begin{aligned}
c &= \tfrac{1}{2} \ \{ \ \min_{x \in A_1} \ [\, \langle w \cdot x \rangle + b \,] + \ \min_{x \in A_2} \ [-\langle w \cdot x \rangle - b \,] \ \} \\
&= \tfrac{1}{2} \ \{ \ \min_{x \in A_1} \ \langle w \cdot x \rangle - \max_{x \in A_2} \ \langle w \cdot x \rangle \ \}
\end{aligned} \tag{20.2}$$

between the hyperplane and the two groups in the training set determines the *maximal margin separating hyperplane*. The situation is illustrated in Figures 20.1 and 20.2. The figures show a set of samples containing two classes plotted in the original input space, which happens to be two-dimensional. The two groups of samples are linearly separable in the original input space. Figure 20.1 shows a hyperplane that separates the two groups. Figure 20.2 shows the maximal margin hyperplane and the separating span referred to as the maximal margin c. It can be seen that the maximal margin hyperplane is a special case of a separating hyperplane.

Hence, for linearly separable groups A_1 and A_2, one can find a separating hyperplane $\mathcal{D}$ defined by a unit weight vector w and a bias b such that each of the two groups has a maximal margin, or distance, of c units from $\mathcal{D}$ and that the two groups are $2c$ units from each other. That is,

$$\begin{aligned}
\langle w \cdot x_i \rangle + b \ &\geq \ c, \quad \text{if } x_i \in A_1, \\
\langle w \cdot x_i \rangle + b \ &\leq \ -c, \quad \text{if } x_i \in A_2.
\end{aligned} \tag{20.3}$$

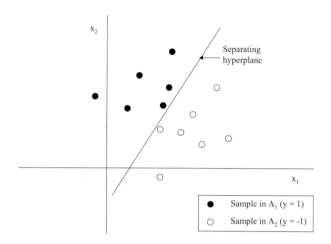

Figure 20.1. The figure shows a set of samples containing two classes that are linearly separable in the original input space. The figure shows a hyperplane that separates the two class groups.

Mathematically, the above optimization problem is to find unit weight vector $\boldsymbol{w}$ and bias b such that the margin c is maximized, i.e.,

$$\max_{\boldsymbol{w},b} \; c \qquad (20.4)$$

subject to the conditions

$$
\begin{aligned}
\langle \boldsymbol{w} \cdot \boldsymbol{x}_i \rangle + b &\geq c, && \text{if } y_i = 1; \\
\langle \boldsymbol{w} \cdot \boldsymbol{x}_i \rangle + b &\leq -c, && \text{if } y_i = -1; \\
\text{and } \| \boldsymbol{w} \| &= 1.
\end{aligned}
\qquad (20.5)
$$

Notice that if we multiply both sides of the above inequalities by class label y_i, the two leading constraints in (20.5) can be combined into one.

$$y_i \left[\langle \boldsymbol{w} \cdot \boldsymbol{x}_i \rangle + b \right] \geq c, \quad \text{for all } i. \qquad (20.6)$$

On the other hand, the unit norm constraint $\| \boldsymbol{w} \| = 1$ in equation (20.5) can be removed if we divide any weight vector $\boldsymbol{w}$ by its norm. Hence the optimization problem in (20.3) can be restated as follows.

$$\max_{\boldsymbol{w},b} \; c \qquad (20.7)$$

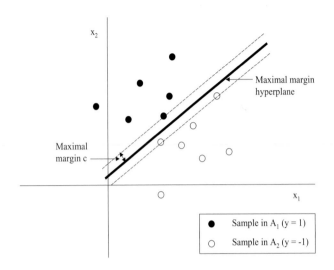

Figure 20.2. The figure shows a set of samples containing two classes that are linearly separable in the original input space. The figure shows the maximal margin hyperplane and the maximal margin c.

subject to the condition that

$$\frac{1}{\| \, w \, \|} \, y_i \, [\, \langle w \cdot x_i \rangle + b \,] \; \geq c, \quad \text{for all } i. \tag{20.8}$$

Furthermore, for any weight w and bias b satisfying (20.8), any positive multiple of them will satisfy the equation as well. Hence we can let $\| \, w \, \| = 1/c$. As a result, for linearly separable groups in the training set, the optimization problem in (20.6) is equivalent to finding w (and b) that solves

$$\min_{w,b} \| \, w \, \|, \tag{20.9}$$

subject to the condition that

$$y_i \, [\langle w \cdot x_i \rangle + b] \; \geq 1, \quad \text{for all } i. \tag{20.10}$$

Equation (20.10) implies that the value of $\langle w \cdot x_i \rangle + b$ has the same sign as the true class label y_i.

20.2.   Convex Optimization in the Dual Space

In this section we show that the quadratic optimization problem in (20.9) can be transformed into a convex optimization problem in the dual space of Lagrange multipliers.

For linearly separable groups, finding the weight vector $\boldsymbol{w}$ with minimal norm subject to inequality constraints (20.10) is equivalent to finding the saddle point of the Lagrange function

$$\mathcal{L}(\boldsymbol{w}, b, \boldsymbol{\alpha}) = \frac{1}{2}\langle \boldsymbol{w} \cdot \boldsymbol{w}\rangle - \sum_{i=1}^{n} \alpha_i\{\ y_i[\ \langle \boldsymbol{w} \cdot \boldsymbol{x}_i\rangle + b\ ] - 1\ \}, \qquad (20.11)$$

subject to the condition that the Lagrange multipliers, $\boldsymbol{\alpha} = (\alpha_1, \ldots, \alpha_n)$, satisfy $\alpha_i \geq 0$, for all i. The scalar of $1/2$ is added for convenience. To find the saddlepoint, one has to (1) minimize the function $\mathcal{L}(\boldsymbol{w}, b, \boldsymbol{\alpha})$ over $\boldsymbol{w}$ and b and (2) maximize it over the nonnegative Lagrange multipliers $\alpha_i \geq 0$. The minimum point of $\mathcal{L}(\boldsymbol{w}, b, \boldsymbol{\alpha})$ has to satisfy the conditions

$$\frac{\partial \mathcal{L}(\boldsymbol{w}, b, \boldsymbol{\alpha})}{\partial \boldsymbol{w}} = \boldsymbol{w} - \sum_{i=1}^{n} y_i\alpha_i\boldsymbol{x}_i = 0,$$

$$\frac{\partial \mathcal{L}(\boldsymbol{w}, b, \boldsymbol{\alpha})}{\partial b} = -\sum_{i=1}^{n} y_i\alpha_i = 0. \qquad (20.12)$$

The above conditions imply that, for the weight vector $\boldsymbol{w}$ that defines the optimal hyperplane, the following equalities hold true.

$$\boldsymbol{w} = \sum_{i=1}^{n} y_i\alpha_i\boldsymbol{x}_i \quad \text{and} \quad \sum_{i=1}^{n} y_i\alpha_i = 0. \qquad (20.13)$$

Substituting (20.13) into (20.11), and taking into account the equalities in (20.12), the function $\mathcal{L}$ can be written in terms of Lagrange multipliers as follows.

$$\mathcal{L}(\alpha_1, \ldots, \alpha_n) = \sum_{i=1}^{n} \alpha_i - \frac{1}{2} \sum_{i=1}^{n}\sum_{j=1}^{n} y_i y_j \alpha_i \alpha_j \langle \boldsymbol{x}_i \cdot \boldsymbol{x}_j\rangle \qquad (20.14)$$

To maximize the function $\mathcal{L}(\alpha_1, \ldots, \alpha_n)$, subject to the inequality constraints

$$\alpha_i \geq 0, \quad \text{for } i = 1, \ldots, n, \qquad (20.15)$$

the optimal solution $\boldsymbol{w}^*$ and b^* must satisfy the Kuhn-Tucker conditions that

$$\alpha_i{}^* \{\ y_i[\ \langle \boldsymbol{w}^* \cdot \boldsymbol{x}_i\rangle + b^*\ ] - 1\ \} = 0, \quad \text{for } i = 1, \ldots, n. \qquad (20.16)$$

20.3.   Support Vectors

Recall that the constraints specified in equation (20.10) require that the margins $y_i [\langle \boldsymbol{w}^* \cdot \boldsymbol{x}_i \rangle + b^*] \geq 1$, for all points $i = 1, \ldots, n$. Therefore, the conditions in (20.16) imply that

$$\alpha_i^* = 0 \quad \text{if the margin} \quad y_i [\langle \boldsymbol{w}^* \cdot \boldsymbol{x}_i \rangle + b^*] > 1,$$

and that positive values α_i^* correspond only to those points $\boldsymbol{x}_i$ that have a margin of 1, i.e.,

$$\alpha_i^* > 0 \quad \text{if the margin} \quad y_i [\langle \boldsymbol{w}^* \cdot \boldsymbol{x}_i \rangle + b^*] = 1. \tag{20.17}$$

Geometrically, points $\boldsymbol{x}_i$ that have margins equal to 1 are the closest to the optimal separating hyperplane. Hence they are called the *support vectors*. The support vectors in Figure 20.2 are those sample points **x** that lie on the dashed lines.

Let Ω denote the set of indices i for these support vectors $\boldsymbol{x}_i$. The vector $\boldsymbol{w}^*$ that defines the optimal hyperplane has nonzero weights for these support vectors:

$$\boldsymbol{w}^* = \sum_{i \in \Omega} y_i \, \alpha_i^* \, \boldsymbol{x}_i.$$

The value of b^* is obtained by reference to (20.17), where it can be seen that b^* is the common value that solves $y_i [\langle \boldsymbol{w}^* \cdot \boldsymbol{x}_i \rangle + b^*] = 1$ for each $i \in \Omega$. The optimal separating hyperplane is thus of the form

$$\langle \boldsymbol{w}^* \cdot \boldsymbol{x} \rangle + b^* = \sum_{i \in \Omega} y_i \, \alpha_i^* \langle \boldsymbol{x}_i \cdot \boldsymbol{x} \rangle + b^*. \tag{20.18}$$

Using the optimal separating hyperplane, a classification decision for any point $\boldsymbol{x}$ can be made according to the relative position of $\boldsymbol{x}$ with respect to the separating hyperplane (20.18). Denoting the decision rule by $\varphi(\boldsymbol{x})$, we have the classification decision function

$$\varphi(\boldsymbol{x}) = \text{sign} [\langle \boldsymbol{w}^* \cdot \boldsymbol{x} \rangle + b^*] = \text{sign} [\sum_{i \in \Omega} y_i \, \alpha_i^* \langle \boldsymbol{x}_i \cdot \boldsymbol{x} \rangle + b^*]. \tag{20.19}$$

Note that $\varphi(\boldsymbol{x})$ is either 1 or -1 according to the sign of the margin. Note also that the optimal separating hyperplane defined in (20.18) does not depend explicitly on the dimensionality of the vector $\boldsymbol{x}$ but only on the inner products of $\boldsymbol{x}$ with the support vectors $\boldsymbol{x}_i$, for $i \in \Omega$.

20.4.   Linearly Nonseparable Groups

We now consider the situation where groups A_1 and A_2 are not linearly separable. Using the concept of a *soft margin*, one can allow some training samples to fall on the wrong side of the separating hyperplane. Figure 20.3 illustrates the idea of introducing slack amounts to allow samples to fall on the wrong side of the separating hyperplane.

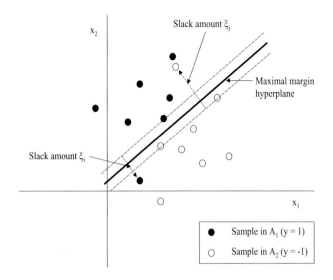

Figure 20.3.   The figure shows slack amounts that allow training samples to fall on the wrong side of the separating hyperplane when two groups are not linearly separable.

Let $\xi_i \geq 0$ denote the *slack* amount by which the margin constraints in (20.10) are violated. The value ξ_i is the amount by which the prediction $\varphi(\boldsymbol{x}_i)$ falls on the wrong side of the margin[3]. Hence, to bound the total amount by which the predictions fall on the wrong side of the margin, one can bound the sum $\sum_{i=1}^{n} \xi_i$ by some constant $\mathcal{E}$, i.e., set $\sum_{i=1}^{n} \xi_i \leq \mathcal{E}$.

For nonlinearly separable groups, we consider the optimization problem

$$\min_{\boldsymbol{w},\, b} \| \boldsymbol{w} \|, \quad \text{subject to} \quad y_i \left[\langle \boldsymbol{w} \cdot \boldsymbol{x}_i \rangle + b \right] \geq 1 - \xi_i, \quad \text{for all } i;$$

$$\text{with } \xi_i \geq 0, \quad \text{for all } i, \quad \text{and} \quad \sum_{i=1}^{n} \xi_i \leq \mathcal{E}. \qquad (20.20)$$

This convex optimization problem can be solved by standard methods, such as the Kuhn-Tucker Theorem.

20.5. Nonlinear Separating Boundary

For real-world problems involving data that are not linearly separable, one solution for the classification problem is to map the data into a higher-dimensional space using a function h, where

$$z = h(x),$$

and to define a separating hyperplane there. The data reside in an *input space* $\mathcal{X} = \mathbf{R}^n$ with dimension $n \leq N$. The higher-dimensional space is called the *feature* space of dimension N and is denoted by $\mathcal{Z} = \mathbf{R}^N$ here. Using kernel representations, the SVM method can locate the separating hyperplane in the feature space without ever representing the space explicitly. This property greatly reduces computational efforts.

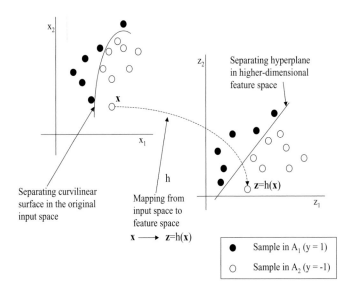

Figure 20.4. An abstract illustration showing the mapping of a curvilinear separating surface in the original input space into a linear separating hyperplane in a higher-dimensional space.

Figure 20.4 illustrates such a mapping. The figure shows a set of samples containing two classes being mapped from the original input space to a higher-dimensional feature space. Linear separability is achieved in the higher-dimensional feature space. Only two of the dimensions of each space can be shown in this diagram. The diagram is an abstraction and does not show a true representation.

20.5.1. Kernel Functions

A *kernel* is a function $\mathcal{K}$ from the product space $\mathcal{X} \times \mathcal{X}$ to $\boldsymbol{R}$ such that there exists a mapping h from input space $\mathcal{X}$ to feature space $\mathcal{Z}$ with the following equality holding for any two points $\boldsymbol{x}_1, \boldsymbol{x}_2 \in \mathcal{X}$.

$$\mathcal{K}(\boldsymbol{x}_1, \boldsymbol{x}_2) = \langle h(\boldsymbol{x}_1) \cdot h(\boldsymbol{x}_2) \rangle$$

That is, the value of the kernel function $\mathcal{K}$ on $(\boldsymbol{x}_1, \boldsymbol{x}_2)$ can be evaluated by the inner product of $h(\boldsymbol{x}_1)$, and $h(\boldsymbol{x}_2)$ in the feature space $\mathcal{Z}$. Using the kernel representation, points $\boldsymbol{x}_1, \boldsymbol{x}_2$ in the input space $\mathcal{X}$ are related to points $\boldsymbol{z}_1 = h(\boldsymbol{x}_1)$ and $\boldsymbol{z}_2 = h(\boldsymbol{x}_2)$ in the feature space $\mathcal{Z}$ via the inner product.

As $\mathcal{K}(\boldsymbol{x}_1, \boldsymbol{x}_2) = \langle h(\boldsymbol{x}_1) \cdot h(\boldsymbol{x}_2) \rangle = \langle h(\boldsymbol{x}_2) \cdot h(\boldsymbol{x}_1) \rangle = \mathcal{K}(\boldsymbol{x}_2, \boldsymbol{x}_1)$,

the kernel function is symmetric. The kernel function also satisfies the Cauchy-Schwarz inequality,

$$
\begin{aligned}
\mathcal{K}(\boldsymbol{x}_1, \boldsymbol{x}_2)^2 &= \langle h(\boldsymbol{x}_1) \cdot h(\boldsymbol{x}_2) \rangle^2 \\
&\leq \| h(\boldsymbol{x}_1) \|^2 \| h(\boldsymbol{x}_2) \|^2 \\
&= \langle h(\boldsymbol{x}_1) \cdot h(\boldsymbol{x}_1) \rangle \langle h(\boldsymbol{x}_2) \cdot h(\boldsymbol{x}_2) \rangle \\
&= \mathcal{K}(\boldsymbol{x}_1, \boldsymbol{x}_1) \mathcal{K}(\boldsymbol{x}_2, \boldsymbol{x}_2).
\end{aligned}
$$

Using the dual representation one does not represent the feature vectors $\boldsymbol{z} \in \mathcal{Z}$ explicitly. The use of kernels makes it possible to map the data implicitly into a feature space and to train a linear SVM in such a space. Hence, the computational problems inherent in evaluating the feature map can be avoided.

20.5.2. Kernels Defined by Symmetric Functions

Consider a finite set of n sample points $\boldsymbol{x}_1, \ldots, \boldsymbol{x}_n$ in input space $\mathcal{X}$. If $\mathcal{K}(\boldsymbol{x}, \boldsymbol{x}_2)$ is a symmetric function, the only information used about the training samples is their *kernel matrix* defined by

$$\boldsymbol{K} = (\mathcal{K}(\boldsymbol{x}_i, \boldsymbol{x}_j); i, j = 1, .., n).$$

Since the kernel matrix $\boldsymbol{K}$ is symmetric, there exists an orthogonal matrix $\boldsymbol{V}$ such that

$$\boldsymbol{K} = \boldsymbol{V} \boldsymbol{\Lambda} \boldsymbol{V}',$$

where $\boldsymbol{\Lambda}$ is a diagonal matrix containing the eigenvalues λ_t of $\boldsymbol{K}$, with corresponding eigenvectors $\boldsymbol{v}_t = (v_{t1}, \ldots, v_{tn})'$, $t = 1, \ldots, n$.

If we further assume that the kernel matrix $\boldsymbol{K}$ is positive semi-definite, then all the eigenvalues λ_t are non-negative. We can define a feature mapping

$$h : \ \boldsymbol{x}_i \rightarrow (\sqrt{\lambda_1} v_{1i}, \ldots, \sqrt{\lambda_n} v_{ni}), \quad \text{such that}$$

$$\langle h(\boldsymbol{x}_i) \cdot h(\boldsymbol{x}_j) \rangle = \sum_{t=1}^{n} \lambda_t \, v_{ti} \, v_{tj} = (\boldsymbol{V}\boldsymbol{\Lambda}\boldsymbol{V}')_{ij} = \mathcal{K}(\boldsymbol{x}_i, \boldsymbol{x}_j).$$

The above result is a special case of Mercer's Theorem which implies that, for any continuous positive definite function $\mathcal{K}(\boldsymbol{x}_1, \boldsymbol{x}_2)$, there exists a mapping h such that

$$\mathcal{K}(\boldsymbol{x}_1, \boldsymbol{x}_2) \ = \ \langle h(\boldsymbol{x}_1) \cdot h(\boldsymbol{x}_2) \rangle$$

for all $\boldsymbol{x}_1, \boldsymbol{x}_2 \in \boldsymbol{R}^n$. See Vapnik (1998) and Cristianini and Shawe-Taylor (2000) for Mercer's Theorem and its extension to Hilbert space.

Two popular choices for kernel function $\mathcal{K}$ are

- dth degree polynomial:

$$\mathcal{K}(\boldsymbol{x}_1, \boldsymbol{x}_2) = (1 + \langle \boldsymbol{x}_1 \cdot \boldsymbol{x}_2 \rangle)^d \ \text{ and}$$

- radial basis:

$$\mathcal{K}(\boldsymbol{x}_1, \boldsymbol{x}_2) = \exp(- \parallel \boldsymbol{x}_1 - \boldsymbol{x}_2 \parallel^2 / s), \quad \text{where } s > 0 \text{ is a scalar.}$$

20.5.3.   Use of SVM for Classifying Genes

In previous sections in this chapter, we have considered how to classify a new specimen into one of two groups in a training set of specimens. In this section, we will demonstrate via an example that the method of SVM can also be used to classify genes into two groups, say, two functional categories.

Using the same notation we adopted previously in Chapter 15, given n specimens, the expression level of each gene $g = 1, \ldots, G$ can be represented as an n-dimensional vector $\boldsymbol{x}^{(g)} = (x_{g1}, \ldots, x_{gn})'$ in the input space $\mathcal{X}$. To construct a decision surface corresponding to, say, a second-order polynomial in the input space, one can create a feature space $\mathcal{Z}$ that has a total of $N = n(n+3)/2$ coordinates of the form

$$z_1 = x_{g1}, \quad \ldots, \quad z_n = x_{gn}, \qquad (n \text{ coordinates}),$$
$$z_{n+1} = x_{g1}{}^2, \quad \ldots, \quad z_{2n} = x_{gn}{}^2, \qquad (n \text{ coordinates}),$$
$$z_{2n+1} = x_{g1}x_{g2}, \quad \ldots, \quad z_N = x_{gn-1}x_{gn}, \qquad (n(n-1)/2 \text{ coordinates}).$$

This set of coordinates defines the mapping $z = h(x)$ from input space $\mathcal{X}$ to feature space $\mathcal{Z}$, where $z = (z_1, \ldots, z_N)'$. A separating hyperplane constructed in this feature space $\mathcal{Z}$ is a second-degree polynomial in the input space $\mathcal{X}$.

Given any gene g, $g = 1, \ldots, G$, consider the expression vector $x^{(g)}$ in the input space and the corresponding point $z^{(g)} = h(x^{(g)})$ in the higher dimensional feature space. Following the earlier reasoning, let

$$\{z^{(g)}, g \in \Omega\}$$

denote the set of support vectors for a separating hyperplane in the feature space. A linear decision function in the feature space $\mathcal{Z}$ will take the form

$$\varphi(z) = \text{sign}\left(\sum_{g \in \Omega} y_g\, \alpha_g{}^* \, \langle z^{(g)} \cdot z \rangle + b^* \right). \qquad (20.21)$$

This decision function in the feature space corresponds to the following nonlinear decision function in the input space (a second-order polynomial in this case).

$$\varphi(x) = \text{sign}\left(\sum_{g \in \Omega} y_g\, \alpha_g{}^* \, \mathcal{K}(x^{(g)}, x) + b^* \right). \qquad (20.22)$$

In conclusion, we see that the specification of a support vector machine requires setting only two parameters: the kernel function and the magnitude of the penalty for violating the soft margin. The settings of these parameters depend on the specific data at hand. We are now ready to look at some case examples.

20.6. Examples

20.6.1. Functional Classification of Genes

Brown, Grundy, Lin *et al.*[4] (2000) use the SVM method to classify genes by function. To begin, a set of genes in a functional class (i.e., sharing a common function) is selected. A second set of genes that are known not to share this function is also specified. Together, these two sets of genes form a training set in which the genes are labeled as group A_1 if they are in the functional class and as group A_2 if they are not in the functional class.

Given any gene $g = 1, \ldots, G$, let $\boldsymbol{x}^{(g)}$ denote the vector of normalized logarithms of ratios of expression level E_{gi} for the ith experimental specimen to the expression level R_{gi} for the reference sample. The expression vector is normalized so that $\boldsymbol{x}^{(g)} = (x_{g1}, \ldots, x_{gn})$ has Euclidean length 1. The normalized log-ratio measurement x_{gi}

$$x_{gi} = \frac{\log(E_{gi}/R_{gi})}{\sqrt{\sum_{j=1}^{n} \log^2(E_{gj}/R_{gj})}} \tag{20.23}$$

is positive if gene g is up-regulated with respect to the reference sample and negative if it is down-regulated. The authors use a training set of 2467 yeast genes, each having an expression vector of dimension $n = 79$.

The authors consider polynomial kernels defined by

$$\mathcal{K}(\boldsymbol{x}^{(g_1)}, \boldsymbol{x}^{(g_2)}) = [\langle \boldsymbol{x}^{(g_1)} \cdot \boldsymbol{x}^{(g_2)} \rangle + 1]^d$$

for degrees $d = 1$, 2, and 3. They also consider a radial basis kernel, which has a Gaussian form

$$\mathcal{K}(\boldsymbol{x}^{(g_1)}, \boldsymbol{x}^{(g_2)}) = \exp\left(-\frac{||\boldsymbol{x}^{(g_1)} - \boldsymbol{x}^{(g_2)}||^2}{2\alpha^2}\right).$$

In their experiments, parameter α is set equal to the median of the Euclidean distances from each positive example to the nearest negative example.

Brown *et al.* (2000) compared the performance of the SVM classifiers with that of four standard machine learning algorithms, including Parzen windows, Fisher's linear discriminant analysis, and two decision tree learners. The authors used a three-way cross-validation to test the performance of these methods. They randomly divided the genes into three groups. Classifiers were trained by using two-thirds of the gene expression vectors and were tested on the remaining third. This procedure was then repeated two more times, each time using a different third of the genes as test genes. Their results show that, for all but one of the six functional classes of genes that they considered, the support vector machine using the radial basis or a higher-dimensional inner-product kernel outperformed the other methods. Using five separate tests, they also show that the radial basis SVM has better performance in functional classification than Fisher's linear discriminant analysis.

20.6.2. SVM and One-versus-All Classification Scheme

We show in previous sections that SVM can be used for distinguishing between two classes. Making multiclass distinctions can be a much greater challenge. Ramaswamy, Tamayo, Rifkin *et al.*[5] (2001) devise an analytical scheme for making multiclass distinctions by dividing the multiclass problem into a series of pairwise comparisons. To determine whether the diagnosis of common adult malignancies could be achieved by molecular classification, Ramaswamy *et al.* (2001) used 218 tumor samples representing 14 common tumor types and 90 normal tissue samples in an experiment with oligonucleotide microarrays. Unsupervised clustering results of their analysis were previously discussed in subsection 17.5.3. In this section we consider their results using SVM as a supervised learning method.

On the basis of the expression levels of 16,063 genes and expressed sequence tags, the authors applied a linear SVM algorithm which maximizes the distance between a hyperplane and the samples from two tumor classes that are closest to the hyperplane, with the constraint that the samples lie on opposite sides of the hyperplane[6]. This distance is calculated in the space of 16,063 genes. Given m classes and m trained classifiers, Ramaswamy *et al.* (2001) used a *one-versus-all* (OVA) approach for multiclass classification. This method recursively removes features based on the absolute magnitude of each hyperplane element. The authors divided the multiclass problem into a series of 14 OVA pairwise comparisons. Each test sample is presented sequentially to these 14 pairwise classifiers, each of which either claims or rejects that sample as belonging to a single class. This method results in 14 separate OVA classifications per sample, each with an associated confidence. Each test sample is assigned to the class with the highest OVA classifier confidence. They evaluated several classification algorithms for these OVA pairwise classifiers including weighted voting, k-nearest neighbors, and SVM; all of which yielded significant prediction accuracy. They showed that the SVM algorithm consistently out-performed other algorithms. Figure 20.5 illustrates the multiclass classification scheme.

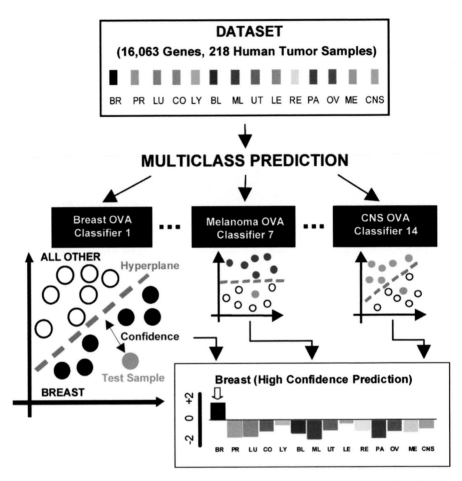

Figure 20.5. Multiclass classification scheme. The multiclass cancer classification problem is divided into a series of 14 OVA problems, and each OVA problem is addressed by a different class-specific classifier (e.g., "breast cancer" vs. "not breast cancer"). Each classifier uses the SVM algorithm to define a hyperplane. In the example shown, a test sample is sequentially presented to each of 14 OVA classifiers and is predicted to be breast cancer, based on the breast OVA classifier having the highest confidence. Source: Proceedings of the National Academy of Sciences (PNAS), volume 98, Copyright (2001) National Academy of Sciences, U.S.A. Reprinted with permission from PNAS.

Notes

1 Vapnik, V. (1998). *Statistical Learning Theory*, John Wiley and Sons, New York.

2 Cristianini, N. and Shawe-Tylor, J. (2000). *Support Vector Machines*, Cambridge University Press, Cambridge, U.K.

3 Hastie, T., Tibshirani, R., and Friedman, J. (2001). *The Elements of Statistical Learning*, Springer, New York.

4 Brown M.P.S., Grundy, W.N., Lin, D., *et al.*. (2000), *Proceedings of the National Academy of Sciences, USA*, **97**, 262-267.

5 Ramaswamy, S., Tamayo, P., Rifkin, R., *et al.* (2001). *Proceedings of the National Academy of Sciences, USA*, **98**, 15149-15154.

6 The implementation of SVM-FU is available at www.ai.mit.edu/projects/cbcl

Appendix A
Sample Size Table for Treatment-control Designs

TABLE 1: Sample Size Table for Treatment-control Designs

This table gives sample sizes for matched-pairs and completely randomized designs in which differential expression between a treatment condition and a control condition is of interest. Refer to subsection 14.11.1 of Chapter 14 for detailed technical background for this table.

The following list summarizes notation for items cross-referenced in the table.

$E(R_0)$: The mean number of false positives.

μ_1: The mean difference in log-expression between treatment and control conditions as postulated under the alternative hypothesis H_1.

σ_d: The anticipated standard deviation of the difference in log-expression between treatment and control conditions.

$|\mu_1|/\sigma_d$: The standardized statistical distance of the difference in gene expression between treatment and control conditions under H_1.

$1 - \beta_1$: The specified power level for an individual gene, which represents the expected proportion of differentially expressed genes that will be declared as such by the tests.

G_0: The anticipated number of undifferentially expressed genes in the experiment.

Use of the Table

The following subsections tell how this table can be used for matched-pairs and completely randomized designs.

Matched-pairs Design

Table 1 gives the number of treatment-control pairs n required to achieve a specified individual power level $1 - \beta_1$ for a matched-pairs design. The table is entered based on the specified mean number of false positives $E(R_0)$, ratio $|\mu_1|/\sigma_d$, anticipated number of undifferentially expressed genes G_0, and desired individual power level $1 - \beta_1$. If G_0 is expected to be similar to the total gene count G, the table could be entered using G without introducing great error. To conserve space, only three individual power levels are offered in the table, 0.90, 0.95, and 0.99. The sample size shown in the table is the smallest whole number that will yield the specified power. The total number of experimental units C is double the entry in the table, i.e., $C = 2n$. Observe that the ratio $|\mu_1|/\sigma_d$ can be interpreted as the statistical distance (i.e., the number of standard deviations) between the treatment and control log-expression levels under the alternative hypothesis H_1.

Example 1. Consider a study where it is anticipated that $G_0 = 2000$ genes will not be differentially expressed. The investigator wishes to control the mean number of false positives at $E(R_0) = 1$ and to detect a two-fold difference between treatment and control conditions with an individual power level of 0.90. Previous studies by the investigator may suggest that the standard deviation of gene expression differences in matched pairs will be about $\sigma_d = 0.5$ on a log-2 scale. The two-fold difference represents a value of $\log_2(2) = 1.00$ for $|\mu_1|$ on a log-2 scale. Thus, the ratio $|\mu_1|/\sigma_d$ equals $1.00/0.5 = 2.0$. Reference to Table 1 for these specifications shows that $n = 6$. Thus, 6 pairs of treatment and control conditions are required in the study. The specified individual power level of 0.90 indicates that 90 percent of the differentially expressed genes are expected to be discovered.

Example 2. Should an investigator assume that the estimated differential expression vectors $\hat{\boldsymbol{I}}_g$ are possibly dependent then the Bonferroni approach is used and we have $E(R_0) = G_0\alpha_0 = \alpha_F$. Thus, the expected number of false positives is necessarily smaller than 1. The table should be entered accordingly. Two values of $E(R_0)$ smaller than 1 are displayed in the table, namely, 0.1 and 0.5. To illustrate this use of the table, consider the situation where α_F is set at 0.1, $G_0 = 2000$, $|\mu_1|/\sigma_d = 2.0$ and an individual power level of 0.90 is desired. In this case, $E(R_0) = \alpha_F = 0.1$ and reference to the table indicates that 8 pairs of treatment and control conditions are required in the study.

Completely Randomized Design

Consider a completely randomized design with an equal number n of treatment and control conditions. The standard deviation of the difference in log-expression between treatment and control is given by $\sigma_d = \sqrt{2}\sigma$, where σ is the experimental error standard deviation of gene log-expression.

Example. To illustrate, suppose σ is anticipated to be 0.40 in a completely randomized design. Furthermore, suppose that $\mu_1 = 1.00$, $E(R_0) = 1$, $G_0 = 2000$ and the desired individual power level is 0.90 as specified before. As $\sigma_d = \sqrt{2}\sigma = \sqrt{2}(0.40) = 0.566$, the ratio $|\mu_1|/\sigma_d = 1.00/0.566 = 1.77$. Making reference to ratio 1.75 in the table (the closest value), the required sample size can be seen to be $n = 8$ and, hence, a total of $C = 2(8) = 16$ experimental units are required.

Table 1. Sample Sizes for Treatment-control Designs
Mean Number of False Positives $E(R_0) = 0.1$

The table gives the group sample size n for treatment and control groups in matched-pairs or completely randomized designs. $E(R_0)$ denotes the mean number of false positives, $|\mu_1|/\sigma_d$ the statistical distance between treatment and control conditions under H_1, G_0 the anticipated number of undifferentially expressed genes.

	Distance $|\mu_1|/\sigma_d$										
	0.50	0.75	1.00	1.25	1.50	1.75	2.00	2.25	2.50	2.75	3.00
Genes G_0	Power $1 - \beta_1$ =Proportion correctly declared as differentially expressed = 0.80										
100	69	31	18	11	8	6	5	4	3	3	2
200	75	34	19	12	9	7	5	4	3	3	3
500	84	37	21	14	10	7	6	5	4	3	3
1000	90	40	23	15	10	8	6	5	4	3	3
2000	96	43	24	16	11	8	6	5	4	4	3
5000	105	47	27	17	12	9	7	6	5	4	3
10000	111	50	28	18	13	10	7	6	5	4	4
20000	117	52	30	19	13	10	8	6	5	4	4
Genes G_0	Power $1 - \beta_1$ =Proportion correctly declared as differentially expressed = 0.90										
100	84	38	21	14	10	7	6	5	4	3	3
200	91	41	23	15	11	8	6	5	4	3	3
500	101	45	26	17	12	9	7	5	5	4	3
1000	108	48	27	18	12	9	7	6	5	4	3
2000	114	51	29	19	13	10	8	6	5	4	4
5000	124	55	31	20	14	11	8	7	5	5	4
10000	130	58	33	21	15	11	9	7	6	5	4
20000	137	61	35	22	16	12	9	7	6	5	4
Genes G_0	Power $1 - \beta_1$ =Proportion correctly declared as differentially expressed = 0.95										
100	98	44	25	16	11	8	7	5	4	4	3
200	106	47	27	17	12	9	7	6	5	4	3
500	116	52	29	19	13	10	8	6	5	4	4
1000	123	55	31	20	14	11	8	7	5	5	4
2000	130	58	33	21	15	11	9	7	6	5	4
5000	140	63	35	23	16	12	9	7	6	5	4
10000	147	66	37	24	17	12	10	8	6	5	5
20000	155	69	39	25	18	13	10	8	7	6	5
Genes G_0	Power $1 - \beta_1$ =Proportion correctly declared as differentially expressed = 0.99										
100	127	57	32	21	15	11	8	7	6	5	4
200	135	60	34	22	15	12	9	7	6	5	4
500	147	65	37	24	17	12	10	8	6	5	5
1000	155	69	39	25	18	13	10	8	7	6	5
2000	163	73	41	27	19	14	11	9	7	6	5
5000	174	78	44	28	20	15	11	9	7	6	5
10000	182	81	46	30	21	15	12	9	8	7	6
20000	190	85	48	31	22	16	12	10	8	7	6

Table 1. Sample Sizes for Treatment-control Designs
Mean Number of False Positives $E(R_0) = 0.5$

The table gives the group sample size n for treatment and control groups in matched-pairs or completely randomized designs. $E(R_0)$ denotes the mean number of false positives, $|\mu_1|/\sigma_d$ the statistical distance between treatment and control conditions under H_1, G_0 the anticipated number of undifferentially expressed genes.

	Distance $|\mu_1|/\sigma_d$										
	0.50	0.75	1.00	1.25	1.50	1.75	2.00	2.25	2.50	2.75	3.00
Genes G_0	Power $1-\beta_1$ =Proportion correctly declared as differentially expressed = 0.80										
100	54	24	14	9	6	5	4	3	3	2	2
200	60	27	15	10	7	5	4	3	3	2	2
500	69	31	18	11	8	6	5	4	3	3	2
1000	75	34	19	12	9	7	5	4	3	3	3
2000	82	37	21	13	10	7	6	5	4	3	3
5000	90	40	23	15	10	8	6	5	4	3	3
10000	96	43	24	16	11	8	6	5	4	4	3
20000	103	46	26	17	12	9	7	6	5	4	3
Genes G_0	Power $1-\beta_1$ =Proportion correctly declared as differentially expressed = 0.90										
100	67	30	17	11	8	6	5	4	3	3	2
200	75	33	19	12	9	7	5	4	3	3	3
500	84	38	21	14	10	7	6	5	4	3	3
1000	91	41	23	15	11	8	6	5	4	3	3
2000	98	44	25	16	11	8	7	5	4	4	3
5000	108	48	27	18	12	9	7	6	5	4	3
10000	114	51	29	19	13	10	8	6	5	4	4
20000	121	54	31	20	14	10	8	6	5	4	4
Genes G_0	Power $1-\beta_1$ =Proportion correctly declared as differentially expressed = 0.95										
100	80	36	20	13	9	7	5	4	4	3	3
200	88	39	22	14	10	8	6	5	4	3	3
500	98	44	25	16	11	8	7	5	4	4	3
1000	106	47	27	17	12	9	7	6	5	4	3
2000	113	51	29	19	13	10	8	6	5	4	4
5000	123	55	31	20	14	11	8	7	5	5	4
10000	130	58	33	21	15	11	9	7	6	5	4
20000	138	62	35	22	16	12	9	7	6	5	4
Genes G_0	Power $1-\beta_1$ =Proportion correctly declared as differentially expressed = 0.99										
100	106	47	27	17	12	9	7	6	5	4	3
200	115	51	29	19	13	10	8	6	5	4	4
500	127	57	32	21	15	11	8	7	6	5	4
1000	135	60	34	22	15	12	9	7	6	5	4
2000	144	64	36	23	16	12	9	8	6	5	4
5000	155	69	39	25	18	13	10	8	7	6	5
10000	163	73	41	27	19	14	11	9	7	6	5
20000	172	77	43	28	20	14	11	9	7	6	5

Table 1. Sample Sizes for Treatment-control Designs
Mean Number of False Positives $E(R_0) = 1$

The table gives the group sample size n for treatment and control groups in matched-pairs or completely randomized designs. $E(R_0)$ denotes the mean number of false positives, $|\mu_1|/\sigma_d$ the statistical distance between treatment and control conditions under H_1, G_0 the anticipated number of undifferentially expressed genes.

| | Distance $|\mu_1|/\sigma_d$ | | | | | | | | | | |
	0.50	0.75	1.00	1.25	1.50	1.75	2.00	2.25	2.50	2.75	3.00
Genes G_0	Power $1 - \beta_1$ =Proportion correctly declared as differentially expressed = 0.80										
100	47	21	12	8	6	4	3	3	2	2	2
200	54	24	14	9	6	5	4	3	3	2	2
500	62	28	16	10	7	6	4	4	3	3	2
1000	69	31	18	11	8	6	5	4	3	3	2
2000	75	34	19	12	9	7	5	4	3	3	3
5000	84	37	21	14	10	7	6	5	4	3	3
10000	90	40	23	15	10	8	6	5	4	3	3
20000	96	43	24	16	11	8	6	5	4	4	3
Genes G_0	Power $1 - \beta_1$ =Proportion correctly declared as differentially expressed = 0.90										
100	60	27	15	10	7	5	4	3	3	2	2
200	67	30	17	11	8	6	5	4	3	3	2
500	77	34	20	13	9	7	5	4	4	3	3
1000	84	38	21	14	10	7	6	5	4	3	3
2000	91	41	23	15	11	8	6	5	4	3	3
5000	101	45	26	17	12	9	7	5	5	4	3
10000	108	48	27	18	12	9	7	6	5	4	3
20000	114	51	29	19	13	10	8	6	5	4	4
Genes G_0	Power $1 - \beta_1$ =Proportion correctly declared as differentially expressed = 0.95										
100	72	32	18	12	8	6	5	4	3	3	2
200	80	36	20	13	9	7	5	4	4	3	3
500	90	40	23	15	10	8	6	5	4	3	3
1000	98	44	25	16	11	8	7	5	4	4	3
2000	106	47	27	17	12	9	7	6	5	4	3
5000	116	52	29	19	13	10	8	6	5	4	4
10000	123	55	31	20	14	11	8	7	5	5	4
20000	130	58	33	21	15	11	9	7	6	5	4
Genes G_0	Power $1 - \beta_1$ =Proportion correctly declared as differentially expressed = 0.99										
100	97	43	25	16	11	8	7	5	4	4	3
200	106	47	27	17	12	9	7	6	5	4	3
500	118	53	30	19	14	10	8	6	5	4	4
1000	127	57	32	21	15	11	8	7	6	5	4
2000	135	60	34	22	15	12	9	7	6	5	4
5000	147	65	37	24	17	12	10	8	6	5	5
10000	155	69	39	25	18	13	10	8	7	6	5
20000	163	73	41	27	19	14	11	9	7	6	5

Table 1. Sample Sizes for Treatment-control Designs
Mean Number of False Positives $E(R_0) = 2$

The table gives the group sample size n for treatment and control groups in matched-pairs or completely randomized designs. $E(R_0)$ denotes the mean number of false positives, $|\mu_1|/\sigma_d$ the statistical distance between treatment and control conditions under H_1, G_0 the anticipated number of undifferentially expressed genes.

	\multicolumn{11}{c}{Distance $|\mu_1|/\sigma_d$}										
	0.50	0.75	1.00	1.25	1.50	1.75	2.00	2.25	2.50	2.75	3.00
Genes G_0	\multicolumn{11}{l}{Power $1 - \beta_1$ =Proportion correctly declared as differentially expressed = 0.80}										
100	41	18	11	7	5	4	3	2	2	2	2
200	47	21	12	8	6	4	3	3	2	2	2
500	56	25	14	9	7	5	4	3	3	2	2
1000	62	28	16	10	7	6	4	4	3	3	2
2000	69	31	18	11	8	6	5	4	3	3	2
5000	77	35	20	13	9	7	5	4	4	3	3
10000	84	37	21	14	10	7	6	5	4	3	3
20000	90	40	23	15	10	8	6	5	4	3	3
Genes G_0	\multicolumn{11}{l}{Power $1 - \beta_1$ =Proportion correctly declared as differentially expressed = 0.90}										
100	53	24	14	9	6	5	4	3	3	2	2
200	60	27	15	10	7	5	4	3	3	2	2
500	70	31	18	12	8	6	5	4	3	3	2
1000	77	34	20	13	9	7	5	4	4	3	3
2000	84	38	21	14	10	7	6	5	4	3	3
5000	93	42	24	15	11	8	6	5	4	4	3
10000	101	45	26	17	12	9	7	5	5	4	3
20000	108	48	27	18	12	9	7	6	5	4	3
Genes G_0	\multicolumn{11}{l}{Power $1 - \beta_1$ =Proportion correctly declared as differentially expressed = 0.95}										
100	64	29	16	11	8	6	4	4	3	3	2
200	72	32	18	12	8	6	5	4	3	3	2
500	82	37	21	14	10	7	6	5	4	3	3
1000	90	40	23	15	10	8	6	5	4	3	3
2000	98	44	25	16	11	8	7	5	4	4	3
5000	108	48	27	18	12	9	7	6	5	4	3
10000	116	52	29	19	13	10	8	6	5	4	4
20000	123	55	31	20	14	11	8	7	5	5	4
Genes G_0	\multicolumn{11}{l}{Power $1 - \beta_1$ =Proportion correctly declared as differentially expressed = 0.99}										
100	87	39	22	14	10	8	6	5	4	3	3
200	97	43	25	16	11	8	7	5	4	4	3
500	109	49	28	18	13	9	7	6	5	4	4
1000	118	53	30	19	14	10	8	6	5	4	4
2000	127	57	32	21	15	11	8	7	6	5	4
5000	138	62	35	23	16	12	9	7	6	5	4
10000	147	65	37	24	17	12	10	8	6	5	5
20000	155	69	39	25	18	13	10	8	7	6	5

Table 1. Sample Sizes for Treatment-control Designs
Mean Number of False Positives $E(R_0) = 3$

The table gives the group sample size n for treatment and control groups in matched-pairs or completely randomized designs. $E(R_0)$ denotes the mean number of false positives, $|\mu_1|/\sigma_d$ the statistical distance between treatment and control conditions under H_1, G_0 the anticipated number of undifferentially expressed genes.

	Distance $|\mu_1|/\sigma_d$										
	0.50	0.75	1.00	1.25	1.50	1.75	2.00	2.25	2.50	2.75	3.00
Genes G_0	Power $1 - \beta_1$ =Proportion correctly declared as differentially expressed = 0.80										
100	37	17	10	6	5	3	3	2	2	2	2
200	43	20	11	7	5	4	3	3	2	2	2
500	52	23	13	9	6	5	4	3	3	2	2
1000	59	26	15	10	7	5	4	3	3	2	2
2000	65	29	17	11	8	6	5	4	3	3	2
5000	74	33	19	12	9	6	5	4	3	3	3
10000	80	36	20	13	9	7	5	4	4	3	3
20000	86	39	22	14	10	8	6	5	4	3	3
Genes G_0	Power $1 - \beta_1$ =Proportion correctly declared as differentially expressed = 0.90										
100	48	22	12	8	6	4	3	3	2	2	2
200	56	25	14	9	7	5	4	3	3	2	2
500	65	29	17	11	8	6	5	4	3	3	2
1000	73	33	19	12	9	6	5	4	3	3	3
2000	80	36	20	13	9	7	5	4	4	3	3
5000	89	40	23	15	10	8	6	5	4	3	3
10000	96	43	24	16	11	8	6	5	4	4	3
20000	103	46	26	17	12	9	7	6	5	4	3
Genes G_0	Power $1 - \beta_1$ =Proportion correctly declared as differentially expressed = 0.95										
100	59	26	15	10	7	5	4	3	3	2	2
200	67	30	17	11	8	6	5	4	3	3	2
500	78	35	20	13	9	7	5	4	4	3	3
1000	86	38	22	14	10	7	6	5	4	3	3
2000	93	42	24	15	11	8	6	5	4	4	3
5000	104	46	26	17	12	9	7	6	5	4	3
10000	111	50	28	18	13	10	7	6	5	4	4
20000	119	53	30	19	14	10	8	6	5	4	4
Genes G_0	Power $1 - \beta_1$ =Proportion correctly declared as differentially expressed = 0.99										
100	81	36	21	13	9	7	6	4	4	3	3
200	91	41	23	15	11	8	6	5	4	3	3
500	103	46	26	17	12	9	7	6	5	4	3
1000	113	50	29	18	13	10	8	6	5	4	4
2000	122	54	31	20	14	10	8	6	5	5	4
5000	133	59	34	22	15	11	9	7	6	5	4
10000	142	63	36	23	16	12	9	7	6	5	4
20000	150	67	38	24	17	13	10	8	6	5	5

Table 1. Sample Sizes for Treatment-control Designs
Mean Number of False Positives $E(R_0) = 4$

The table gives the group sample size n for treatment and control groups in matched-pairs or completely randomized designs. $E(R_0)$ denotes the mean number of false positives, $|\mu_1|/\sigma_d$ the statistical distance between treatment and control conditions under H_1, G_0 the anticipated number of undifferentially expressed genes.

	Distance $|\mu_1|/\sigma_d$										
	0.50	0.75	1.00	1.25	1.50	1.75	2.00	2.25	2.50	2.75	3.00
Genes G_0	Power $1 - \beta_1$ =Proportion correctly declared as differentially expressed = 0.80										
100	34	15	9	6	4	3	3	2	2	2	1
200	41	18	11	7	5	4	3	2	2	2	2
500	49	22	13	8	6	4	4	3	2	2	2
1000	56	25	14	9	7	5	4	3	3	2	2
2000	62	28	16	10	7	6	4	4	3	3	2
5000	71	32	18	12	8	6	5	4	3	3	2
10000	77	35	20	13	9	7	5	4	4	3	3
20000	84	37	21	14	10	7	6	5	4	3	3
Genes G_0	Power $1 - \beta_1$ =Proportion correctly declared as differentially expressed = 0.90										
100	45	20	12	8	5	4	3	3	2	2	2
200	53	24	14	9	6	5	4	3	3	2	2
500	62	28	16	10	7	6	4	4	3	3	2
1000	70	31	18	12	8	6	5	4	3	3	2
2000	77	34	20	13	9	7	5	4	4	3	3
5000	86	39	22	14	10	8	6	5	4	3	3
10000	93	42	24	15	11	8	6	5	4	4	3
20000	101	45	26	17	12	9	7	5	5	4	3
Genes G_0	Power $1 - \beta_1$ =Proportion correctly declared as differentially expressed = 0.95										
100	55	25	14	9	7	5	4	3	3	2	2
200	64	29	16	11	8	6	4	4	3	3	2
500	74	33	19	12	9	7	5	4	3	3	3
1000	82	37	21	14	10	7	6	5	4	3	3
2000	90	40	23	15	10	8	6	5	4	3	3
5000	100	45	25	16	12	9	7	5	4	4	3
10000	108	48	27	18	12	9	7	6	5	4	3
20000	116	52	29	19	13	10	8	6	5	4	4
Genes G_0	Power $1 - \beta_1$ =Proportion correctly declared as differentially expressed = 0.99										
100	77	35	20	13	9	7	5	4	4	3	3
200	87	39	22	14	10	8	6	5	4	3	3
500	100	45	25	16	12	9	7	5	4	4	3
1000	109	49	28	18	13	9	7	6	5	4	4
2000	118	53	30	19	14	10	8	6	5	4	4
5000	130	58	33	21	15	11	9	7	6	5	4
10000	138	62	35	23	16	12	9	7	6	5	4
20000	147	65	37	24	17	12	10	8	6	5	5

Appendix B
Power Table for Multiple-treatment Designs

TABLE 2: Power Table for Multiple-treatment Designs

This table provides power values for multiple-treatment designs in which differential gene expression for each gene is summarized by the quadratic statistic defined in section 14.9.3 of Chapter 14.

The following list summarizes the notation for items cross-referenced in the table.

$E(R_0)$: The mean number of false positives.

ψ_1: The non-centrality parameter for the design.

T: The number of treatments.

G_0: The anticipated number of undifferentially expressed genes in the experiment.

The number listed in each cell is the individual power level $1-\beta_1$. This power value is the expected fraction of truly differentially expressed genes that will be correctly declared as differentially expressed by the tests.

Use of the Table

The following examples show how this table can be used for several kinds of designs for which the quadratic statistic is appropriate.

Multiple-treatment Design Having an Isolated Treatment Effect

The experiment has either a completely randomized design or a randomized block design. Under the alternative hypothesis H_1, one treatment is distinguished from the other $T-1$ treatments by exhibiting differential expression for the gene. The difference in expression between the distinguished treatment and the other treatments is μ_1 on the log-intensity scale. The number of replicates or blocks, as the case may be, is n. Hence, there are nT readings on each gene. If the error variance (expected mean square error) of the associated ANOVA model is denoted by σ^2, the non-centrality parameter previously defined in equation (14.31) in Chapter 14 has the following form.

$$\psi_1 = \frac{n(T-1)}{T}\left(\frac{|\mu_1|}{\sigma}\right)^2 \tag{B.1}$$

Example 1. Consider a randomized block design involving $T = 6$ treatments and $G_0 = 10,000$ undifferentially expressed genes. The investigator wishes to control the mean number of false positives at $E(R_0) = 2$ and to detect an isolated effect that amounts to a 1.5-fold difference between the distinguished treatment and the other treatments. The experimental error standard deviation σ is anticipated to be 0.30 on a log-2 scale. The 1.5-fold difference represents a value of $|\mu_1| = \log_2(1.5) = 0.5850$ on a log-2 scale. Thus, the ratio $|\mu_1|/\sigma$ equals $0.5850/0.30 = 1.950$. Eight blocks ($n = 8$) are to be used. For these specifications, the non-centrality parameter (B.1) equals

$$\psi_1 = 8\frac{(6-1)}{6}(1.950)^2 = 25.3.$$

Without interpolating, reference to the closest cell of the table corresponding to $E(R_0) = 2$, $T = 6$, $\psi_1 = 26$ and $G_0 = 10,000$, shows an individual power level of $1 - \beta_1 = .72$. Thus, about 72 percent of genes that exhibit an isolated 1.5-fold differential expression in one of the $T = 6$ treatments are expected to be discovered with this study design. The table can be used iteratively to explore the effect on power of specific design changes. For example, if $n = 10$ blocks are used in lieu of $n = 8$, then recalculation of the non-centrality parameter gives $\psi_1 = 31.7$. Reference to the closest cell of the table corresponding to $E(R_0) = 2$, $T = 6$, $\psi_1 = 32$ and $G_0 = 10,000$ gives an individual power level of .87, which is somewhat larger than previously.

Example 2. Should an investigator assume that the estimated differential expression vectors $\hat{\mathcal{I}}_g$ are possibly dependent then the Bonferroni approach is used. As shown in equation (14.19), we have $E(R_0) = G_0\alpha_0 = \alpha_F$ in the Bonferroni approach. Thus, the expected number of false positives is necessarily smaller than 1 and the table should be entered accordingly. Two values of $E(R_0)$ smaller than 1 are displayed in the table, namely, 0.1 and 0.5. To illustrate this use of the table, consider the same situation as in *Example 1* but now we set $E(R_0) = \alpha_F = 0.5$. Reference to the closest cell of the table corresponding to $E(R_0) = 0.5$, $T = 6$, $\psi_1 = 26$ and $G_0 = 10,000$ gives an individual power level of .62.

Treatment-control Design

A treatment-control design corresponds to a multiple-treatment design with two treatments (i.e., $T = 2$). The design has equal sample sizes for all treatments and may be of the completely randomized or randomized block variety. The non-centrality parameter for this special case has the following form.

$$\psi_1 = \frac{n}{2}\left(\frac{\mu_1}{\sigma}\right)^2 \tag{B.2}$$

For both a completely randomized design and matched-pairs design, we have the identity $\sigma_d^2 = 2\sigma^2$, where σ_d^2 denotes the variance of the difference in log-expression between treatment and control conditions and σ^2 the error variance of the associated ANOVA model.

Example 1. Consider a treatment-control design of the completely randomized variety involving $G_0 = 5000$ undifferentially expressed genes. The investigator wishes to control the mean number of false positives at $E(R_0) = 2$ and to detect a two-fold differential expression between the treatment and control conditions. The experimental error standard deviation is anticipated to be about $\sigma = 0.40$ on a log-2 scale. The two-fold difference represents a value of $|\mu_1| = \log_2(2.0) = 1.000$ on a log-2 scale. Thus, the ratio $|\mu_1|/\sigma$ equals $1.000/0.40 = 2.500$. Eight replications are to be used ($n = 8$). For these specifications, the non-centrality parameter in (B.2) equals

$$\psi_1 = \frac{8}{2}(2.500)^2 = 25$$

Without interpolating, reference to the closest cell of the table corresponding to $E(R_0) = 2$, $T = 2$, $\psi_1 = 24$ and $G_0 = 5000$ shows an individual power level of $1 - \beta_1 = 0.91$. Thus, about 91 percent of genes that exhibit a two-fold differential expression between treatment and control (whether up- or down-regulated) are expected to be discovered with this study design.

Example 2. Consider a matched-pairs design involving $G_0 = 5000$ undifferentially expressed genes. The investigator wishes to control the mean number of false positives at $E(R_0) = 2$ and to detect a two-fold differential expression between the treatment and control conditions. The standard deviation of the experimental error is anticipated to be $\sigma = 0.30$. (This corresponds to a standard deviation of $\sigma_d = \sqrt{(2)}(0.30) = 0.424$ on a log-2 scale for the difference between treatment and control expression levels). The two-fold difference represents a value of $|\mu_1| = \log_2(2.0) = 1.000$ on a log-2 scale. Thus, the ratio $|\mu_1|/\sigma$ equals $1.000/0.30 = 3.333$. Four matched pairs are to be used ($n = 4$). For these specifications, the non-centrality parameter in (B.2) equals

$$\psi_1 = \frac{4}{2}(3.333)^2 = 22.2$$

Without interpolation, reference to the closest cell of the table corresponding to $E(R_0) = 2$, $T = 2$, $\psi_1 = 22$ and $G_0 = 5000$ shows an individual power level of $1 - \beta_1 = 0.87$. Thus, about 87 percent of genes that exhibit a two-fold differential expression between matched treatment and control conditions (whether up- or down-regulated) are expected to be discovered with this study design.

Table 2. Power Table for Multiple-treatment Designs
Mean Number of False Positives $E(R_0) = 0.1$

This table lists power values for multiple-treatment designs. $E(R_0)$ denotes the mean number of false positives, ψ_1 the non-centrality parameter for the design, T the number of treatment conditions in the experiment, G_0 the anticipated number of undifferentially expressed genes in the experiment.

	Number of treatments $T = 2$								
	Non-centrality ψ_1								
Genes G_0	12	14	16	18	20	22	24	26	28	30
100	.57	.67	.76	.83	.88	.92	.95	.96	.98	.99
200	.49	.60	.70	.78	.84	.89	.92	.95	.96	.98
500	.40	.51	.61	.70	.77	.83	.88	.92	.94	.96
1000	.33	.44	.54	.64	.72	.79	.84	.89	.92	.94
	Non-centrality ψ_1								
Genes G_0	22	24	26	28	30	32	34	36	38	40
2000	.74	.80	.85	.89	.92	.95	.96	.97	.98	.99
5000	.66	.74	.80	.85	.89	.92	.94	.96	.97	.98
10000	.61	.69	.75	.81	.86	.89	.92	.94	.96	.97
20000	.55	.63	.70	.77	.82	.86	.90	.92	.95	.96

	Number of treatments $T = 3$								
	Non-centrality ψ_1								
Genes G_0	12	14	16	18	20	22	24	26	28	30
100	.46	.56	.66	.74	.81	.86	.90	.93	.95	.97
200	.38	.49	.59	.68	.76	.82	.87	.91	.93	.95
500	.30	.40	.50	.59	.68	.75	.81	.86	.90	.93
1000	.24	.34	.43	.53	.62	.70	.76	.82	.87	.90
	Non-centrality ψ_1								
Genes G_0	22	24	26	28	30	32	34	36	38	40
2000	.64	.71	.77	.83	.87	.90	.93	.95	.96	.98
5000	.56	.64	.71	.77	.82	.86	.90	.93	.95	.96
10000	.50	.58	.66	.72	.78	.83	.87	.90	.93	.95
20000	.44	.52	.60	.67	.74	.79	.84	.88	.91	.93

	Number of treatments $T = 4$								
	Non-centrality ψ_1								
Genes G_0	22	24	26	28	30	32	34	36	38	40
100	.81	.86	.90	.93	.95	.97	.98	.99	.99	.99
200	.76	.82	.87	.90	.93	.95	.97	.98	.98	.99
500	.68	.75	.81	.86	.89	.92	.94	.96	.97	.98
1000	.62	.70	.76	.82	.86	.90	.92	.94	.96	.97
	Non-centrality ψ_1								
Genes G_0	22	24	26	28	30	32	34	36	38	40
2000	.56	.64	.71	.77	.82	.86	.90	.92	.95	.96
5000	.48	.56	.64	.71	.77	.82	.86	.89	.92	.94
10000	.42	.50	.58	.65	.72	.77	.82	.86	.90	.92
20000	.37	.45	.53	.60	.67	.73	.78	.83	.87	.90

Table 2: Power Table for Multiple-treatment Designs
Mean Number of False Positives $E(R_0) = 0.1$

This table lists power values for multiple-treatment designs. $E(R_0)$ denotes the mean number of false positives, ψ_1 the non-centrality parameter for the design, T the number of treatment conditions in the experiment, G_0 the anticipated number of undifferentially expressed genes in the experiment.

Number of treatments $T = 5$									
Non-centrality ψ_1									
Genes G_0	22	24	26	28	30	32	34	36	38	40
100	.77	.82	.87	.91	.93	.95	.97	.98	.98	.99
200	.71	.77	.83	.87	.90	.93	.95	.97	.98	.98
500	.63	.70	.76	.82	.86	.90	.92	.94	.96	.97
1000	.56	.64	.71	.77	.82	.86	.90	.92	.94	.96
Non-centrality ψ_1									
Genes G_0	32	34	36	38	40	42	44	46	48	50
2000	.83	.87	.90	.92	.94	.96	.97	.98	.99	.99
5000	.77	.82	.86	.89	.92	.94	.95	.97	.98	.98
10000	.72	.78	.82	.86	.89	.92	.94	.96	.97	.98
20000	.68	.73	.79	.83	.87	.90	.92	.94	.96	.97
Number of treatments $T = 6$									
Non-centrality ψ_1									
Genes G_0	22	24	26	28	30	32	34	36	38	40
100	.73	.79	.84	.88	.91	.94	.95	.97	.98	.98
200	.66	.73	.79	.84	.88	.91	.93	.95	.97	.98
500	.58	.65	.72	.78	.83	.87	.90	.93	.95	.96
1000	.51	.59	.66	.73	.78	.83	.87	.90	.93	.95
Non-centrality ψ_1									
Genes G_0	32	34	36	38	40	42	44	46	48	50
2000	.79	.83	.87	.90	.93	.95	.96	.97	.98	.99
5000	.73	.78	.83	.86	.89	.92	.94	.96	.97	.98
10000	.68	.74	.79	.83	.87	.90	.92	.94	.96	.97
20000	.63	.69	.74	.79	.83	.87	.90	.92	.94	.96
Number of treatments $T = 7$									
Non-centrality ψ_1									
Genes G_0	22	24	26	28	30	32	34	36	38	40
100	.69	.75	.81	.85	.89	.92	.94	.96	.97	.98
200	.62	.69	.76	.81	.85	.89	.92	.94	.96	.97
500	.53	.61	.68	.74	.80	.84	.88	.91	.93	.95
1000	.47	.55	.62	.69	.75	.80	.84	.88	.91	.93
Non-centrality ψ_1									
Genes G_0	32	34	36	38	40	42	44	46	48	50
2000	.75	.80	.84	.88	.91	.93	.95	.96	.97	.98
5000	.69	.74	.79	.84	.87	.90	.92	.94	.96	.97
10000	.63	.70	.75	.80	.84	.87	.90	.92	.94	.96
20000	.58	.64	.70	.76	.80	.84	.87	.90	.92	.94

Table 2: Power Table for Multiple-treatment Designs
Mean Number of False Positives $E(R_0) = 0.1$

This table lists power values for multiple-treatment designs. $E(R_0)$ denotes the mean number of false positives, ψ_1 the non-centrality parameter for the design, T the number of treatment conditions in the experiment, G_0 the anticipated number of undifferentially expressed genes in the experiment.

	Number of treatments $T = 8$								
	Non-centrality ψ_1								
Genes G_0	22	24	26	28	30	32	34	36	38	40
100	.65	.72	.78	.83	.87	.90	.93	.95	.96	.97
200	.58	.66	.72	.78	.83	.87	.90	.93	.95	.96
500	.49	.57	.64	.71	.77	.81	.86	.89	.92	.94
1000	.43	.51	.58	.65	.71	.77	.82	.86	.89	.91
	Non-centrality ψ_1								
Genes G_0	32	34	36	38	40	42	44	46	48	50
2000	.72	.77	.82	.86	.89	.91	.93	.95	.96	.97
5000	.65	.71	.76	.81	.85	.88	.91	.93	.95	.96
10000	.59	.66	.71	.77	.81	.85	.88	.91	.93	.94
20000	.54	.61	.67	.72	.77	.81	.85	.88	.91	.93

	Number of treatments $T = 9$								
	Non-centrality ψ_1								
Genes G_0	22	24	26	28	30	32	34	36	38	40
100	.62	.69	.75	.80	.85	.88	.91	.94	.95	.97
200	.55	.62	.69	.75	.80	.85	.88	.91	.93	.95
500	.46	.53	.61	.68	.74	.79	.83	.87	.90	.92
1000	.39	.47	.54	.62	.68	.74	.79	.83	.87	.90
	Non-centrality ψ_1								
Genes G_0	32	34	36	38	40	42	44	46	48	50
2000	.69	.74	.79	.83	.87	.90	.92	.94	.95	.97
5000	.61	.67	.73	.78	.82	.86	.89	.91	.93	.95
10000	.56	.62	.68	.73	.78	.82	.86	.89	.91	.93
20000	.50	.57	.63	.69	.74	.79	.83	.86	.89	.91

	Number of treatments $T = 10$								
	Non-centrality ψ_1								
Genes G_0	22	24	26	28	30	32	34	36	38	40
100	.59	.66	.72	.78	.83	.87	.90	.92	.94	.96
200	.52	.59	.66	.72	.78	.82	.86	.90	.92	.94
500	.43	.50	.58	.64	.71	.76	.81	.85	.88	.91
1000	.36	.44	.51	.58	.65	.71	.76	.81	.85	.88
	Non-centrality ψ_1								
Genes G_0	32	34	36	38	40	42	44	46	48	50
2000	.65	.71	.76	.81	.85	.88	.91	.93	.95	.96
5000	.58	.64	.70	.75	.80	.84	.87	.90	.92	.94
10000	.52	.59	.65	.70	.75	.80	.84	.87	.90	.92
20000	.47	.53	.60	.66	.71	.76	.80	.84	.87	.90

Table 2: Power Table for Multiple-treatment Designs
Mean Number of False Positives $E(R_0) = 0.5$

This table lists power values for multiple-treatment designs. $E(R_0)$ denotes the mean number of false positives, ψ_1 the non-centrality parameter for the design, T the number of treatment conditions in the experiment, G_0 the anticipated number of undifferentially expressed genes in the experiment.

	Number of treatments $T = 2$									
	Non-centrality ψ_1									
Genes G_0	2	4	6	8	10	12	14	16	18	20
100	.08	.21	.36	.51	.64	.74	.83	.88	.92	.95
200	.05	.15	.28	.42	.56	.67	.76	.84	.89	.93
500	.03	.10	.20	.32	.45	.57	.67	.76	.83	.88
1000	.02	.07	.15	.26	.38	.49	.60	.70	.78	.84
	Non-centrality ψ_1									
Genes G_0	12	14	16	18	20	22	24	26	28	30
2000	.42	.53	.63	.72	.79	.85	.89	.92	.95	.97
5000	.33	.44	.54	.64	.72	.79	.84	.89	.92	.94
10000	.28	.38	.48	.57	.66	.74	.80	.85	.89	.92
20000	.23	.32	.41	.51	.60	.68	.75	.81	.86	.90
	Number of treatments $T = 3$									
	Non-centrality ψ_1									
Genes G_0	12	14	16	18	20	22	24	26	28	30
100	.64	.74	.81	.87	.91	.94	.96	.98	.98	.99
200	.56	.66	.75	.82	.87	.91	.94	.96	.97	.98
500	.46	.56	.66	.74	.81	.86	.90	.93	.95	.97
1000	.38	.49	.59	.68	.76	.82	.87	.91	.93	.95
	Non-centrality ψ_1									
Genes G_0	22	24	26	28	30	32	34	36	38	40
2000	.77	.83	.87	.91	.93	.95	.97	.98	.99	.99
5000	.70	.76	.82	.87	.90	.93	.95	.96	.98	.98
10000	.64	.71	.77	.83	.87	.90	.93	.95	.96	.98
20000	.58	.66	.73	.79	.83	.88	.91	.93	.95	.97
	Number of treatments $T = 4$									
	Non-centrality ψ_1									
Genes G_0	12	14	16	18	20	22	24	26	28	30
100	.57	.67	.75	.82	.87	.91	.94	.96	.97	.98
200	.48	.59	.68	.76	.82	.87	.91	.94	.96	.97
500	.38	.49	.59	.67	.75	.81	.86	.90	.93	.95
1000	.31	.42	.51	.61	.69	.76	.82	.87	.90	.93
	Non-centrality ψ_1									
Genes G_0	22	24	26	28	30	32	34	36	38	40
2000	.70	.77	.82	.87	.90	.93	.95	.97	.98	.98
5000	.62	.70	.76	.82	.86	.90	.92	.94	.96	.97
10000	.56	.64	.71	.77	.82	.86	.90	.92	.95	.96
20000	.50	.58	.66	.72	.78	.83	.87	.90	.93	.95

Table 2: Power Table for Multiple-treatment Designs
Mean Number of False Positives $E(R_0) = 0.5$

This table lists power values for multiple-treatment designs. $E(R_0)$ denotes the mean number of false positives, ψ_1 the non-centrality parameter for the design, T the number of treatment conditions in the experiment, G_0 the anticipated number of undifferentially expressed genes in the experiment.

	Number of treatments $T = 5$								
	Non-centrality ψ_1								
Genes G_0	12	14	16	18	20	22	24	26	28	30
100	.51	.61	.70	.78	.84	.88	.92	.94	.96	.97
200	.43	.53	.63	.71	.78	.84	.88	.92	.94	.96
500	.33	.43	.53	.62	.70	.77	.82	.87	.91	.93
1000	.27	.36	.45	.55	.63	.71	.77	.83	.87	.90
	Non-centrality ψ_1								
Genes G_0	22	24	26	28	30	32	34	36	38	40
2000	.65	.72	.78	.83	.87	.91	.93	.95	.96	.98
5000	.56	.64	.71	.77	.82	.86	.90	.92	.94	.96
10000	.50	.58	.65	.72	.78	.83	.87	.90	.92	.94
20000	.44	.52	.60	.67	.73	.78	.83	.87	.90	.93
	Number of treatments $T = 6$								
	Non-centrality ψ_1								
Genes G_0	12	14	16	18	20	22	24	26	28	30
100	.46	.57	.66	.74	.80	.85	.90	.93	.95	.96
200	.38	.48	.58	.67	.74	.80	.85	.89	.92	.95
500	.29	.38	.48	.57	.65	.73	.79	.84	.88	.91
1000	.23	.32	.41	.50	.58	.66	.73	.79	.84	.88
	Non-centrality ψ_1								
Genes G_0	22	24	26	28	30	32	34	36	38	40
2000	.60	.67	.74	.79	.84	.88	.91	.93	.95	.97
5000	.51	.59	.66	.73	.78	.83	.87	.90	.93	.95
10000	.45	.53	.60	.67	.74	.79	.83	.87	.90	.93
20000	.39	.47	.55	.62	.68	.74	.79	.84	.87	.90
	Number of treatments $T = 7$								
	Non-centrality ψ_1								
Genes G_0	22	24	26	28	30	32	34	36	38	40
100	.83	.87	.91	.93	.95	.97	.98	.99	.99	.99
200	.77	.83	.87	.90	.93	.95	.97	.98	.98	.99
500	.69	.75	.81	.85	.89	.92	.94	.96	.97	.98
1000	.62	.69	.76	.81	.85	.89	.92	.94	.96	.97
	Non-centrality ψ_1								
Genes G_0	22	24	26	28	30	32	34	36	38	40
2000	.55	.63	.70	.76	.81	.85	.89	.92	.94	.95
5000	.47	.55	.62	.69	.75	.80	.84	.88	.91	.93
10000	.41	.48	.56	.63	.70	.75	.80	.84	.88	.91
20000	.35	.42	.50	.57	.64	.70	.76	.81	.85	.88

Table 2: Power Table for Multiple-treatment Designs
Mean Number of False Positives $E(R_0) = 0.5$

This table lists power values for multiple-treatment designs. $E(R_0)$ denotes the mean number of false positives, ψ_1 the non-centrality parameter for the design, T the number of treatment conditions in the experiment, G_0 the anticipated number of undifferentially expressed genes in the experiment.

	Number of treatments $T = 8$								
	Non-centrality ψ_1								
Genes G_0	22	24	26	28	30	32	34	36	38	40
100	.80	.85	.89	.92	.94	.96	.97	.98	.99	.99
200	.74	.80	.85	.89	.92	.94	.96	.97	.98	.99
500	.65	.72	.78	.83	.87	.90	.93	.95	.96	.97
1000	.58	.66	.72	.78	.83	.87	.90	.93	.95	.96
	Non-centrality ψ_1								
Genes G_0	32	34	36	38	40	42	44	46	48	50
2000	.83	.87	.90	.92	.94	.96	.97	.98	.98	.99
5000	.77	.82	.86	.89	.91	.94	.95	.96	.97	.98
10000	.72	.77	.82	.86	.89	.91	.93	.95	.96	.97
20000	.67	.72	.78	.82	.86	.89	.91	.93	.95	.96
	Number of treatments $T = 9$								
	Non-centrality ψ_1								
Genes G_0	22	24	26	28	30	32	34	36	38	40
100	.78	.83	.87	.91	.93	.95	.97	.98	.98	.99
200	.71	.77	.82	.87	.90	.93	.95	.96	.97	.98
500	.62	.69	.75	.80	.85	.88	.91	.94	.95	.97
1000	.55	.62	.69	.75	.80	.85	.88	.91	.93	.95
	Non-centrality ψ_1								
Genes G_0	32	34	36	38	40	42	44	46	48	50
2000	.80	.85	.88	.91	.93	.95	.96	.97	.98	.99
5000	.74	.79	.83	.87	.90	.92	.94	.96	.97	.98
10000	.69	.74	.79	.83	.87	.90	.92	.94	.95	.97
20000	.63	.69	.75	.79	.83	.87	.90	.92	.94	.95
	Number of treatments $T = 10$								
	Non-centrality ψ_1								
Genes G_0	22	24	26	28	30	32	34	36	38	40
100	.75	.81	.85	.89	.92	.94	.96	.97	.98	.99
200	.68	.75	.80	.85	.88	.91	.94	.95	.97	.98
500	.59	.66	.72	.78	.83	.87	.90	.92	.94	.96
1000	.52	.59	.66	.72	.78	.82	.86	.90	.92	.94
	Non-centrality ψ_1								
Genes G_0	32	34	36	38	40	42	44	46	48	50
2000	.78	.82	.86	.89	.92	.94	.95	.97	.98	.98
5000	.71	.76	.81	.85	.88	.91	.93	.95	.96	.97
10000	.65	.71	.76	.81	.85	.88	.91	.93	.95	.96
20000	.60	.66	.72	.77	.81	.85	.88	.91	.93	.94

Table 2: Power Table for Multiple-treatment Designs
Mean Number of False Positives $E(R_0) = 1$

This table lists power values for multiple-treatment designs. $E(R_0)$ denotes the mean number of false positives, ψ_1 the non-centrality parameter for the design, T the number of treatment conditions in the experiment, G_0 the anticipated number of undifferentially expressed genes in the experiment.

	Number of treatments $T = 2$								
	Non-centrality ψ_1								
Genes G_0	2	4	6	8	10	12	14	16	18	20
100	.12	.28	.45	.60	.72	.81	.88	.92	.95	.97
200	.08	.21	.36	.51	.64	.74	.83	.88	.92	.95
500	.05	.14	.26	.40	.53	.65	.74	.82	.88	.92
1000	.03	.10	.20	.32	.45	.57	.67	.76	.83	.88
	Non-centrality ψ_1								
Genes G_0	12	14	16	18	20	22	24	26	28	30
2000	.49	.60	.70	.78	.84	.89	.92	.95	.96	.98
5000	.40	.51	.61	.70	.77	.83	.88	.92	.94	.96
10000	.33	.44	.54	.64	.72	.79	.84	.89	.92	.94
20000	.28	.38	.48	.57	.66	.74	.80	.85	.89	.92
	Number of treatments $T = 3$								
	Non-centrality ψ_1								
Genes G_0	12	14	16	18	20	22	24	26	28	30
100	.72	.80	.87	.91	.94	.96	.98	.99	.99	.99
200	.64	.74	.81	.87	.91	.94	.96	.98	.98	.99
500	.53	.64	.73	.80	.86	.90	.93	.95	.97	.98
1000	.46	.56	.66	.74	.81	.86	.90	.93	.95	.97
	Non-centrality ψ_1								
Genes G_0	22	24	26	28	30	32	34	36	38	40
2000	.82	.87	.91	.93	.95	.97	.98	.99	.99	.99
5000	.75	.81	.86	.90	.93	.95	.96	.98	.98	.99
10000	.70	.76	.82	.87	.90	.93	.95	.96	.98	.98
20000	.64	.71	.77	.83	.87	.90	.93	.95	.96	.98
	Number of treatments $T = 4$								
	Non-centrality ψ_1								
Genes G_0	12	14	16	18	20	22	24	26	28	30
100	.65	.75	.82	.87	.91	.94	.96	.98	.98	.99
200	.57	.67	.75	.82	.87	.91	.94	.96	.97	.98
500	.46	.56	.66	.74	.81	.86	.90	.93	.95	.97
1000	.38	.49	.59	.67	.75	.81	.86	.90	.93	.95
	Non-centrality ψ_1								
Genes G_0	22	24	26	28	30	32	34	36	38	40
2000	.76	.82	.87	.90	.93	.95	.97	.98	.98	.99
5000	.68	.75	.81	.86	.89	.92	.94	.96	.97	.98
10000	.62	.70	.76	.82	.86	.90	.92	.94	.96	.97
20000	.56	.64	.71	.77	.82	.86	.90	.92	.95	.96

Table 2: Power Table for Multiple-treatment Designs
Mean Number of False Positives $E(R_0) = 1$

This table lists power values for multiple-treatment designs. $E(R_0)$ denotes the mean number of false positives, ψ_1 the non-centrality parameter for the design, T the number of treatment conditions in the experiment, G_0 the anticipated number of undifferentially expressed genes in the experiment.

	Number of treatments $T = 5$									
	Non-centrality ψ_1									
Genes G_0	12	14	16	18	20	22	24	26	28	30
100	.60	.69	.77	.84	.89	.92	.95	.96	.98	.99
200	.51	.61	.70	.78	.84	.88	.92	.94	.96	.97
500	.40	.51	.60	.69	.76	.82	.87	.91	.93	.95
1000	.33	.43	.53	.62	.70	.77	.82	.87	.91	.93
	Non-centrality ψ_1									
Genes G_0	22	24	26	28	30	32	34	36	38	40
2000	.71	.77	.83	.87	.90	.93	.95	.97	.98	.98
5000	.63	.70	.76	.82	.86	.90	.92	.94	.96	.97
10000	.56	.64	.71	.77	.82	.86	.90	.92	.94	.96
20000	.50	.58	.65	.72	.78	.83	.87	.90	.92	.94

	Number of treatments $T = 6$									
	Non-centrality ψ_1									
Genes G_0	12	14	16	18	20	22	24	26	28	30
100	.55	.65	.74	.80	.86	.90	.93	.95	.97	.98
200	.46	.57	.66	.74	.80	.85	.90	.93	.95	.96
500	.36	.46	.55	.64	.72	.78	.84	.88	.91	.94
1000	.29	.38	.48	.57	.65	.73	.79	.84	.88	.91
	Non-centrality ψ_1									
Genes G_0	22	24	26	28	30	32	34	36	38	40
2000	.66	.73	.79	.84	.88	.91	.93	.95	.97	.98
5000	.58	.65	.72	.78	.83	.87	.90	.93	.95	.96
10000	.51	.59	.66	.73	.78	.83	.87	.90	.93	.95
20000	.45	.53	.60	.67	.74	.79	.83	.87	.90	.93

	Number of treatments $T = 7$									
	Non-centrality ψ_1									
Genes G_0	12	14	16	18	20	22	24	26	28	30
100	.51	.61	.70	.77	.83	.88	.91	.94	.96	.97
200	.42	.52	.62	.70	.77	.83	.87	.91	.93	.95
500	.32	.42	.51	.60	.68	.75	.81	.86	.89	.92
1000	.26	.34	.44	.53	.61	.69	.75	.81	.85	.89
	Non-centrality ψ_1									
Genes G_0	22	24	26	28	30	32	34	36	38	40
2000	.62	.69	.76	.81	.85	.89	.92	.94	.96	.97
5000	.53	.61	.68	.74	.80	.84	.88	.91	.93	.95
10000	.47	.55	.62	.69	.75	.80	.84	.88	.91	.93
20000	.41	.48	.56	.63	.70	.75	.80	.84	.88	.91

Table 2: Power Table for Multiple-treatment Designs
Mean Number of False Positives $E(R_0) = 1$

This table lists power values for multiple-treatment designs. $E(R_0)$ denotes the mean number of false positives, ψ_1 the non-centrality parameter for the design, T the number of treatment conditions in the experiment, G_0 the anticipated number of undifferentially expressed genes in the experiment.

	Number of treatments $T = 8$								
	Non-centrality ψ_1								
Genes G_0	12	14	16	18	20	22	24	26	28	30
100	.48	.58	.67	.74	.81	.86	.90	.93	.95	.96
200	.39	.49	.58	.67	.74	.80	.85	.89	.92	.94
500	.29	.38	.48	.56	.65	.72	.78	.83	.87	.91
1000	.23	.31	.40	.49	.57	.65	.72	.78	.83	.87
	Non-centrality ψ_1								
Genes G_0	22	24	26	28	30	32	34	36	38	40
2000	.58	.66	.72	.78	.83	.87	.90	.93	.95	.96
5000	.49	.57	.64	.71	.77	.81	.86	.89	.92	.94
10000	.43	.51	.58	.65	.71	.77	.82	.86	.89	.91
20000	.37	.44	.52	.59	.66	.72	.77	.82	.86	.89

	Number of treatments $T = 9$								
	Non-centrality ψ_1								
Genes G_0	22	24	26	28	30	32	34	36	38	40
100	.84	.88	.91	.94	.96	.97	.98	.99	.99	.99
200	.78	.83	.87	.91	.93	.95	.97	.98	.98	.99
500	.69	.75	.81	.85	.89	.92	.94	.96	.97	.98
1000	.62	.69	.75	.80	.85	.88	.91	.94	.95	.97
	Non-centrality ψ_1								
Genes G_0	32	34	36	38	40	42	44	46	48	50
2000	.85	.88	.91	.93	.95	.96	.97	.98	.99	.99
5000	.79	.83	.87	.90	.92	.94	.96	.97	.98	.98
10000	.74	.79	.83	.87	.90	.92	.94	.96	.97	.98
20000	.69	.74	.79	.83	.87	.90	.92	.94	.95	.97

	Number of treatments $T = 10$								
	Non-centrality ψ_1								
Genes G_0	22	24	26	28	30	32	34	36	38	40
100	.82	.86	.90	.93	.95	.96	.97	.98	.99	.99
200	.75	.81	.85	.89	.92	.94	.96	.97	.98	.99
500	.66	.73	.78	.83	.87	.90	.93	.95	.96	.97
1000	.59	.66	.72	.78	.83	.87	.90	.92	.94	.96
	Non-centrality ψ_1								
Genes G_0	32	34	36	38	40	42	44	46	48	50
2000	.82	.86	.90	.92	.94	.96	.97	.98	.98	.99
5000	.76	.81	.85	.88	.91	.93	.95	.96	.97	.98
10000	.71	.76	.81	.85	.88	.91	.93	.95	.96	.97
20000	.65	.71	.76	.81	.85	.88	.91	.93	.95	.96

Table 2: Power Table for Multiple-treatment Designs
Mean Number of False Positives $E(R_0) = 2$

This table lists power values for multiple-treatment designs. $E(R_0)$ denotes the mean number of false positives, ψ_1 the non-centrality parameter for the design, T the number of treatment conditions in the experiment, G_0 the anticipated number of undifferentially expressed genes in the experiment.

	Number of treatments $T = 2$								
	Non-centrality ψ_1								
Genes G_0	2	4	6	8	10	12	14	16	18	20
100	.18	.37	.55	.69	.80	.87	.92	.95	.97	.98
200	.12	.28	.45	.60	.72	.81	.88	.92	.95	.97
500	.07	.19	.33	.48	.61	.72	.81	.87	.91	.94
1000	.05	.14	.26	.40	.53	.65	.74	.82	.88	.92

	Non-centrality ψ_1								
Genes G_0	12	14	16	18	20	22	24	26	28	30
2000	.57	.67	.76	.83	.88	.92	.95	.96	.98	.99
5000	.47	.58	.68	.76	.82	.87	.91	.94	.96	.97
10000	.40	.51	.61	.70	.77	.83	.88	.92	.94	.96
20000	.33	.44	.54	.64	.72	.79	.84	.89	.92	.94

	Number of treatments $T = 3$								
	Non-centrality ψ_1								
Genes G_0	2	4	6	8	10	12	14	16	18	20
100	.13	.28	.44	.58	.70	.80	.86	.91	.94	.97
200	.08	.20	.35	.49	.61	.72	.80	.87	.91	.94
500	.05	.13	.24	.37	.50	.61	.71	.79	.85	.90
1000	.03	.09	.18	.30	.42	.53	.64	.73	.80	.86

	Non-centrality ψ_1								
Genes G_0	12	14	16	18	20	22	24	26	28	30
2000	.46	.56	.66	.74	.81	.86	.90	.93	.95	.97
5000	.36	.47	.57	.66	.74	.80	.85	.90	.93	.95
10000	.30	.40	.50	.59	.68	.75	.81	.86	.90	.93
20000	.24	.34	.43	.53	.62	.70	.76	.82	.87	.90

	Number of treatments $T = 4$								
	Non-centrality ψ_1								
Genes G_0	2	4	6	8	10	12	14	16	18	20
100	.11	.23	.37	.51	.64	.74	.82	.87	.92	.95
200	.07	.16	.29	.42	.54	.65	.75	.82	.87	.91
500	.04	.10	.19	.31	.43	.54	.64	.73	.80	.86
1000	.02	.07	.14	.24	.35	.46	.56	.66	.74	.81

	Non-centrality ψ_1								
Genes G_0	12	14	16	18	20	22	24	26	28	30
2000	.38	.49	.59	.67	.75	.81	.86	.90	.93	.95
5000	.29	.39	.49	.58	.67	.74	.80	.85	.89	.92
10000	.24	.33	.42	.52	.60	.68	.75	.81	.86	.89
20000	.19	.27	.36	.45	.54	.62	.70	.76	.82	.86

Table 2: Power Table for Multiple-treatment Designs
Mean Number of False Positives $E(R_0) = 2$

This table lists power values for multiple-treatment designs. $E(R_0)$ denotes the mean number of false positives, ψ_1 the non-centrality parameter for the design, T the number of treatment conditions in the experiment, G_0 the anticipated number of undifferentially expressed genes in the experiment.

	Number of treatments $T = 5$								
	Non-centrality ψ_1								
Genes G_0	12	14	16	18	20	22	24	26	28	30
100	.69	.77	.84	.89	.93	.95	.97	.98	.99	.99
200	.60	.69	.77	.84	.89	.92	.95	.96	.98	.99
500	.48	.59	.68	.76	.82	.87	.91	.94	.96	.97
1000	.40	.51	.60	.69	.76	.82	.87	.91	.93	.95

	Non-centrality ψ_1								
Genes G_0	22	24	26	28	30	32	34	36	38	40
2000	.77	.82	.87	.91	.93	.95	.97	.98	.98	.99
5000	.69	.76	.81	.86	.90	.92	.94	.96	.97	.98
10000	.63	.70	.76	.82	.86	.90	.92	.94	.96	.97
20000	.56	.64	.71	.77	.82	.86	.90	.92	.94	.96

	Number of treatments $T = 6$								
	Non-centrality ψ_1								
Genes G_0	12	14	16	18	20	22	24	26	28	30
100	.65	.74	.81	.86	.91	.94	.96	.97	.98	.99
200	.55	.65	.74	.80	.86	.90	.93	.95	.97	.98
500	.44	.54	.63	.71	.78	.84	.88	.92	.94	.96
1000	.36	.46	.55	.64	.72	.78	.84	.88	.91	.94

	Non-centrality ψ_1								
Genes G_0	22	24	26	28	30	32	34	36	38	40
2000	.73	.79	.84	.88	.91	.94	.95	.97	.98	.98
5000	.64	.71	.77	.83	.87	.90	.93	.95	.96	.97
10000	.58	.65	.72	.78	.83	.87	.90	.93	.95	.96
20000	.51	.59	.66	.73	.78	.83	.87	.90	.93	.95

	Number of treatments $T = 7$								
	Non-centrality ψ_1								
Genes G_0	12	14	16	18	20	22	24	26	28	30
100	.61	.70	.78	.84	.89	.92	.95	.96	.98	.98
200	.51	.61	.70	.77	.83	.88	.91	.94	.96	.97
500	.40	.50	.59	.68	.75	.81	.86	.90	.93	.95
1000	.32	.42	.51	.60	.68	.75	.81	.86	.89	.92

	Non-centrality ψ_1								
Genes G_0	22	24	26	28	30	32	34	36	38	40
2000	.69	.75	.81	.85	.89	.92	.94	.96	.97	.98
5000	.60	.67	.74	.79	.84	.88	.91	.93	.95	.96
10000	.53	.61	.68	.74	.80	.84	.88	.91	.93	.95
20000	.47	.55	.62	.69	.75	.80	.84	.88	.91	.93

Table 2: Power Table for Multiple-treatment Designs
Mean Number of False Positives $E(R_0) = 2$

This table lists power values for multiple-treatment designs. $E(R_0)$ denotes the mean number of false positives, ψ_1 the non-centrality parameter for the design, T the number of treatment conditions in the experiment, G_0 the anticipated number of undifferentially expressed genes in the experiment.

	Number of treatments $T = 8$								
	Non-centrality ψ_1								
Genes G_0	12	14	16	18	20	22	24	26	28	30
100	.58	.67	.75	.81	.87	.90	.93	.95	.97	.98
200	.48	.58	.67	.74	.81	.86	.90	.93	.95	.96
500	.37	.46	.56	.64	.72	.78	.83	.88	.91	.94
1000	.29	.38	.48	.56	.65	.72	.78	.83	.87	.91

	Non-centrality ψ_1								
Genes G_0	22	24	26	28	30	32	34	36	38	40
2000	.65	.72	.78	.83	.87	.90	.93	.95	.96	.97
5000	.56	.64	.70	.76	.81	.86	.89	.92	.94	.96
10000	.49	.57	.64	.71	.77	.81	.86	.89	.92	.94
20000	.43	.51	.58	.65	.71	.77	.82	.86	.89	.91

	Number of treatments $T = 9$								
	Non-centrality ψ_1								
Genes G_0	12	14	16	18	20	22	24	26	28	30
100	.55	.64	.72	.79	.85	.89	.92	.94	.96	.97
200	.45	.55	.64	.72	.78	.84	.88	.91	.94	.96
500	.34	.43	.52	.61	.69	.75	.81	.86	.89	.92
1000	.27	.35	.44	.53	.61	.69	.75	.81	.85	.89

	Non-centrality ψ_1								
Genes G_0	22	24	26	28	30	32	34	36	38	40
2000	.62	.69	.75	.80	.85	.88	.91	.94	.95	.97
5000	.52	.60	.67	.73	.79	.83	.87	.90	.93	.95
10000	.46	.53	.61	.68	.74	.79	.83	.87	.90	.92
20000	.39	.47	.54	.62	.68	.74	.79	.83	.87	.90

	Number of treatments $T = 10$								
	Non-centrality ψ_1								
Genes G_0	12	14	16	18	20	22	24	26	28	30
100	.52	.61	.70	.77	.83	.87	.91	.93	.95	.97
200	.42	.52	.61	.69	.76	.82	.86	.90	.93	.95
500	.31	.40	.49	.58	.66	.73	.79	.84	.88	.91
1000	.24	.33	.41	.50	.58	.66	.73	.78	.83	.87

	Non-centrality ψ_1								
Genes G_0	22	24	26	28	30	32	34	36	38	40
2000	.59	.66	.72	.78	.83	.87	.90	.92	.94	.96
5000	.49	.57	.64	.70	.76	.81	.85	.89	.91	.93
10000	.43	.50	.58	.64	.71	.76	.81	.85	.88	.91
20000	.36	.44	.51	.58	.65	.71	.76	.81	.85	.88

Table 2: Power Table for Multiple-treatment Designs
Mean Number of False Positives $E(R_0) = 3$

This table lists power values for multiple-treatment designs. $E(R_0)$ denotes the mean number of false positives, ψ_1 the non-centrality parameter for the design, T the number of treatment conditions in the experiment, G_0 the anticipated number of undifferentially expressed genes in the experiment.

	Number of treatments $T = 2$									
	Non-centrality ψ_1									
Genes G_0	2	4	6	8	10	12	14	16	18	20
100	.23	.43	.61	.74	.84	.90	.94	.97	.98	.99
200	.15	.33	.51	.65	.77	.85	.90	.94	.96	.98
500	.09	.23	.38	.53	.66	.76	.84	.89	.93	.96
1000	.06	.17	.30	.44	.58	.69	.78	.85	.90	.93
	Non-centrality ψ_1									
Genes G_0	12	14	16	18	20	22	24	26	28	30
2000	.61	.71	.80	.86	.90	.94	.96	.97	.98	.99
5000	.51	.62	.72	.79	.85	.90	.93	.95	.97	.98
10000	.44	.55	.65	.73	.80	.86	.90	.93	.95	.97
20000	.37	.48	.58	.67	.75	.82	.87	.90	.93	.95

	Number of treatments $T = 3$									
	Non-centrality ψ_1									
Genes G_0	2	4	6	8	10	12	14	16	18	20
100	.17	.34	.50	.64	.76	.84	.90	.93	.96	.98
200	.11	.25	.40	.54	.67	.77	.84	.89	.93	.96
500	.06	.16	.29	.42	.55	.66	.75	.83	.88	.92
1000	.04	.11	.22	.34	.46	.58	.68	.77	.83	.88
	Non-centrality ψ_1									
Genes G_0	12	14	16	18	20	22	24	26	28	30
2000	.50	.61	.70	.78	.84	.89	.92	.95	.96	.98
5000	.40	.51	.61	.70	.77	.83	.88	.91	.94	.96
10000	.33	.44	.54	.63	.71	.78	.84	.88	.92	.94
20000	.27	.37	.47	.57	.65	.73	.79	.84	.89	.92

	Number of treatments $T = 4$									
	Non-centrality ψ_1									
Genes G_0	2	4	6	8	10	12	14	16	18	20
100	.14	.28	.43	.57	.69	.78	.85	.90	.94	.96
200	.09	.20	.34	.47	.60	.70	.79	.85	.90	.93
500	.05	.13	.23	.35	.48	.59	.69	.77	.84	.88
1000	.03	.09	.17	.28	.39	.51	.61	.70	.78	.84
	Non-centrality ψ_1									
Genes G_0	12	14	16	18	20	22	24	26	28	30
2000	.43	.53	.63	.71	.78	.84	.89	.92	.94	.96
5000	.33	.43	.53	.62	.71	.77	.83	.88	.91	.94
10000	.27	.37	.46	.56	.64	.72	.78	.84	.88	.91
20000	.22	.30	.40	.49	.58	.66	.73	.79	.84	.88

Table 2: Power Table for Multiple-treatment Designs
Mean Number of False Positives $E(R_0) = 3$

This table lists power values for multiple-treatment designs. $E(R_0)$ denotes the mean number of false positives, ψ_1 the non-centrality parameter for the design, T the number of treatment conditions in the experiment, G_0 the anticipated number of undifferentially expressed genes in the experiment.

	Number of treatments $T = 5$								
	Non-centrality ψ_1								
Genes G_0	12	14	16	18	20	22	24	26	28	30
100	.74	.82	.87	.92	.94	.96	.98	.99	.99	.99
200	.65	.74	.81	.87	.91	.94	.96	.97	.98	.99
500	.53	.63	.72	.79	.85	.89	.93	.95	.97	.98
1000	.45	.55	.65	.73	.80	.85	.89	.92	.95	.96
	Non-centrality ψ_1								
Genes G_0	22	24	26	28	30	32	34	36	38	40
2000	.80	.85	.89	.92	.95	.96	.97	.98	.99	.99
5000	.72	.79	.84	.88	.91	.94	.96	.97	.98	.99
10000	.66	.73	.79	.84	.88	.91	.94	.95	.97	.98
20000	.60	.68	.74	.80	.85	.88	.91	.94	.95	.97

	Number of treatments $T = 6$								
	Non-centrality ψ_1								
Genes G_0	12	14	16	18	20	22	24	26	28	30
100	.70	.78	.85	.89	.93	.95	.97	.98	.99	.99
200	.61	.70	.78	.84	.89	.92	.95	.96	.98	.98
500	.49	.59	.68	.76	.82	.87	.91	.93	.95	.97
1000	.40	.50	.60	.68	.76	.82	.87	.90	.93	.95
	Non-centrality ψ_1								
Genes G_0	22	24	26	28	30	32	34	36	38	40
2000	.76	.82	.86	.90	.93	.95	.96	.98	.98	.99
5000	.68	.75	.80	.85	.89	.92	.94	.96	.97	.98
10000	.61	.69	.75	.81	.85	.89	.92	.94	.96	.97
20000	.55	.63	.70	.76	.81	.85	.89	.92	.94	.96

	Number of treatments $T = 7$								
	Non-centrality ψ_1								
Genes G_0	12	14	16	18	20	22	24	26	28	30
100	.67	.75	.82	.87	.91	.94	.96	.97	.98	.99
200	.57	.66	.75	.81	.86	.90	.93	.95	.97	.98
500	.45	.55	.64	.72	.79	.84	.88	.92	.94	.96
1000	.36	.46	.56	.65	.72	.79	.84	.88	.91	.94
	Non-centrality ψ_1								
Genes G_0	22	24	26	28	30	32	34	36	38	40
2000	.72	.79	.84	.88	.91	.93	.95	.97	.98	.98
5000	.64	.71	.77	.82	.86	.90	.92	.95	.96	.97
10000	.57	.65	.71	.77	.82	.86	.90	.92	.94	.96
20000	.50	.58	.66	.72	.78	.82	.86	.90	.92	.94

Table 2: Power Table for Multiple-treatment Designs
Mean Number of False Positives $E(R_0) = 3$

This table lists power values for multiple-treatment designs. $E(R_0)$ denotes the mean number of false positives, ψ_1 the non-centrality parameter for the design, T the number of treatment conditions in the experiment, G_0 the anticipated number of undifferentially expressed genes in the experiment.

	Number of treatments $T = 8$								
	Non-centrality ψ_1								
Genes G_0	12	14	16	18	20	22	24	26	28	30
100	.64	.72	.80	.85	.90	.93	.95	.97	.98	.99
200	.53	.63	.72	.79	.84	.89	.92	.94	.96	.97
500	.41	.51	.60	.69	.76	.82	.86	.90	.93	.95
1000	.33	.43	.52	.61	.69	.76	.81	.86	.90	.92
	Non-centrality ψ_1								
Genes G_0	22	24	26	28	30	32	34	36	38	40
2000	.69	.76	.81	.86	.89	.92	.94	.96	.97	.98
5000	.60	.67	.74	.79	.84	.88	.91	.93	.95	.96
10000	.53	.61	.68	.74	.79	.84	.88	.91	.93	.95
20000	.47	.54	.62	.68	.74	.80	.84	.88	.91	.93

	Number of treatments $T = 9$								
	Non-centrality ψ_1								
Genes G_0	12	14	16	18	20	22	24	26	28	30
100	.61	.70	.77	.83	.88	.91	.94	.96	.97	.98
200	.51	.60	.69	.76	.82	.87	.90	.93	.95	.97
500	.38	.48	.57	.66	.73	.79	.84	.88	.91	.94
1000	.31	.40	.49	.58	.66	.73	.79	.84	.88	.91
	Non-centrality ψ_1								
Genes G_0	22	24	26	28	30	32	34	36	38	40
2000	.66	.73	.78	.83	.87	.90	.93	.95	.96	.97
5000	.57	.64	.71	.77	.82	.86	.89	.92	.94	.96
10000	.50	.57	.64	.71	.77	.81	.86	.89	.92	.94
20000	.43	.51	.58	.65	.71	.77	.81	.85	.89	.91

	Number of treatments $T = 10$								
	Non-centrality ψ_1								
Genes G_0	12	14	16	18	20	22	24	26	28	30
100	.58	.67	.75	.81	.86	.90	.93	.95	.97	.98
200	.48	.57	.66	.74	.80	.85	.89	.92	.94	.96
500	.36	.45	.54	.63	.70	.77	.82	.87	.90	.93
1000	.28	.37	.46	.55	.63	.70	.76	.82	.86	.89
	Non-centrality ψ_1								
Genes G_0	22	24	26	28	30	32	34	36	38	40
2000	.63	.70	.76	.81	.85	.89	.92	.94	.96	.97
5000	.53	.61	.68	.74	.79	.84	.87	.90	.93	.95
10000	.46	.54	.61	.68	.74	.79	.83	.87	.90	.92
20000	.40	.48	.55	.62	.68	.74	.79	.83	.87	.90

Table 2: Power Table for Multiple-treatment Designs
Mean Number of False Positives $E(R_0) = 4$

This table lists power values for multiple-treatment designs. $E(R_0)$ denotes the mean number of false positives, ψ_1 the non-centrality parameter for the design, T the number of treatment conditions in the experiment, G_0 the anticipated number of undifferentially expressed genes in the experiment.

Number of treatments $T = 2$										
Non-centrality ψ_1										
Genes G_0	2	4	6	8	10	12	14	16	18	20
100	.26	.48	.65	.78	.87	.92	.95	.97	.99	.99
200	.18	.37	.55	.69	.80	.87	.92	.95	.97	.98
500	.11	.26	.42	.57	.70	.79	.86	.91	.94	.97
1000	.07	.19	.33	.48	.61	.72	.81	.87	.91	.94
Non-centrality ψ_1										
Genes G_0	12	14	16	18	20	22	24	26	28	30
2000	.65	.74	.82	.88	.92	.95	.96	.98	.99	.99
5000	.54	.65	.74	.81	.87	.91	.94	.96	.97	.98
10000	.47	.58	.68	.76	.82	.87	.91	.94	.96	.97
20000	.40	.51	.61	.70	.77	.83	.88	.92	.94	.96
Number of treatments $T = 3$										
Non-centrality ψ_1										
Genes G_0	2	4	6	8	10	12	14	16	18	20
100	.20	.38	.55	.69	.79	.86	.91	.95	.97	.98
200	.13	.28	.44	.58	.70	.80	.86	.91	.94	.97
500	.07	.18	.32	.46	.59	.70	.78	.85	.90	.93
1000	.05	.13	.24	.37	.50	.61	.71	.79	.85	.90
Non-centrality ψ_1										
Genes G_0	12	14	16	18	20	22	24	26	28	30
2000	.53	.64	.73	.80	.86	.90	.93	.95	.97	.98
5000	.43	.54	.64	.72	.79	.85	.89	.92	.95	.96
10000	.36	.47	.57	.66	.74	.80	.85	.90	.93	.95
20000	.30	.40	.50	.59	.68	.75	.81	.86	.90	.93
Number of treatments $T = 4$										
Non-centrality ψ_1										
Genes G_0	2	4	6	8	10	12	14	16	18	20
100	.17	.32	.48	.62	.73	.82	.88	.92	.95	.97
200	.11	.23	.37	.51	.64	.74	.82	.87	.92	.95
500	.06	.15	.26	.39	.51	.63	.72	.80	.86	.90
1000	.04	.10	.19	.31	.43	.54	.64	.73	.80	.86
Non-centrality ψ_1										
Genes G_0	12	14	16	18	20	22	24	26	28	30
2000	.46	.56	.66	.74	.81	.86	.90	.93	.95	.97
5000	.36	.46	.56	.65	.73	.80	.85	.89	.92	.95
10000	.29	.39	.49	.58	.67	.74	.80	.85	.89	.92
20000	.24	.33	.42	.52	.60	.68	.75	.81	.86	.89

Table 2: Power Table for Multiple-treatment Designs
Mean Number of False Positives $E(R_0) = 4$

This table lists power values for multiple-treatment designs. $E(R_0)$ denotes the mean number of false positives, ψ_1 the non-centrality parameter for the design, T the number of treatment conditions in the experiment, G_0 the anticipated number of undifferentially expressed genes in the experiment.

	Number of treatments $T = 5$									
	Non-centrality ψ_1									
Genes G_0	2	4	6	8	10	12	14	16	18	20
100	.15	.29	.43	.57	.68	.78	.85	.90	.93	.96
200	.09	.20	.33	.46	.58	.69	.77	.84	.89	.93
500	.05	.12	.22	.34	.46	.57	.67	.75	.82	.87
1000	.03	.08	.16	.26	.37	.48	.59	.68	.76	.82
	Non-centrality ψ_1									
Genes G_0	12	14	16	18	20	22	24	26	28	30
2000	.40	.51	.60	.69	.76	.82	.87	.91	.93	.95
5000	.31	.41	.50	.59	.68	.75	.81	.86	.89	.92
10000	.25	.34	.43	.52	.61	.69	.76	.81	.86	.90
20000	.20	.28	.37	.46	.54	.63	.70	.76	.82	.86
	Number of treatments $T = 6$									
	Non-centrality ψ_1									
Genes G_0	12	14	16	18	20	22	24	26	28	30
100	.74	.81	.87	.91	.94	.96	.98	.98	.99	.99
200	.65	.74	.81	.86	.91	.94	.96	.97	.98	.99
500	.52	.62	.71	.78	.84	.89	.92	.94	.96	.97
1000	.44	.54	.63	.71	.78	.84	.88	.92	.94	.96
	Non-centrality ψ_1									
Genes G_0	22	24	26	28	30	32	34	36	38	40
2000	.78	.84	.88	.91	.94	.96	.97	.98	.99	.99
5000	.71	.77	.82	.87	.90	.93	.95	.96	.97	.98
10000	.64	.71	.77	.83	.87	.90	.93	.95	.96	.97
20000	.58	.65	.72	.78	.83	.87	.90	.93	.95	.96
	Number of treatments $T = 7$									
	Non-centrality ψ_1									
Genes G_0	12	14	16	18	20	22	24	26	28	30
100	.71	.79	.85	.89	.93	.95	.97	.98	.99	.99
200	.61	.70	.78	.84	.89	.92	.95	.96	.98	.98
500	.48	.58	.67	.75	.81	.86	.90	.93	.95	.97
1000	.40	.50	.59	.68	.75	.81	.86	.90	.93	.95
	Non-centrality ψ_1									
Genes G_0	22	24	26	28	30	32	34	36	38	40
2000	.75	.81	.86	.89	.92	.94	.96	.97	.98	.99
5000	.67	.73	.79	.84	.88	.91	.93	.95	.97	.98
10000	.60	.67	.74	.79	.84	.88	.91	.93	.95	.96
20000	.53	.61	.68	.74	.80	.84	.88	.91	.93	.95

Table 2: Power Table for Multiple-treatment Designs
Mean Number of False Positives $E(R_0) = 4$

This table lists power values for multiple-treatment designs. $E(R_0)$ denotes the mean number of false positives, ψ_1 the non-centrality parameter for the design, T the number of treatment conditions in the experiment, G_0 the anticipated number of undifferentially expressed genes in the experiment.

	Number of treatments $T = 8$									
	Non-centrality ψ_1									
Genes G_0	12	14	16	18	20	22	24	26	28	30
100	.68	.76	.83	.88	.91	.94	.96	.97	.98	.99
200	.58	.67	.75	.81	.87	.90	.93	.95	.97	.98
500	.45	.55	.64	.72	.79	.84	.88	.92	.94	.96
1000	.37	.46	.56	.64	.72	.78	.83	.88	.91	.94
	Non-centrality ψ_1									
Genes G_0	22	24	26	28	30	32	34	36	38	40
2000	.72	.78	.83	.87	.91	.93	.95	.96	.98	.98
5000	.63	.70	.76	.81	.86	.89	.92	.94	.96	.97
10000	.56	.64	.70	.76	.81	.86	.89	.92	.94	.96
20000	.49	.57	.64	.71	.77	.81	.86	.89	.92	.94

	Number of treatments $T = 9$									
	Non-centrality ψ_1									
Genes G_0	12	14	16	18	20	22	24	26	28	30
100	.65	.74	.80	.86	.90	.93	.95	.97	.98	.99
200	.55	.64	.72	.79	.85	.89	.92	.94	.96	.97
500	.42	.52	.61	.69	.76	.82	.86	.90	.93	.95
1000	.34	.43	.52	.61	.69	.75	.81	.86	.89	.92
	Non-centrality ψ_1									
Genes G_0	22	24	26	28	30	32	34	36	38	40
2000	.69	.75	.81	.85	.89	.92	.94	.96	.97	.98
5000	.59	.67	.73	.79	.83	.87	.90	.93	.95	.96
10000	.52	.60	.67	.73	.79	.83	.87	.90	.93	.95
20000	.46	.53	.61	.68	.74	.79	.83	.87	.90	.92

	Number of treatments $T = 10$									
	Non-centrality ψ_1									
Genes G_0	12	14	16	18	20	22	24	26	28	30
100	.63	.71	.78	.84	.89	.92	.94	.96	.97	.98
200	.52	.61	.70	.77	.83	.87	.91	.93	.95	.97
500	.39	.49	.58	.66	.73	.80	.84	.88	.92	.94
1000	.31	.40	.49	.58	.66	.73	.79	.84	.88	.91
	Non-centrality ψ_1									
Genes G_0	22	24	26	28	30	32	34	36	38	40
2000	.66	.73	.78	.83	.87	.90	.93	.95	.96	.97
5000	.56	.64	.70	.76	.81	.85	.89	.92	.94	.95
10000	.49	.57	.64	.70	.76	.81	.85	.89	.91	.93
20000	.43	.50	.58	.64	.71	.76	.81	.85	.88	.91

Glossary of Notation

α_0 Probability of incorrectly declaring an undifferentially expressed gene as differentially expressed (individual type I error probability)

α_F Familywise type I error probability

β_1 Probability of incorrectly declaring a differentially expressed gene as undifferentially expressed (individual type II error probability)

β_F Familywise type II error probability

B_{gc} Additive background noise component of instrument measurement w_{gc}.

$\mathcal{C}$ Set of all biological specimens or experimental conditions and their replicates, indexed by $c = 1, 2, \ldots, C$ and $n = 1, 2, \ldots, N$, respectively.

$\boldsymbol{\Delta}$ A non-zero vector $\boldsymbol{\mathcal{I}}_g = \boldsymbol{\Delta} = (\Delta_1, \Delta_2, \ldots, \Delta_C)'$ of interaction terms specified in the alternative hypothesis H_1 representing differential expression among conditions in set $\mathcal{C}$.

ϵ Statistical error term.

$\mathcal{G}_P$ Population set of genes in the biological specimen under study.

$\mathcal{G}_A$ Designated set of genes under study, indexed by $g = 1, 2, \ldots, G$.

G Total number of genes considered in the experiment.

$\mathcal{G}_0$ Set of truly undifferentiated genes under study.

G_0 Number of genes that are truly undifferentially expressed.

$\mathcal{G}_1$ Set of truly differentiated genes under study.

G_1 Numbers of genes that are truly differentially expressed.

$\hat{\mathbf{\mathcal{I}}}_g$ Estimated differential expression vector $\hat{\mathbf{\mathcal{I}}}_g = (\hat{\mathcal{I}}_{gc}, c = 1, \ldots, C)'$, where the $\mathcal{I}_{gc} = (\gamma\tau)_{gc}$ denote the interaction terms in an ANOVA model.

ζ_{gc} Population concentration of genetic material in specimen c attributable to gene g.

x_{gc} True gene expression component of instrument measurement w_{gc}.

v_g Statistic $v_g = h(\hat{\mathbf{\mathcal{I}}}_g)$ denotes a summary measure of the estimated differential expression vector $\hat{\mathbf{\mathcal{I}}}_g$ for gene g, where h is a function specified by the investigator.

$\mathbf{x}_n$ $\mathbf{x}_n = (x_{1n}, \ldots, x_{Gn})'$ denotes a column vector of observed expression levels across G genes in specimen n, where x_{gn} represents the intensity of gene g measured from specimen n and the prime denotes a vector or matrix transpose.

$\mathbf{x}^{(g)}$ $\mathbf{x}^{(g)} = (x_{g1}, \ldots, x_{gN})$ denotes a row vector of observed expression levels across N specimens for gene g where x_{gn} represents the intensity of gene g measured from specimen n.

y_{gc} A variable representing a transformation of w_{gc}, the instrument measurement for gene expression. The transformation is selected to give y_{gc} desired statistical properties.

w_{gc} A variable representing the instrument measurement of expression for gene g under condition c, possibly having undergone various adjustments (calibration, background correction, etc.) in the internal software of the instrument.

References

Affymetrix (2001). *Affymetrix Microarray Suite 5.0 User Guide*, Affymetrix Inc., Santa Clara, CA.

Affymetrix GeneChip, 700228 rev.2.

Afifi, A.A. and Clark, V. (1990). *Computer-aided Multivariate Analysis*. 2nd edition, Chapman and Hall, New York.

Alberts, B., Johnson, A. Lewis, J., Raff, M., Roberts, K., Walter, P. (2002). *Molecular Biology of the Cell*, 4th edition, Garland Publishing.

Alizadeh, A., Eisen, M.B., Davis, R.E., Ma, C., Lossos, I.S., Rosenwald, A., Boldrick, J.C., Sabet, H., Tran, T., Yu, X., Powell, J.I., Yang, L., Marti, G.E., Moore, T., Hudson, J., Lu, L., Lewish, D.B., Tibshrani, R., Sherlock, G., Chan, W.C., Greiner, T.C., Weisenburger D.D., Armitage, J.O., Warnke, R., Levy, R., Wilson, W., Grever, M.R., Byrd, J.C., Botstein, D., Brown, P.O., and Staudt, L.M. (2000). Distinct types of diffuse large B-cell lymphoma identified by gene expression profiling. *Nature*, **403**, 503-511.

Alon, U., Barkai, N., Notterman, D.A., Gish, K., Ybarra, S., Mack, D., Levine, A.J., (1999). Broad patterns of gene expression revealed by clustering analysis of tumor and normal colon tissues probed by oligonucleotide arrays. *Proceedings of the National Academy of Sciences, USA*, **96**, 6745-6750.

Alter, O. , Brown, P. O. and Botstein, D. (2000). Singular value decomposition for genome-wide expression data processing and modeling, *Proceedings of the National Academy of Sciences, USA*, **97**, 10101-10106.

Alter, O., Brown, P.O., and Botstein, D. (2003). Generalized singular value decomposition for comparative analysis of genome-scale expression data sets of two different organisms, *Proceedings of the National Academy of Sciences, USA*, **100**, 3351-3356.

Anderson, T.W. (1958). *An Introduction to Multivariate Statistical Analysis*. John Wiley & Sons, Inc., New York.

Ashburner, M., Ball, C.A., Blake, J.A., Botstein, D., Butler H., Cherry, J.M., Davis, A.P., Dolinski, K., Dwight, S.S., Eppig, J.T., Harris, M.A., Hill, D.P., Issel-Tarver, L., Kasarskis, A., Lewis, S., Matese, J.C., Richardson, J.E., Ringwald, M., Rubin, G.M., Sherlock, G. (2000). Gene ontology: tool for the unification of biology. The Gene Ontology Consortium. *Nature Genetics*, **25**, 25-29.

Ausubel, F. M. et al. (editors) (1993). *Current Protocols in Molecular Biology*, John Wiley & Sons, Inc., New York.

Baggerly, K.A., Coombes, K.R., Hess, K.R., Stivers, D.N., Abruzzo, L.V., and Zhang, W. (2001). Identifying differentially expressed genes in cDNA microarray experiments. *Journal of Computational Biology*, **8**, 639-659.

Baldi, P., Hatfield, G.W. (2002). *DNA Microarrays and Gene Expression*, Cambridge University Press, Cambridge, U.K.

Baldi, P., Long, A.D. (2001). A Bayesian framework for the analysis of microarray expression data: regularized t-test and statistical inferences of gene changes. *Bioinformatics*, **17**, 509-519.

Benjamini, Y., Hochberg, Y. (1995). Controlling the false discovery rate: A practical and powerful approach to multiple testing, *Journal of Royal Statistical Society*, **B 57**, 289-300.

Beran, R. (1988). Balanced simultaneous confidence sets, *Journal of the American Statistical Association*, **83**, 679-686.

Bernardo, J.M., Giron, J. (1988). A Bayesian approach to cluster analysis, *Questiio*, **12**, 97-112.

Binder, D.A. (1978). Bayesian cluster analysis, *Biometrika*, **65**, 31-38.

Bishop, C.M. (1995). *Neural Networks for Pattern Recognition*. Clarendon Press, Oxford.

Bittner, M., Chen, Y., Amundson, S.A., Khan, J., Forance, A.J., Dougherty, E.R., Meltzer, P.S., and Trent, J.M., (2000). Obtaining and evaluating gene expression profiles with cDNA microarrays. In *Genomics and Proteomics: Functional and Computational Aspects*, ed. Suhai, S., pp 5-25. Kluwer Academic, Plenum Publisher, New York.

Bittner M., Meltzer P., Chen Y., Jiang Y., Seftor E., Hendrix M., Radmacher M., Simon R., Yakhini Z., Ben-Dor A., Sampas N., Dougherty E., Wang E., Marincola F., Gooden C., Lueders J., Glatfelter A., Pollock P., Carpten J., Gillanders E., Leja D., Dietrich K., Beaudry C., Berens M., Alberts D., Sondak V. (2000). Molecular classification of cutaneous malignant melanoma by gene expression profiling. *Nature*, **406**, 536-540.

Björkbacka, H., personal communication, 2003.

Bowtell, D.D. (1999). Options available–from start to finish–for obtaining expression data by microarray. *Nature Genetics*, **21**, 25-32.

Bowtell, D., Sambrook, J. (Editors) (2002). *DNA Microarrays: A Molecular Cloning Manual.*, Cold Spring Harbor Laboratory.

Brazma, A., Hingamp, P., Quackenbush, J., Sherlock, G., Spelllman, P., Stoeckert C., Aach, J., Ansorge, W., Ball, C.A., Causton, H.C., Gaasterland, T., Glenisson, P., Holstege, F.C.P., Kim, I.F., Markowitz, V., Matese, J.C., Parkinson, H., Robinson, A., Sarkans, U., Schulze-Kremer, S., Steward, J., Taylor, R., Vilo, J., and Vingron, M. (2001). Minimum information about a microarray experiment (MIAME): toward standards for microarray data. *Nature Genetics*, **29**, 365-371.

Breiman, L. (1996a). Bagging predictors. *Machine Learning*, **24**: 123-140.

Breiman, L. (1996b). Out-of-bag estimation. Technical report, Statistics Department, U.C. Berkeley, http://www.stat.Berkeley.EDU/users/breiman/OOBestimation.ps.

Breiman, L. (1998). Arcing classifiers. *Annals of Statistics*, **26**: 801-824.

Breiman, L. (2001a). Random forests, random features, Technical report, Statistics Department, U.C. Berkeley, http://www.stat.Berkeley.EDU/users/breiman/.

Breiman, L. (2001b). Notes on setting up, using, and understanding random forests. Technical report, Statistics Department, University of California, Berkeley, http://www.stat.Berkeley.EDU/users/breiman/.

Breiman, L. , Friedman, J. , Stone, C. and Olshen, R. (1984). *Classification and Regression Trees*, Wadsworth, Monterey, CA.

Broët, P. Richardson, S., Radvanyi, F. (2002). Bayesian hierarchical model for identifying changes in gene expression from microarray experiments. *Journal of Computational Biology*, **9**, 671-683.

Brown, C.S., Goodwin, P.C., and Sorger, P.K. (2001). Image metrics in the statistical analysis of DNA microarray data, *Proc. Natl. Acad. Sci.* USA, Vol. 98, 8944-8949.

Brown, M.P.S. , Grundy, W.N. , Lin, D. , Cristianini, N. , Sugnet, C.W. , Furey, T.S. , Ares, M., Jr., Haussler, D. (2000). Knowledge-based analysis of microarray gene expression data by using support vector machines, *Proc. Natl. Acad. Sci.*, **97**, 262-267.

Brown, P.O., Botstein, D. (1999). Exploring the new world of the genome with DNA microarrays. *Nature Genetics*, **21**, 33-37.

Bulyk, M.L., Huang X., Choo Y. and Church G.M. (2001). Exploring the DNA-binding specificities of zinc fingers with DNA microarrays. *Proceedings of the National Academy of Sciences, USA*, **98**, 7158-7163.

Bulyk, M.L., Johnson, P.L.F., and Church, G.M. (2002). Nucleotides of transcription factor binding sites exert interdependent effects on the binding affinities of transcription factors, *Nucleic Acids Res.*, **30**,(5):1255-1261.

Burge, C.B. (2001). Chipping away at the transcriptome. *Nature Genetics*, **27**, 232-234.

Burges, C.J.C. (1998). *Data Mining and Knowledge Discovery*, **2**, 121-167.

Bustin, S.A. (2002). Quantification of mRNA using real-time RT-PCR: trends and problems. *Journal of Molecular Endocrinology*, **29**, 23-39.

Carlin, B.P., Louis, T.A. (1996). *Bayes and Empirical Bayes Methods for Data Analysis*, Chapman and Hall, New York.

Chakravarti, A. (1999). Population genetics–making sense out of sequence. *Nature Genetics*, **21**, 56-60.

Chen, Y., Dougherty, E.R., and Bittner, M.L. (1997). Ratio-based decisions and the quantitative analysis of cDNA microarray images, *Journal of Biomedical Optics*, **2**, 364-374.

Cheung, V.G., Morley M., Aguilar F., Massimi A., Kucherlapati, R., Childs, G. (1999). Making and reading microarrays. *Nature Genetics*, **21**, 15-19.

Cho, R.J., Campbell, J.J., Winzeler, E.A., Steinmetz, L., Conway, A., Wodicka, L., Wolfsberg, T.G., Gabrielian, A.E., Landsman, D., Lockart, D.J., and Davis, R.W. (1998). A Genome-wide transcriptional analysis of the mitotic cell cycle. *Mol. Cell*, **2**, 65-73.

Choo, Y. and Klug, A. (1994a). Toward a code for the interactions of zinc fingers with DNA: Selection of a randomized fingers displayed on phage. *Proceedings of the National Academy of Sciences, USA*, **91**, 11163-11167.

Choo, Y. and Klug, A. (1994b). Selection of DNA binding sites for zinc fingers using rationally randomized DNA reveals coded interactions. *Proceedings of the National Academy of Sciences, USA*, **91**, 11168-11172.

Chu, S., DeRisi, J., Eisen, M., Mulholland, J., Botstein, D., Brown, P.O., Herskowitz, I. (1998). The transcriptional program of sporulation in budding yeast, *Science*, **282**, 699-705.

Chu, G., Narasimhan, B., Tibshirani, R., Tusher, V. (2002). Significant analysis of microarrays: users guide and technical document.
http://www-stat.stanford.edu/~ tibs/clickwrap/sam

Claverie, J.-M. (1999). Computational methods for the identification of differential and coordinated gene expression. *Human Molecular Genetics*, **8**, 1821-1832.

Cleveland, W.S. (1979). Robust locally weighted regression and smoothing scatterplots. *Journal of the American Statistical Association*, **74**, 829-836.

Cleveland, W.S. (1981). LOWESS: A program for smoothing scatterplots by robust locally weighted regression. *The American Statistician*, **35**, 54.

Cochran, W.G., and Cox, G.M., (1992), *Experimental Designs*, John Wiley & Sons, New York.

Cohen, B.A., Mitra, R.D., Hughes, J.D., Church, G.M. (2000). A computational analysis of whole-genome expression data reveals chromosomal domains of gene expression. *Nature Genetics*, **26**, 183-186.

Cole, K.A., Krizman, D.B., Emmert-Buck, M.R. (1999). The genetics of cancer–a 3D model. *Nature Genetics*, **21**, 38-41.

Comstock, R.E., and Winters, L.M. (1942). Design of experimental comparisons between lines of breeding in livesotck. *Journal of Agricultural Research*, **64**, 523-532.

Cooper, G.M. (2000). *The Cell - A Molecular Approach*, 2nd ed., Sinauer Associates Inc, Sunderland, Massachusetts.

Cover, T. and Hart, P. (1967). Nearest neighbor pattern classification. *Proc. IEEE Trans. Inform. Theory*, **11**, 21-27.

Cristianini, N. and Shawe-Tylor, J. (2000). *Support Vector Machines*, Cambridge University Press, Cambridge, U.K.

Croux, C., Filzmoser, P., Pison, G. and Rousseeuw, P. J. (2002) *Statistical Computations*, **13**, 23-36.

Dale, J.W. and von Schantz, M. (2002). *From Genes to Genomes: Concepts and Applications of DNA Technology.* John Wiley & Sons, Ltd, England.

Debouck, C., Goodfellow, P.N. (1999). DNA microarrays in drug discovery and development. *Nature Genetics*, **21**, 48-50.

Dellaportas, P. (1998). Bayesian classification of Neolithic tools. *Applied Statistics*, **47**, 279-297.

Delongchamp, R. R., C. Velasco, R. Evans, A. Harris and D. Casciano. Adjusting cDNA array data for nuisance effects, Division of Biometry and Risk Assessment, HFT-20, 2001, National Center for Toxicological Research, Jefferson, Arkansas.

Denison, D.G.T., Holmes, C.C., Mallick, B.K., and Smith, A.F.M. (2002). *Bayesian Methods for Nonlinear Classification and Regression*, John Wiley & Sons, Ltd., West Sussex, U.K.

DeRisi, J.L., Iyer, V.R., and Brown, P.O. (1997). Exploring the metabolic and genetic control of gene expression on a genomic scale, *Science*, **278**: 680-865.

DeRisi, J.L., Penland, L., Brown, P.O., Bittner, M.L., Meltzer, P.S., Ray, M., Chen, Y., Su, Y.A., and Trent, J.M. (1996). Use of a cDNA microarray to analyse gene expression patterns in human cancer, *Nature Genetics*, **14**, 457-460.

Diehn, M., Eisen, M.B., Botstein, D., Brown, P.O. (2000). Large-scale identification of secreted and membrane-associated gene products using DNA microarrays. *Nature Genetics*, **25**, 58-62.

Dobbin, K., Shih, J.H., Simon, R. (2003). Statistical design of reverse dye microarrays. *Bioinformatics*, **19**, 803-810.

Drobyshev, A.L., Machka C., Horsch, M., Seltmann, M., Liebscher, V., Hrabe de Angelis, M., Beckers, J. (2003). Specificity assessment from fractionation experiments (SAFE): a novel method to evaluate microarray probe specificity based on hybridisation stringencies. *Nucleic Acids Research*, **31**, e1.

Dudley, A.M., Aach, J., Steffen, M.A. and Church, G.M. (2002). Measuring absolute expression with microarrays using a calibrated reference sample and an extended signal intensity range. *Proceedings of the National Academy of Sciences, USA,* **99**,7554-7559.

Dudoit, S., Fridlyand, J., and Speed, T.P. (2002). Comparison of discrimination methods for the classification of tumors using gene expression data. *Journal of the American Statistical Association,* **97**, 77-87.

Dudoit, S., Shaffer, J.P., Boldrick, J.C. (2003). Multiple hypothesis testing in microrray experiments. Technical Report No. 110, Division of Biostatistics, University of California, Berkeley.

Dudoit, S., Shaffer, J.P., Boldrick, J.C. (2003). Multiple hypothesis testing in microarray experiments. *Statistical Sciences,* **18**, 71-103.

Dudoit, S., Yang Y.H., Callow, M.J., Speed, T.P. (2002). Statistical methods for identifying differentially expressed genes in replicated cDNA microarray experiments. *Statistica Sinica,* **12**, 111-139.

Duggan D.J., Bittner M., Chen Y., Meltzer P., Trent J.M. (1999). Expression profiling using cDNA microarrays. *Nature Genetics,* **21**,10-14.

Dunn, G. and Everitt, B.S. (1982). *An Introduction to Mathematical Taxonomy,* Cambridge University Press, Cambridge.

Dunteman, G.H. (1989), *Principal Component Analysis,* Sage University Papers, Sage, Newbury Park, California.

Durbin, B.P., Hardin, J.S., Hawkins, D.M., and Rocke, D.M. (2002). A variance-stabilizing transformation for gene-expression microarray data. *Bioinformatics,* **18**, S105-S110.

Efron, B. (2001). Robbins, empirical Bayes, and microarrays. Technical Report No.2001-30B/219, Department of Statistics, Stanford University, Stanford, California.

Efron, B., Morris, C. (1973). Stein's estimation rule and its competitors - an empirical Bayes approach. *Journal of American Statistical Association,* **68**, 117-130.

Efron, B., Morris, C. (1975). Data analysis using Stein's estimator and its generalizations. *Journal of American Statistical Association,* **70**, 311-319.

Efron, B., Storey, J.D., Tibshirani, R. (2001). Microarrays empirical Bayes methods and false discovery rates. Technical Report No.2001-23B/217, Department of Statistics, Stanford University, Stanford, California.

Efron, B., Tibshirani, R., Storey, J.D., Tusher, V. (2001). Empirical Bayes analysis of a microarray experiment. *Journal of the American Statistical Association,* **96**, 1151-1160.

Eisen, M.B. (1999). *ScanAlyze User Manual,* Version 2.32; Stanford University: Stanford, CA.

Eisen, M.B., Brown, P.O. (1999). DNA arrays for analysis of gene expression. *Methods in Enzymology,* **303**, 179-205.

Eisen, M., Spellman, P.T., Brown, P.O. and Botstein, D. (1998). Cluster analysis and display of genome-wide expression patterns, *Proceedings of the National Academy of Sciences, USA,* **95**, 14863-14868.

Emptage, M.R., Hudson-Curtis, B., Sen, K. (2003). Treatment of microarray experiments as split-plot designs. *Journal of Biopharmaceutical Statistics,* **13**, 159-178.

Everitt, B.S. (1993). *Cluster Analysis.* Edward Arnold, New York.

Fisher, R.A., (1936). The use of multiple measurements in taxonomic problems, *Annal of Eugenics,* **7**, 179-188.

Fisher, R.A., (1947). *The Design of Experiments,* Oliver and Boyd, Edinburgh, 4th ed.

Fix, E, and Hodges, J. (1951). Discriminatory analysis, nonparametric discrimination: consistency properties. Technical report, Randloph Field, Texas, USAF School of Aviation Medicine.

Flury, B. (1988), *Common Principal Components and Related Multivariate Models*, John Wiley & Sons, New York.

Fodor, S.P., Rava, R.P., Huang, X.C., Pease, A.C., Holmes, C.P., Adams, C.L. (1993). Multiplexed biochemical assays with biological chips, *Nature*, **364**, 555-556.

Gabriel, K.R. and Zamir, S. (1979) *Technometrics*, **21**, 489-498.

Geller, S.C., Gregg, J.P., Hagerman P., Rocke, D.M. (2003). Transformation and normalization of oligonucleotide microarray data. http://handel.cipic.ucdavis.edu/ dm-rocke/gellernormpaper.pdf, (in pess at *Bioinformatics*).

Getz, G., Levine, E., Domany, E. (2000). Coupled two-way clustering analysis of gene microarray data. *Proceedings of the National Academy of Sciences, USA*, **97**, 12079-12084.

Ginzinger, D.G. (2002). Gene quantification using real-time quantitative PCR: an emerging technology hits the mainstream. *Experimental Hematology*, **30**, 503-512.

Golub, T.R., Slonim, D.K., Tamayo, P., Huard, C., Gaasenbeek, M., Mesirov, J.P., Coller, H., Loh, M.L., Downing, J.R., Caligiuri, M.A., Bloomfield, C.D., Lander, E.S. (1999). Molecular classification of cancer: class discovery and class prediction by gene expression monitoring. *Science*, **286**, 531-537.

Good, P. (2000). *Permutation Tests: A Practical Guide to Resampling Methods for Testing Hypotheses*, 2nd edition, Springer, New York.

Gordon, A.E. (1981). *Classification: Methods for the Exploratory Analysis of Multivariate Data*, Chapman and Hall, New York.

Gower, J.C. (1967). A comparison of some methods of cluster analysis. *Biometrics*, **23**, 623-628.

Gower, J.C. (1967). Multivariate analysis and multidimensional geometry. *The Statistician*, **17**, 13-25.

Gower, J.C. (1971). A general coefficient of similarity and some of its properties. *Biometrics*, **27**, 857-872.

Gower, J.C. (1985). Measures of similarity, dissimilarity and distance. In *Encyclopedia of Statistical Sciences*, **V.5**, (S. Kotz, N.L. Johnson and C.B. Read, eds.), John Wiley & Sons, New York.

Griffiths, A.J.F., Gelbart, W.M., Miller, J.H., and Lewontin R.C. (1999). *Modern Genetic Analysis*, Freeman, New York.

Griffiths, A.J.F., Miller, J.H., Suzuki D.T., Lewontin R.C., and Gelbart, W.M. (2000). *An Introduction to Genetic Analysis*, Freeman, New York.

Grigorenko, E.V., Editor. (2002). *DNA Arrays: Technologies and Experimental Strategies*, CRC Press, New York.

Hacia, J.G. (1999). Resequencing and mutational analysis using oligonucleotide microarrays. *Nature Genetics*, **21**, 42-47.

Handran, S. and Zhai, J.Y. (2001). Biological Relevance of GenePix Results, Axon Instruments, Inc., Technical Report.

Hartigan, J. (1975). *Clustering Algorithms*. John Wiley & Sons, New York.

Hastie, T., Tibshirani, R., and Friedman, J. (2001). *The Elements of Statistical Learning*, Springer, New York.

Hawkins, D.M. (2002). Diagnostics for conformity of paired quantitative measurements. *Statistics in Medicine*, **21**, 1913-1935.

Hedayat, A.S., Majumdar, D. (1984). Optimal incomplete block designs for control-test treatment comparisons. *Technometrics*, **26**, 363-370.

Heden, B., Ohlin, H., Rittner, R., and Edenbrandt, L. (1997). Acute myocardial infartion detected in the 120lead ECG by artificial neural network. *Circulation*, **96**, 1798-1802.

Hedenfalk, I., Duggan, D., Chen, Y., Radmacher, M., Bittner, M., Simon, R., Meltzer, P., Gusterson, B., Esteller, M., Raffeld, M. , *et al.* (2001). Gene-expression profiles in hereditary breast cancer. *New England Journal of Medicine*, **344**, 539-548.

Hedenfalk, I., Ringner, M., Ben-Dor, A., Yakhini, Z., Chen, Y., Chebil, G., Ach, R., Loman, N., Olsson, H., Meltzer, P., Borg, A., Trent, J. (2003). Molecular classification of familial non-*BRCA1/BRCA2* breast cancer. *Proceedings of the National Academy of Sciences, USA*, **100**, 2532-2537.

Held, G.A., Grinstein, G., Tu, Y. (2003). Modeling of DNA microarray data by using physical properties of hybridization. *Proceedings of the National Academy of Sciences, USA*, **100**, 7575-7580.

Heller, M.J. (2002). DNA microarray technology: devices, systems, and applications. *Annu. Rev. Biomed. Eng.*, **4**, 129-153.

Hertz, J., Krogh, A., Palmer, R.G. (1991). *Introduction to the Theory of Neural Computation*, Addison-Wesley, Redwood City, CA.

Herzel, H., Beule, D., Kielbasa, S., Korbel, J., Sers, C., Malik, A., Eickhoff, H., Lehrach, H., Schuchhardt, J. (2001). Extracting information from cDNA arrays. *CHAOS*, **11**.

Hessner, M.J., Wang, X., Hulse, K., Meyer, L., Wu, Y., Nye, S., Guo, S.-W., Ghosh, S. (2003). Three color cDNA microarrays: quantitative assessment through the use of fluorescein-labeled probes. *Nucleic Acids Research*, **31**, No. 4, e14.

Hochberg, Y. (1988). A sharper Bonferroni procedure for multiple tests of significance, *Biometrika*, **75**, 800-802.

Hochberg, Y. and Tamhane A.C. (1987). *Multiple Comparison Procedures*. John Wiley & Sons, New York.

Holm, S. (1979). A simple sequentially rejective multiple test procedure, *Scandinavian Journal of Statistics*, **6**, 65-70.

Holter, N.S., Maritan, A., Cieplak, M., Fedoroff, N.V., Banavar, J.R. (2001). Dynamic modeling of gene expression data. *Proceedings of the National Academy of Sciences, USA*, **98**, 1693-1698.

Hotelling, H. (1933). Analysis of a complex of statistical variables into principal components. *Journal of Educational Psychology*, **24**, 417-441.

Huber, W., von Heydebreck, A., Sültmann, H., Poustka, A., and Vingron, M. (2002). Variance stabilization applied to microarray data calibration and to the quantification of differential expression. *Bioinformatics*, **18**, *Suppl. S-96-S104.*

Hudson, D.L. and Cohen, M.E. (2000). *Neural Networks and Artificial Intelligence for Biomedical Engineering*, The Institute of Electric and Electronics Engineers, Inc., New York.

Hughes, T.R., Roberts, C.J., Dai, H., Jones, A.R., Meyer, M.R., Slade, D., Burchard, J., Dow, S., Ward, T.R., Kidd, M.J., Friend, S.H., Marton, M.J. (2000). Widespread aneuploidy revealed by DNA microarray expression profiling. *Nature Genetics*, **25**, 333-337.

Ibrahim, J.G., Chen M.-H., Gray, R.J. (2002). Bayesian models for gene expression with DNA microarray data. *Journal of the American Statistical Association*, **97**, 88-99.

Irizarry, R.A., Bolstad, B.M., Collin, F., Cope, L.M., Hobbs, B., Speed, T.P. (2003). Summaries of Affymetrix GeneChip probe level data. *Nucleic Acids Research*, **31**, No.4, 1-8.

358

Irizarry, K., Kustanovich, V., Li, C., Brown, N., Nelson, S., Wong, W., Lee, C.J. (2000). Genome-wide analysis of single-nucleotide polymorphisms in human expressed sequences. *Nature Genetics*, **26**, 233-236.

Jackson, J.E. (1991). *A User's Guide to Principal Components*, John Wiley & Sons, New York.

Jain, A.K. and Dubes, R.C. (1988). *Algorithms for Clustering Data*, Prentice Hall, Englewood Cliff, New Jersey.

Jenssen, T.-K., Langaas, M., Kuo, W.P., Smith-Sorensen, B., Myklebost, O., Hovig, E. (2002). Analysis of repeatability in spotted cDNA microarrays. *Nucleic Acids Research*, **30**, 3235-3244.

Jin, W., Riley, R.M., Wolfinger, R.D., White, K.P., Passador-Gurgel, G., and Gibson, G. (2001). The contributions of sex, genotype and age to transcriptional variance in *Drosophila melanogaster*, *Nature Genetics*, **29**, 389 - 395.

Jobson, J.D. (1992). *Applied Multivariate Data Analysis*, Springer-Verlag, New York.

Johnson, R.A., Wichern, D.W. (1998). *Applied Multivariate Statistical Analysis*, 4th ed., Prentice-Hall, Inc., New Jersey.

Johnson, S.C. (1967). Hierarachical clustering schemes. *Psychometrika*, **32**, 241-254.

Joliffe, I.T. (2002). *Principal Component Analysis*, 2nd ed., Springer-Verlag, New York.

Kaufman, L. and Rousseeuw, P.J. (1990). *Finding Groups in Data*, John Wiley & Sons, New York.

Kendziorski, C.M., Newton, M.A., Lan, H., and Gould, M.N. (2003). On parametric empirical Bayes methods for comparing multiple groups using replicated gene expression profiles. Technical Report #166, Department of Biostatistics and Medical Informatics, University of Wisconsin - Madison (submitted). *Statistics in Medicine*. (In press).

Kerr, M.K. and Churchill, G.A. (2001). Statistical design and the analysis of gene expression microarrays, *Genetical Research*, **77**:123-128.

Kerr, M.K., and Churchill, G.A. (2001a). Bootstrapping cluster analysis; assessing the reliability of conclusions from microarray experiments. *Proceedings of the National Academy of Sciences, USA*, **98**, 8961-8965.

Kerr, M.K., and Churchill, G.A. (2001b). Experimental design for gene expression microarrays. *Biostatistics*, **2**, 183-201.

Kerr, M.K., Martin, M., and Churchill, G.A. (2001c). Analysis of variance for gene expression microaray data. *Journal of Computational Biology*, **7**, 819-837.

Kerr, M.K., Afshari, C.A., Bennett, L., Bushel, P., Martinez, J., Walker, N.J., and Churchill, G.A. (2002). Statistical analysis of a gene expression microarray experiment with replication. *Statistics Sinica*, **12**, 203-218.

Khan, J., Wei, J.S., Ringner, M., Saal, L.H., Ladanyi, M., Westermann, F., Berthold, F., Schwab, M., Antonescu, C.R., Peterson, C., and Meltzer, P.S. (2001). Classification and diagnostic prediction of cancers using gene expression profiling and artificial neural networks. *Nature Medicine*, **7**, 673-679.

Kohonen, T. (1997). *Self-Organizing Maps*, Springer, Berlin.

Kong, C.F., Bowtell, D. (2002). Genome-wide gene expression analysis using cDNA microarrays. *Methods in Molecular Medicine*, **68**, 195-204.

Kruglyak, L., Nickerson, D.A. (2001). Variation is the spice of life. *Nature Genetics*, **27**, 234-236.

Krzanowski, W.J. (1988). *Principles of Multivariate Analysis: A User's Perspective*, Oxford University Press, Oxford.

Lance, G.N. and Williams, W.T. (1967). A general theory of classificatory sorting strategies: 1. Hierarchical systems. *Comp. J.*, **9**, 373-380.

Lander, E.S. (1996). The new genomics: global views of biology. *Science*, **274**, 536-539.

Lander, E.S. (1999). Array of hope, *Nature Genetics*, **21**, 3-4.

Lee, M.-L.T., Kuo, F.C., Whitmore, G.A., Sklar, J. (2000). The importance of replication in microarray gene expression studies. Statistical methods and evidence from repetitive cDNA hybridizations. *Proceedings of the National Academy of Sciences, USA*, **97**, 9834-9839.

Lee, M.-L.T., Bulyk, M.L., Whitmore, G.A., Church, G.M. (2002a). A statistical model for investigating binding probabilities of DNA nucleotide sequences using microarrays. *Biometrics*, **58**, 129-136.

Lee, M.-L.T., Lu, W., Whitmore, G.A., Beier, D. (2002b). Models for microarray gene expression data. *Journal of Biopharmaceutical Statistics*, **12(1)**, 1-19.

Lee, M.-L.T., Whitmore, G.A. (2002c). Power and sample size for microarray studies. *Statistics in Medicine*, **21**, 3543-3570.

Lee, M.-L.T., Whitmore, G.A., Yukhananov, R.Y. (2003). Analysis of unbalanced microarray data. *Journal of Data Science*, **1**, 103-121.

Lehninger, A.L., Nelson, D.L., and Cox, M.M. (2000). *Principles of Biochemistry*, Worth Publishing, 3rd ed.

Lewin, B. (2000). *Genes VII*, Oxford University Press, New York.

Li, C. and Wong, W.H. (2001). Model-based analysis of oligonucleotide arrays: model validation, design issues and standard error application, *Genome Biology* 2(8): research0032.1-0032.11. www.genomebiology.com.

Li, C., Wong, W.H. (2001). Model-based analysis of oligonucleotide arrays: expression index computation and outlier detection. *Proceedings of the National Academy of Sciences, USA*, **98**, 31-36.

Lipshutz, R.J., Fodor, S.P., Gingeras, T.R., Lockhart, D.J. (1999). High density synthetic oligonucleotide arrays. *Nature Genetics*, **21**, 20-24.

Little, R.J.A. and Rubin, D.B. (1987). *Statistical Analysis with Missing Data*, John Wiley & Sons, New York.

Liu, L., Hawkins, D.M., Ghosh, S., Young, S.S. (2003). Robust singular value decomposition analysis of microarray data, *Proceedings of the National Academy of Sciences, USA*, **100**, 13167-13172.

Lockhart, D.J., Dong, H., Bryne, M.C., Follettie, M.T., Gallo, M.V., Chee, M.S., Mittmann, M., Wang, C., Kobayashi, M., Horton, H., Brown, E.L. (1996). Expression monitoring by hybridization to high-density oligonucleotide arrays. *Nature Biotechnology*, **14**, 1675-1680.

Ludbrook, J. and Dudley, H. (1998). Why permutation tests are superior to t and F tests in biomedical research, *The American Statistician*, **52**, 127-132.

MacQueen, J.B. (1967). Some methods for classification and analysis of multivariate observations. In: *Proceedings of 5th Berkeley Symposium on Mathematical Statistics and Probability*, **1**, 281-297, University of California Press, Berkeley, California.

Mardia, K.V., Kent, J.T., Bibby, J.M. (1979). *Multivariate Analysis*, Academic Press, Inc., San Diego.

Martin, K.J., Graner, E., Li, Y., Price, L.M., Kritzman, B.M., Fournier, M.V., Rhei, E., and Pardee, A.B. (2001). High-sensitivity array analysis of gene expression for the early detection of disseminated breast tumor cells in peripheral blood. *Proceedings of the National Academy of Sciences, USA*, **98**, 2646-2651.

McShane, L.M., Radmacher, M.D., Freidlin, B., Yu, R., Li, M.-C., and Simon, R. (2002). Methods for assessing reproducibility of clustering patterns observed in analyses of microarray data. *Bioinformatics*, **18**, 1462-1469.

McCulloch, W.S. and Pitts, W. (1943). A logical calculus of the ideas immanent in neural nets. *Bull. Math. Biophys*, **5**, 115-137.

McLachlan, G.J. (1992). *Discriminant Analysis and Statistical Pattern Recognition*. John Wiley & Sons, New York.

Miesfeld, R. L. (1999). *Applied Molecular Genetics*. Wiley-Liss, New York.

Milligan, G.W. (1980). An examination of the effect of six types of error perturbation on fifteen clustering algorithms. *Psychometrika*, **45**, 325-342.

Milligan, G.W. and Cooper, M.C. (1985). An examination of procedures for determining the number of clusters in a data set. *Psychometrika*, **50**, 159-179.

Milligan, G.W. and Cooper, M.C. (1988). A study of standardization of variables in cluster analysis. *Journal of Classification*. **5**, 181-204.

Milliken, G.A., Johnson, D.E. (1992). *Analysis of Messy Data*, Chapman and Hall/CRC, Boca Raton.

Moler, E.J., Chow, M. L., and Mian, I.S. (2000). *Physiol. Genomics*, **4**, 109-126.

Morgan, B.J.T. and Ray, A.P.G. (1995). Non-uniqueness and inversions in cluster analysis. *Applied Statistics*, **44**, 117-134.

Morrison, D.F. (1976). *Multivariate Statistical Methods*, 2nd ed., New York, McGraw-Hill.

Morton, C.C. (2000). personal communication.

Nadeau, J.H., Frankel, W.N. (2000). The roads from phenotypic variation to gene discovery: mutagenesis versus QTLs. *Nature Genetics*, **25**, 381-384.

Naef, F., Hacker, C.R., Patil, N. and Magnasco, M. (2002a). Empirical characterization of the expression ratio noise structure in high density oligonucleotide arrays. *Genome Biology*, **3**; research0018.1-0018.11.

Naef, F., Lim D.A., Patil, N. and Magnasco, M. (2002b). DNA hybridization to mismatched templates: a chip study. *Phys. Rev. E.*, **65**, 040902.

Naef, F., Socci, N.D., Magnasco, M. (2003). A study of accuracy and precision in oligonucleotide arrays: extracting more signal at large concentrations, *Bioinformatics*, **19**; 178-184.

Neter, J., Kutner, M.H., Nachtscheim, C.J., Wasserman, W. (1996). *Applied Linear Statistical Models*, 4th edition, Richard D. Irwin.

Newton, M.A., Kendziorski, C.M., Richmond, C.S., Blattner, F.R., Tsui, K.W. (2001). On differential variability of expression ratios, improving statistical inference about gene expression changes from microarray data. *Journal of Computational Biology*, **8**, 37-52.

Newton, M.A., Kendziorski, C.M. (2003). Parametric empirical Bayes methods for microarrays. In: *The Analysis of Gene Expression Data: Methods and Software*, Parmigiani, G., Garrett, E.S., Irizarray, R.A., and Zeger, S.L., eds. 255-271. Springer, New York.

Newton, M.A., Noueiry, A., Sarkar, D., Ahlquist, P. (2003). Detecting differential gene expression with a semiparametric hierarchical mixture method. Technical Report No. 1074, Department of Statistics, University of Wisconsin, Medison.

Nguyen, D.V., Arpat, A.B., Wang, N, Carroll, R.J. (2002). DNA microarray experiments: biological and technical aspects. *Biometrics*, **58**, 701-717.

Nguyen, D.V., Rocke, D.M. (2002). Partial least squares proportional hazard regression for application to DNA microarray survival data. *Bioinformatics*, **18**, 1625-1632.

Parmigiani, G., Garrett, E.S., Anbazhagan, R., Gabrielson, E. (2002). A statistical framework for expression-based molecular classification in cancer. *Journal of the Royal Statistical Society, Series B*, **64**, 717-736.

Parmigiani, G., Garrett, E.S., Irizarry, R.A., Zeger, S.L., Editors (2003). *The Analysis of Gene Expression Data*. Springer, New York.

Pearson, K. (1901). On lines and planes of closet fit to systems of points in space. *Phil. Mag.* (6), **2**, 559-572.

Perou, C.M., Jeffrey, S.S., van de Rijn, M., Rees, C.A., Eisen, M.B., Ross, D.T., Pergamenschikov, A., Williams, C.F., Zhu, S.X., Lee, J.C.F. , *et al.* (1999). Distinctive gene expression patterns in human mammary epithelial cells and breast cancers. *Proceedings of the National Academy of Sciences, USA*, **96**, 9212-9217.

Phimister, B. (1999). Going global, *Nature Genetics*, **21**, 1.

Pollack, J.R., Perou, C.M., Alizadeh, A.A., Eisen, M.B., Pergamenschikov, A., Williams, C.F., Jeffery, S.S., Botstein, D., Brown, P.O. (1999). Genome-wide analysis of DNA copy-number changes using cDNA microarrays, *Nature Genetics*, **23**, 41-46.

Quackenbush, J. (2001). Computational analysis of microarray data. *Nature Reviews Genetics*, **2**, 418-427.

Ramsey, P.H. (1978). Power differences between pairwise multiple comparisons, *Journal of the American Statistical Association*, **73**, 479-485.

Ramaswamy, S., Tamayo, P.,Rifkin, R., Mukherjee, S., Yeang, C.-H, Angelo, M., Ladd, C., Reich, M., Latulippe, E., Mesirov, J.P., Poggio, T., Gerald, W., Loda, M., Lander, E.S., and Golub, T.R. (2001). Multiclass cancer diagnosis using tumor gene expression signatures, *Proceedings of the National Academy of Sciences, USA*, **98**, 15149-15154.

Rhodes, D. R., T. R. Barrette, M. A. Rubin, D. Ghosh and A. M. Chinnaiyan (2002). Meta-analysis of microarrays: Interstudy validation of gene expression profiles reveals pathway dysregulation in prostate cancer, *Cancer Research*, **62**, 4427-4433.

Ripley, R.B. (1996). *Pattern Recognition and Neural Networks*, Cambridge University Press, Cambridge, U.K.

Ritter, H., Kohonen, T. (1989). *Biological Cybernatics*, **61**, 241.

Robbins, H. (1951). Asymptotically sub-minimax solutions of compound statistical decision problems. *Proc. Second Berkeley Symposium*, **1**, 131-148. Univ. Calif. Press.

Robbins, H. (1964). The empirical Bayes approach to statistical decision problems. *Annals of Mathematical Statistics*, **35**, 1-20.

Robbins, H., Hannan, J. (1955). Asymptotic solutions of the compound decision problem for two completely specified distributions. *Annals of Mathematical Statistics*, **26**, 37-51.

Rocke, D.M., Durbin, B. (2001). A model for measurement error for gene expression arrays. *Journal of Computational Biology*, **8**, 557-569.

Rocke, D.M., Lorenzato, S. (1995). A two-component model for measurement error in analytical chemistry. *Technometrics*, **37**, 176-184.

Rose, K. (1998). *Proc. IEEE*, **96**, 2210-2239.

Rose, K., Gurewitz, E., and Fox, G. (1990). *Phys. Rev. Lett.*, **65**, 945-948.

Rose S.D. (1998). Application of a novel microarraying system in genomics research and drug discovery. *Journal of the Association for Laboratory Automation*, **3**, 53-56.

Rosenblatt, F. (1958). The perceptron: a probabilistic model for information storage and organization in the brain. *Psychological Review*, **65**, 386-408. Reprinted in Shavlik & Dietterick (1990).

362

Rosenblatt, F. (1962). *Principles of Neurodynamics: Perceptrons and the Theory of Brain Mechanisms.* Spartan Books, Washington D.C.

Rosenwald, A., Alizadeh, A.A., Widhopf, G., Simon, R., Davis, R.E., Yu, X., Yang, L., PIckeral, O.K., Rassenti, L.Z., Powell, J., Botstein, D., Byrd, J.C., Grever, M.R., Cheson1, B.R., Chiorazzi1, N., Wilson1, W.H., Kipps, T.J., Brown, P.O., and Staudt, L.M. (2001) Relation of gene expression phenotype to immunoglobulin mutation genotype in B cell chronic lymphocytic leukemia. *Journal of Experimental Medicine*, **194**, 1639-1648.

Ross, D.T., Scherf, U., Eisen, M.B., Perou, C.M., Rees, C., Spellman, P., Iyer, V., Jeffrey, S.S., Van de Rijn, M., Waltham, M., Pergamenschikov, A., Lee, J.C., Lashkari, D., Shalon, D., Myers, T.G., Weinstein, J.N., Botstein, D., Brown P.O. (2000). Systematic variation in gene expression patterns in human cancer cell lines. *Nature Genetics*, **24**, 227-235.

Roweis, S.T. and Saul, L.K. (2000). Nonlinear dimensionality reduction by locally linear embedding. *Science*, **290**, 2323-2326.

Sambrook, J., Fritsch, E. F. and Maniatis T. (1989). *Molecular Cloning: A Laboratory Manual*, 2nd edition, Cold Spring Harbor Laboratory Press.

Šášik, R., Calvo, E., Corbeil, J., (2002). Statistical analysis of high-density oligonucleotide arrays: a multiplicative noise model, *Bioinformatics*, **18**, 1633-1640.

Sapir, M., Churchill, G.A. (2000). Estimating the posterior probability of differential gene expression from microarray data, Poster. The Jackson Laboratory. http://www.jax.org/staff/churchill/labsite/pubs

Schadt, E., Li, C., Ellis, B., and Wong, W.H. (2001). Feature extraction and normalization algorithms for high-density oligonucleotide gene expression array data. *Journal of Cellular Biochemistry*, **84**, S37, 120-125.

Schalon, D., Smith S.J., Brown, P.O. (1996). A DNA microarray system for analyzing complex DNA samples using two-color fluorescent probe hybridization. *Genome Research*, **6**, 639-645.

Schena, M., Editor. (2000). *DNA Microarrays*, Oxford University Press, New York.

Schena, M., Shalon, D., Davis, R.W., Brown, P.O. (1995). Quantitative monitoring of gene expression patterns with a complementary DNA microarray. *Science*, **270**, 467-470.

Scherf, U., Ross, D.T., Waltham, M., Smith, L.H., Lee, J.K., Tanabe, L., Kohn, K.W., Reinhold, W.C., Myers, T.G., Andrews, D.T., Scudiero, D.A., Eisen, M.B., Sausville, E.A., Pommier, Y., Botstein, D., Brown, P.O., Weinstein, J.N. (2000). A gene expression database for the molecular pharmacology of cancer. *Nature Genetics*, **24**, 236-244.

Scholkopf, C., Burges, J.C., Smola, A.J. (1999). *Advances in Kernel Methods*, MIT Press, Cambridge, MA.

Schuchhardt, J., D. Beule, A. Malik, E. Wolski, H. Eickhoff, H. Lehrach and H. Herzel. Normalization strategies for cDNA microarrays, *Nucleic Acids Research* 2000, **28**(10), e47.

Seal, H. (1964). *Multivariate Statistical Analysis for Biologists*, John Wiley & Sons, New York.

Searle, S.R. (1982). *Matrix Algebra Useful for Statistics*. John Wiley & Sons, New York.

Shaffer, J.P. (1986). Modified sequentially rejective multiple test procedures, *Journal of the American Statistical Association*, **81**, 826-831.

Shaffer, J.P. (1995). Multiple hypothesis testing: a review. *Annual Review of Psychology*, **46**, 561-584.

Shavlik, J.W., Dieterich, T.G. (1990). *Readings in Machine Learning*, Morgan Kaufmann, San Mateo, CA.

Shipp, M.A., Ross, K.N., Tamayo, P., Weng, A.P., Kutok, J.L., Aguiar, R.C.T., Gaasenbeek, M., Angelo, M., Reich, M., Pinkus, G.S., Ray, T.S., Koval, M.A., Last, K.W., Norton, A., Lister, T.A., Mesirov, J., Neuberg, D.S., Lander, E.S., Aster, J.C., Golub, T.R. (2002). Diffuse large B-cell lymphoma outcome prediction by gene-expression profiling and supervised machine learning. *Nature Medicine*, **8**, 68-74.

Silipo, R., Gori, M., Taddei, A., Varanini, M., and Marchesi, C. (1995). Classification of arrhythmic events in ambulatory electrocardiogram, using artificial neural networks. *Computational Biomedical Research*, **28**, 305-318.

Simes, R.J. (1986). An improved Bonferroni procedure for multiple tests of significance, *Biometrika*, **73**, 751-754.

Simon, R., Radmacher, M.D., Dobbin, K. (2002). Design of studies using DNA microarrays. *Genetic Epidemiology*, **23**, 21-36.

Soille, P. (1999). *Morphological Image Analysis: Principles and Applications.* Springer.

Somogyi, (1999). *Pharma Informatics*, p.17.

Southern, E., Mir, K., Shchepinov, M. (1999). Molecular interactions on microarrays. *Nature Genetics*, **21**, 5-9.

Speed, T., Editor (2003). *Statistical Analysis of Gene Expression Microarray Data*, Chapman and Hall/CRC Press, New York.

Spellman, P.T., Sherlock, G., Zhang, M.Q., Iyer, V.R., Anders, K. , Eisen, M.B., Brown, P.O., Botstein, D. and Futcher, B. (1998). Comprehensive identification of cell cycle-regulated genes of the yeast *Saccharomyces cerevisiae* by microarray hybridization, *Molecular Biology of the Cell*, **9**, 3273-3297.

Stern, H.S. (1996). Neural networks in applied statistics, (with discussion), *Technometrics*, **38**, 205-220.

Storey, J.D. (2003) The positive false discovery rate: A Bayesian interpretation and the q-value. *Annals of Statistics*, (in press).

Storey, J.D., Taylor, J.E., and Siegmund, D. (2003). Strong control, conservative point estimation, and simultaneous conservative consistency of false discovery rates: A unified approach. *Journal of the Royal Statistical Society*, Series B, (in press).

Storey, J.D. and Tibshirani, R. (2001). Estimating false discovery rates under dependence, with applications to DNA microarrays, Technical Report, Stanford University, 2001.

Storey, J.D. and Tibshirani, R. (2003). Statistical significance for genome-wide studies. *Proceedings of the National Academy of Sciences, USA*, **100**, 9440-9445.

Sudarsanam, P., Vishwanath, R.T., Brown, P.O., and Winston, F. (2000). Whole-genome expression analysis of snf/swi mutants in *Saccharomyces cerevisiae*. *Proceedings of the National Academy of Sciences, USA*, **97**, 3364-3369.

Tamayo, P., Slonim, D., Mesirov, J., Zhu, Q., Kitareewan, S., Dimitrovsky, E., Lander, E.S., Golub, T.R. (1999). Interpreting patterns of gene expression with self-organizing maps: methods and application to hematopoietic differentiation. *Proceedings of the National Academy of Sciences, USA*, **96**, 2907-2912.

Tibshirani, R., Hastie, T., Narasimhan, B., Chu G. (2002). Diagnosis of multiple cancer types by shrunken centroids of gene expression. *Proceedings of the National Academy of Sciences, USA*, **99**, 6567-6572.

Tibshirani, R., Hastie, T., Narasimhan, B., Eisen, M., Sherlock, G., Brown, P., Botstein, D. (2002). Exploratory screening of genes and clusters from microarray experiments. *Statistics Sinica*, **12**, 47-59.

Tavazoie, S., Hughes, J.D., Campbell, M.J., Cho, R.J., Church, G.M. (1999). Systematic determination of genetic network architecture. *Nature Genetics*, **22**, 281-285.

Troendle, J.F. (1996). A permutational step-up method of testing multiple outcomes, *Biometrics*, **52**, 846-859.

Troyanskaya, O., Cantor, M., Sherlock, G., Brown, P., Hastie, T., Tibshirani, R., Botstein, D., Altman, R.B. (2001). Missing value estimation methods for DNA microarrays. *Bioinformatics*, **17 (6)**, 520-525.

Tseng, G.C., Oh, M.-K., Rohlin, L., Liao, J.C., and Wong, W.H. (2001). Issues in cDNA microarray analysis: quality filtering, channel normalization, models of variation and assessment of gene effects. *Nucleid Acids Research*, **29**, 2549-2557.

Tusher, V.G., Tibshirani, R., Chu, G. (2001). Significance analysis of microarrays applied to the ionizing radiation response. *Proceedings of the National Academy of Sciences, USA*, **98**, 5116-5121.

Vapnik, V. (1998). *Statistical Learning Theory*, John Wiley & Sons, New York.

Wang, X., Ghosh, S., and Guo, S. (2001). Quantitative quality control in microarray image processing and data acquisition. *Nucleic Acids Research*, **29**, No. 15, e75.

Wang, X., Hessner, M.J., Pati, N., and Ghosh, S. (2003). Quantitative quality control in microarray experiments and the application in data filtering, normalization and false positive rate prediction. *Bioinformatics*, **19**, 1341-1347.

Ward, J.H. (1963). Hierarchical grouping to optimize an objective function. *Journal of American Statistical Association*, **58**, 236-244.

Watson, J.D., and Crick, F.H.C. (1953). A structure for DNA. *Nature*, **171**, 737-738.

Weiss, K.M., Terwilliger, J.D. (2000). How many diseases does it take to map a gene with SNPs? *Nature Genetics*, **26**, 151-157.

Welsh, J.B., Zarrinkar, P.P., Sapinoso, L.M., Kern, S.G., Behling, C.A., Monk, B.J., Lockhart, D.J., Burger, R.A., Hampton, G.M. (2001). Analysis of gene expression profiles in normal and neoplastic ovarian tissue samples identifies candidate molecular markers of epithelial ovarian cancer. *Proceedings of the National Academy of Sciences, USA*, **98**, 1176-1181.

Wen, X., Fuhrman, S., Michaels, G.S., Carr, D.B., Smith, S., Baker, J.L., and Somogyi, R. (1998). Large-scale temporal gene expression mapping of central nervous system development. *Proceedings of the National Academey of Sciences, USA*, **95**, 334-339.

Wernisch, L., Kendall, S.L., Soneji, S., Wietzorrek, A., Parish, T., Hinds, J., Butcher, P.D., and Stoker, N.G. (2003). Analysis of whole-genome microarray replicates using mixed models. *Bioinformatics*, **19**, 53-61.

Westfall, P.H. and Young, S.S. (1993). *Re-sampling Based Multiple Testing: Examples and Methods for P-value Adjustment*, John Wiley & Sons, New York.

Winer, B.J. (1971). *Statistical Principles in Experimental Design*, 2nd ed., New York, McGraw-Hill.

Wishart, D. (1969). An algorithm for hierarchical classifications. *Biometrics*, **25**, 165-170.

Wit, E., McClure, J. (2003). Statistical adjustment of signal censoring in gene expression experiments. *Bioinformatics*, **19**, 1055-1060.

Wodicka, L., Dong, H., Mittmann, M., Ho, M., and Lockhart, D. (1997). Genome-wide expression monitoring in *Saccharomyces cerevisiae*. *Nature Biotechnology*, **15**, 1359-1367.

Wolfinger, R.D., Gibson,G., Wolfinger, E.D., Bennett, L., Hamadeh, H., Bushel, P., Afshari, C., Paules, R. (2001). Assessing gene significance from cDNA microarray expression data via mixed models. *Journal of Computational Biology*, **8**, 625-637.

Wright, S.P. (1992). Adjusted p-values for simultaneous inference, *Biometrics*, **48**, 1005-1013.

Xiong, M.M., Jin, L., Li, W., and Boerwinkle, E. (2000). *BioTechniques*, **29**, 1264-1270

Yang, Y.H., Buckeley, M.J., Dudoit, S., Speed, T.P. (2000). Comparison of methods for image analysis of cDNA microarray data. *Journal of Computational and Graphical Statistics*, **11**, 108-136.

Yang, Y.H., Dudoit, S., Luu, P., Lin, D.M., Peng, V., Ngai, J., and Speed, T.P. (2002). Normalization for cDNA microarray data: a robust composite method addressing single and multiple slide systematic variation. *Nucleid Acids Research*, **30**, No.4, e15.

Yang, Y.H., Dudoit, S., Luu, P., Speed, T.P. (2000). Normalization for cDNA microarray data. In M. L. Bittner, Y. Chen, A. N. Dorsel, and E. R. Dougherty (eds), *Microarrays: Optical Technologies and Informatics*, Vol. 4266 of Proceedings of SPIE. Society for Optical Engineering, San Jose, CA.

Yates, F. (1935). Complex experiments. *Journal of Royal Statistical Society, Suppl.* **2**, 181-247.

Yekutieli, D., Banjamini, Y. (1999). Resampling-based false discovery rate controlling multiple test procedures for correlated test statistics, *Journal of Statistical Planning and Inference*, **82**, 171-196.

Zhang, H. P. , Holford, T. and Bracken, M. (1995). A tree-based method of analysis for prospective studies. *Statistics in Medicine*, **15**, 37-50.

Zhang, H. P., Singer, B. (1999) *Recursive Partitioning in the Health Sciences*, Springer, New York.

Zhou, X., Kao, M.J., Wong, W.H. (2002) Transitive functional annotation by shortest path analysis of gene exxpression data. *Proceedings of the National Academy of Sciences, USA*, **99**, 12783-12788.

Zhu, H., Klemic, J.F., Chang, S., Bertone, P., Casamayor, A., Klemic, K.G., Smith, D., Gerstein, M., Reed, M.A., Snyder, M. (2000). Analysis of yeast protein kinases using protein chips. *Nature Genetics*, **26**, 283-289.

Author Index

Topic Index